The Moon Year

The Moon Year © 2009 Soul Care Publishing

All rights reserved. No part of this book may be used or reproduced or transmitted in any manner whatsoever, transmitted electronically, or distributed by any means without the written permission of the publisher

Library and Archives Canada Cataloguing in Publication

Bredon, Juliet
 The moon year : a record of Chinese customs and festivals / Juliet Bredon & Igor Mitrophanow.

Includes bibliographical references and index.
ISBN 978-0-9812717-7-4

 1. Festivals--China. 2. Calendar, Chinese. 3. China--Social life and customs. I. Mitrofanov, Igor II. Title.

DS721.B74 2010   390.0951   C2009-907124-X

Photograph by Benjamin March. Used by permission

BURNING INCENSE ON NEW YEAR'S MORNING,

The Moon Year
A Record of Chinese Customs and Festivals
Juliet Bredon & Igor Mitrophanow
(Author and collaborator of *Peking*)

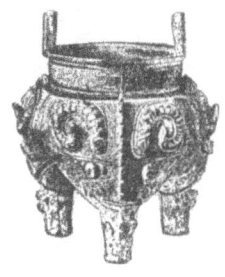

SHANGHAI
KELLY & WALSH, LIMITED
1927

Contents

	PAGE
CONTENTS	v
LIST OF ILLUSTRATIONS	vii
PREFACE	ix
CHAPTER I—THE CHINESE CALENDAR	1
,, II—THE HUNDRED GODS	29
,, III—IMPERIAL CEREMONIES	54
,, IV—THE TWELFTH MOON, OR "BITTER MOON"	69
,, V—THE FIRST MOON, OR "HOLIDAY MOON" .	101
,, VI—THE SECOND MOON, OR "BUDDING MOON"	158
,, VII—THE THIRD MOON, OR "SLEEPY MOON" .	214
,, VIII—THE FOURTH MOON, OR "PEONY MOON" .	250
,, IX—THE FIFTH MOON, OR "DRAGON MOON" .	299
,, X—THE SIXTH MOON, OR "LOTUS MOON" .	334
,, XI—THE SEVENTH MOON, OR "MOON OF HUNGRY GHOSTS"	368
,, XII—THE EIGHTH MOON, OR "HARVEST MOON"	391
,, XIII—THE NINTH MOON, OR "CHRYSANTHEMUM MOON"	425
,, XIV—THE TENTH MOON, OR "KINDLY MOON" .	459
,, XV—THE ELEVENTH MOON, OR "WHITE MOON"	491
BIBLIOGRAPHY	509
INDEX	xiii

List of Illustrations

	FACING PAGE
BURNING INCENSE ON NEW YEAR'S MORNING	*Frontispiece*
THE WU MÊN, OR "MERIDIAN GATE" OF THE PALACE, PEKING	4
ONE OF THE OFFICIAL SUNDIALS IN THE PALACE	5
TAOIST ABBOT	48
BUDDHIST PRIEST IN CEREMONIAL ROBES	49
OFFICIAL ROBE FOR SACRIFICE AT TEMPLE OF HEAVEN, PEKING	62
THE SHÊ CHI T'AN	63
A SHRINE TO THE JADE EMPEROR	146
ALTAR WITH SPIRIT-TABLET OF AN IMPERIAL PRINCE	147
WOMEN ACROBATS	154
VILLAGE STORY-TELLER	155
CONFUCIAN TEMPLE, CH'Ü FU	210
THE "DEW WELL" USED BY CONFUCIUS, CH'Ü FU	211
CARRYING PAPER MONEY TO BURN AT THE TOMBS	226
PILGRIMS AT A TEMPLE	227
BUDDHA SAKYAMUNI, GANDHARA TYPE, CHENG TING FU	258
A BEGGING PRIEST	259
TWO OF THE HEAVENLY MAHARAJAHS	266
IMAGE OF PU TAI, HANGCHOW	267
BRONZE PAGODA, WU T'AI SHAN	276
LAMA PRIEST IN FULL DRESS, PEKING	277
TAOIST MOUNTAIN SHRINE	322
GATHERING FOR A TEMPLE FAIR	323
PEASANTS PRAYING FOR RAIN BESIDE A SPRING	342
FESTIVAL PROCESSION, KANSU	343

List of Illustrations—*Continued*

	FACING PAGE
VILLAGE THEATRICALS	392
LION DANCERS	394
LION DANCERS	395
MOON RABBITS ON TOY-STALL	404
OPEN-AIR ALTAR FOR MOON FESTIVAL	405
A SOOTHSAYER	416
WAYSIDE SHRINE TO A FOX-SPIRIT	417
THEATRICALS AT THE TEMPLE OF A CITY GOD	438
TEMPLE TO A LOCAL GOD	439
SELLER OF CANDIED FRUITS	460
SUMMIT OF MOUNT T'AI SHAN	461
FIVE HUNDRED LOHANS, PI YÜN SSŭ, PEKING	490
BUDDHIST NUNS	491

PREFACE.

IRST impressions of China are invariably confusing. The people look alike to our unaccustomed eyes—just as we do to the Chinese. It takes time and patience to understand even the surface-life of a land where everyone's actions are unfamiliar; where the needle is put over the thread instead of the thread through the needle, and the carpenter pulls his plane towards him instead of pushing it away; where, in fact, most things are done in what we are accustomed to think is the wrong direction.

As for the intimate life of the Chinese, their religions, their superstitions, their ways of thought, the hidden springs by which they move, the customs followed behind the protecting walls of homes and temples,—it is only after many years' residence among them that one can hope to penetrate these inner mysteries.

This book is an attempt to unravel some of the puzzles of an old, old civilisation which, save in the case of a small minority, has not changed for centuries—to describe the everyday beliefs of the Chinese people and the festivals of their "Moon Calendar," used as a diary of daily happenings.

It is a difficult task. The Chinese themselves, fearing that Western sympathy is limited by lack of comprehension, often dislike to disclose the inner meaning of their legends. Moreover, the personalities and adventures of their innumerable divinities and saintly heroes are apt to appear wearyingly similar to strangers whose historical and intellectual background is so dissimilar. Finally, China is such a huge country that traditions kept up in one part of it may have died out in another. True, the "great festivals" described are observed everywhere. But others, no less interesting and no less valuable as a key to the soul of the people, are peculiar to certain provinces and even to certain districts.

Many of the older myths, "comparable for beauty of fancy to those of ancient Greece," are dying out altogether. It seems to us, therefore, important to record these "primitive efforts to find solutions for the Riddle of the Unseen," before they fade beyond recognition. Fox-fairies and flower-sprites will vanish, and many gods desert their altars as Oriental life hardens and coarsens under the shock of contact with modern intellectual and material requirements. Their influence may remain, taking new forms, because Chinese thought is too deeply rooted in them to let them go altogether, but for their poetic personalities we shall soon search in vain.

Much of our material for "The Moon Year" has been gathered at first hand from Chinese friends, some is the result of personal experiences and observations. A great deal more has been gleaned from the literature of four languages, mostly old or rare books now difficult to obtain (*see* "Bibliography").

In conclusion, let us add simply that the Soul of China well repays our study. The time is past, if indeed it ever existed, when foreign residents or visitors could afford to ignore a culture like the Chinese, which is so vast, so important, and so rich in interest. Not to know or even desire to know, not to question strange things seen and heard, not to inquire into the depths of this great nation's heart, is simply to accept an invitation to a banquet and refuse every dish offered, because it is not prepared in a way to which we are accustomed.

The two great civilisations alive in our world to-day are the Western and the Chinese. That we should know as little as we do about the latter —which includes a quarter of the human race— is a matter of wonder—and regret.

J. B. AND I. M.

CHAPTER I.
THE CHINESE CALENDAR.

ALL Nature follows an instinctive calendar. Plants and animals need no almanac to foretell the seasons. The flower unfolds and the leaf falls at the right time. Something beyond our knowledge tells the wild geese when to migrate southwards. The reindeer leave the forest just when the seas unfreeze and open waters provide the salt they crave. Even the oscillations of a barometer are in fairly accurate agreement with the actions of a herd of wild cattle, and scientists have observed that a change of weather is presaged "two days ahead by the behaviour of a colony of caterpillars."

Man alone is less sensitive to Nature's moods. But his defective instinct is compensated by his power of thought and his intelligent observation. As soon as an accurate fore-knowledge of the seasons became essential to primitive human society—as soon as men found that, without it, no crops could be gathered and no successful hunting done—necessity spurred the human mind to the making of the first rude calendar.

The primitive "book of the year" was the sky. Thence, long before the simplest scientific instruments came into being, shepherds and farmers learned to gauge the main divisions of time

Origins of the Calendar

The "First Measurer"

through observing, generation after generation, the position of the sun, moon, and stars.

Now, though light and darkness, or day and night, were obvious enough, more extended time-periods required calculation and a *point de départ,* or guiding planet. The "First Measurer" chosen was the Moon, as the Latin word for month indicates.

But until science developed methods, based on complete consistency between idea and fact, there was no such thing as accurate observation. The human eye is fallible. To observe with precision, and remember without written records the waxing and waning of the Queen of Night was exceedingly difficult. Nor was it easier, without the learned aid of arithmetic, to adjust the time required for the moon to revolve around the earth with the time required for the earth to revolve around the sun. From new moon to new moon is roughly twenty-nine and a half days. Twelve lunar revolutions, therefore, make 354 days only, whereas our planet takes 365¼ days to go round the sun, and on this latter course the changes of the seasons depend. Hence the phenomenon of the return of spring sometimes after twelve, and sometimes after thirteen, complete "moons."

Obviously then, the "First Measurer" alone could not be relied upon to tell the farmer when to sow his seed. The problem of checking her course had to be solved. This riddle was answered by the stars.

Men soon saw that the position of the latter in relation to the principal planets and to one another was in no sense haphazard. There is a definite order in the universe, and each star has its own ecliptic, or path across the heavens. Every evening, just after sunset, one more dazzling than the rest will appear in the west quite close to the horizon. A few weeks later this bright particular star is no longer in its usual place but may be noticed before dawn in the eastern sky, a pale attendant of the rising sun. Furthermore, this journey from one side of the firmament to the other takes place regularly at intervals of several months.

To primitive man this was a wonderful discovery. It was soon followed by another, equally important. The full moon, he observed, rose as the sun set and, when this happened, both were opposite one another. Did the moon keep time in her rising, many calendric complications might be avoided. But no, she appears forty-five minutes later each evening until, finally, she vanishes to be re-born again. Once the position of certain stars above the horizon at dawn and sunset had been established, however, the schedule of her course was easily followed.

The humble mathematician who had learned from studying the skies the simple science of counting up to ten—a science not yet universal —had by this time become a primitive astronomer. Pure guess-work gave place to logic.

If one set of stars could fix the waxing and waning of the moon, it was logical to suppose that another group might well determine the rising and setting of the sun. The finding of this second key for fixing time-periods and seasons was a great advance. In fact, from this knowledge, painfully acquired through the progression of thought, the first lunar and solar calendars were born.

Heirs of the ages, we little realise at what heavy cost of human effort we enjoy the precision of our almanac. Star-gazers on a clear evening, we forget to be grateful for all that our ancestors learned through the slow accumulation of experience, not only about the position of the constellations but, by deduction, about the laws that govern events on our earth, about the mathematical character of natural processes, and the fundamental principles of order and harmony underlying not only the universe but human relations.

* * *

History of the Chinese Calendar

It is to the credit of the Chinese that they early developed a notion of the interdependence of earth and sky and the connection between the spiritual, the animate, and the inanimate worlds. If their scientific theories were largely wrong and, astronomically speaking, their ideas often fantastic,—"the reality of the world-conception which they were trying to express has only been developed in the last three hundred years in Europe." But, ever a practical people, they chiefly concerned themselves with the influence of the stars upon the seasons. Farmers from the earliest ages, dependent upon agriculture for their existence, they were vitally interested in weather conditions, and a reliable calendar meant so much to them that they concentrated their energies upon perfecting one.

"The Book of Records contains some remarkable notices given by the Perfect Emperor Yao (2254

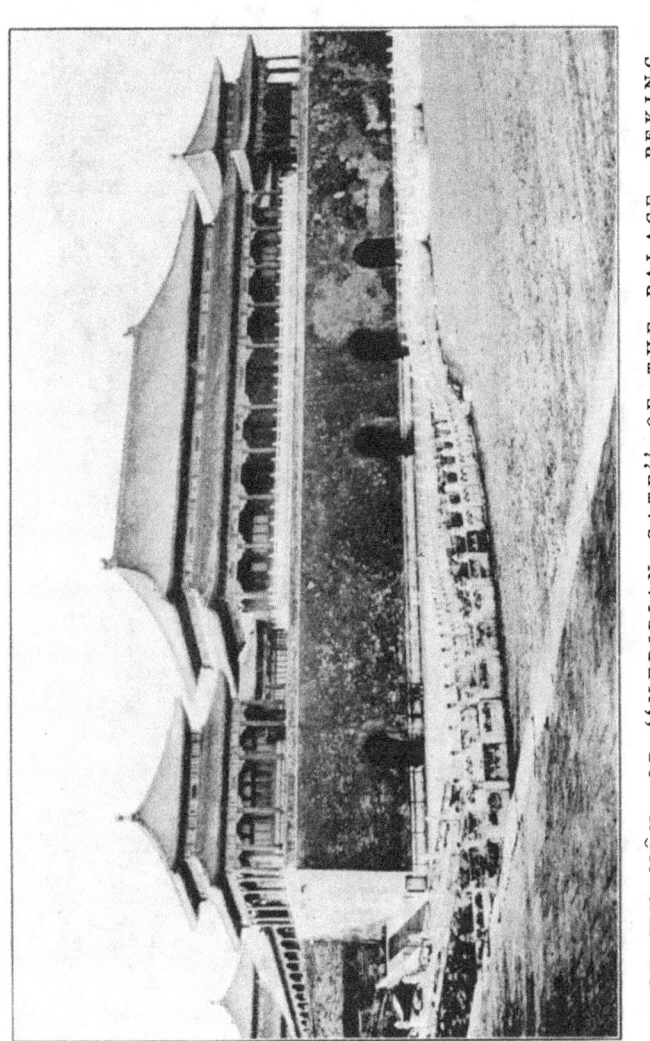

THE WU MÊN, OR "MERIDIAN GATE," OF THE PALACE, PEKING.
(WHERE THE CALENDAR FOR THE NEW YEAR WAS DISTRIBUTED)

ONE OF THE OFFICIAL SUNDIALS IN
THE PALACE, PEKING.

B.C.) to his astronomers to ascertain the solstices and equinoxes, employ intercalary months, and fix the four seasons so that the husbandman might know when to commit his seed to the ground."

For centuries the register of the years was prepared in China by a special Board of Mathematicians whose chief ranked as a Minister of State, and annually submitted his work to the Emperor for approval. When the Sovereign pronounced it good, copies of the new almanac were distributed to the highest officials of the Empire. The precious documents were conveyed and received with Imperial honours. They were carried to their destinations in sedan chairs, placed on pedestals, and greeted with prostrations and a salute of guns. In the capital, calendars were formally distributed to privileged persons outside the chief gate of the Palace, the *Wu Mên*, or "Meridian Gate," on New Year's day.

To issue a counterfeit or pirated edition was a penal offence, and falsification was punished by death, as an infringement of the sovereign Son of Heaven's exclusive right to fathom the laws of Heaven.

All the Chinese dynasties that reigned before Christ—and a few individual Emperors of later date—stressed this right by adopting a different month as the beginning of the civil year. Rebellious feudal lords under the Chou dynasty asserted their independence by issuing a new calendar, and a present of this important publication to friendly countries, or vassal States like Corea and Annam, was considered by the Chinese as one of the highest favours possible to confer.

When the Manchus mounted the Dragon Throne, they carried on the tradition. The young Emperor K'ang Hsi, an enlightened and tolerant man, availed himself of the knowledge and advice of the great Jesuit Father Verbiest, in A.D. 1669, for the correction of errors that had crept into the work of his predecessors. Verbiest revised the calendar on the basis of more accurate Western calculations until the year A.D. 2020. But the present Republican Government set aside the work of the priest-scientist and adopted the Gregorian calendar *in toto*, thus following the precedent of centuries in order to underline the break with monarchic tradition. Officially, the new calendar became general. But the people cling to the Moon almanac used by their forefathers, and their private festivals are still guided by it.

Infinitely old, this system of time-reckoning based on the "First Measurer" was begun with very simple equipment, and perfected only after centuries of stubborn thinking. The lunar month, as we have already observed, does not consist of an even number of days. There is always a fraction left over. A similar difficulty complicates an attempt to fit in the days and months to the year. One solution might be to start every new year at different dates and different hours. But this would be impracticable, even for the purpose of making true time agree with the divisions of the calendar.

Intercalations   So, the mathematicians decided to ignore the overlapping fractions of a day until they totalled a suitable calendric unit. Then they adjusted the difficulty by adding the extra day, or month, or year, required to set the calendar right. Avoiding tedious and technical details, we may say simply that this can be done in different ways and after different lapses of time. The Mohammedans, for example, who follow the only purely lunar calendar, add a whole year at regular intervals, whereas we insert a 29th of February each leap year to make up the difference between the nominal 365 days of the calendar year as compared with the 365¼ days of the true sun year.

The Chinese use still another method. They add an extra moon at stated periods. Thus, sometimes, their years consist of twelve moons only, sometimes of thirteen. The latter are known as "complete years."

The first intercalations seem to have been made at a very early date. Furthermore, the discovery of their first astronomical instrument, a rude sundial known as the gnomon, was an important aid in calculations that resulted in the true harmonising of the lunar and the solar years. With its aid the Chinese learned to date the winter and the summer solstices from the increase and decrease in the length of shadows, and the spring and autumn equinoxes, when the length of day and night are equal. Thus were established the first accurate milestones dividing the year into four seasons. Nevertheless, the Chinese chose to begin their moons not at the solstices and equinoxes, but at a date half-way between, which brings their seasons forty-five days ahead of ours.

It is astonishing how early and how accurately the people of the Middle Kingdom made their calculations "respecting the length of the lunar and the solar year, the obliquity of the ecliptic, and the general motions of the planets."

By the VIIIth century B.C. they could calculate and record eclipses, though they believed them caused by a dog or dragon, trying to eat the darkened planet, and attempted to drive the animal off with gongs and fire-crackers. Even earlier, they discovered what is known as the Metonic Cycle.[1]

With this knowledge, and the principle of intercalation established, they began to work out a calendar. The 19 year periods at whose beginning and end the sun and moon are in the same relative positions was divided into twelve years of twelve moons each, and seven "complete years" of thirteen moons. During the latter, two moons were duplicated. Which moons were chosen, and how the choice varied, is too technical to explain here, and for such details the reader is referred to the work of specialists on this subject (*see* "Bibliography"). Let us add, however, only that the intercalation must so be made that the winter solstice always falls in the eleventh moon, the summer solstice in the fifth, the spring equinox in the second, and the autumn equinox in the eighth. The eleventh, twelfth, and first moons are never duplicated.

---

[1] The Metonic Cycle is based on the fact that 235 "moons" correspond to 19 solar years. With our accumulated knowledge, this discovery may not seem remarkable. But when, later, the Greeks stumbled on this same principle, they considered it so important that the "law" was engraved in golden letters on the Temple of Athena as the formula, or "golden number," permitting astronomy to establish the exact relative position of the sun and moon, the "rank of the year," etc.

THE CHINESE CALENDAR

For the sake of further accuracy, the Chinese make some moons *long,* with thirty days, and some *short,* with twenty-nine days. The double correction results in an astronomical cycle of nineteen solar years divided into 130 long lunar moons, 110 short lunar moons, and seven intercalary moons or "moons made of the remnants of odd days." If we remember the half day short in every short moon, this total practically represents the 235 lunations required for this time-period.

Such, briefly stated, are the fundamental principles used in compiling the Chinese calendar, such its principal variations from other systems of almanac-making of which there are, in all, three: the purely lunar system, the purely solar system (like our own), and a combination of the two, which rivalled and yet borrowed from one another.

*   *   *

Let us now give a brief summary of the terms used by the Chinese to mark the years, months, days, and hours. We date our era from the birth of Christ, the Mohammedans from the Hegira, or flight of Mohammed from Mecca, the Jews from the Creation. The Chinese start computing time from the reign of the "Yellow Emperor," Huang Ti, whose Prime Minister is supposed (2637 B.C.) to have started the "grouping of the years" into *Cyclical* chronological cycles of sixty years each, though *Signs* why this arbitrary number was chosen is not recorded. Instead of numbers, the years of the cycle have each a separate name "formed by taking ten characters called *t'ien kan* 天干, or

'Ten Heavenly Stems,' and joining to them twelve other characters called the *ti chih* 地支, or 'Twelve Earthly Branches,' the latter five times repeated . . . . These hoary ideographs are also applied to hours, days and months."

Solar Zodiac

The Twelve Earthly Branches have the name of as many animals, and correspond to the twelve signs of the Chinese Solar Zodiac, that is to say, the twelve constellations by means of which they fix the position of the sun every month.

They are as follows:

Chinese Zodiac.		Græco-Chaldean Zodiac.
Rat	鼠	Aries
Ox	牛	Taurus
Tiger	虎	Gemini
Hare	兔	Cancer
Dragon	龍	Leo
Snake	蛇	Virgo
Horse	馬	Libra
Sheep	羊	Scorpio
Monkey	猴	Sagittarius
Fowl	雞	Capricornus
Dog	犬	Aquarius
Pig	猪	Pisces

The word "zodiac" comes from the Greek *zodion*, a "small animal," but, concerning the meaning of the symbols used, there are many theories. The Chinese form which, strange to say, scarcely differs from that of the Aztecs, is described by Humboldt as "a zodiac of hunters and shepherds" which spread over all Asia. It includes six wild, or mythical, animals, and six domestic animals used in ancient sacrifices. The original choice is wrapped in mystery. Very probably, there were at first only four symbols to represent the four seasons and the four cardinal directions, the Dragon for the Spring and the East, the Phœnix, or Fowl, for the Summer and the South, the glowing Tiger for the Autumn and the West, and the sluggish Tortoise for the Winter and the North. Each, in addition, belongs either to the *Yang*, active or male element, or to the *Yin*, passive or female element (*see* "The Hundred Gods," and "The Eighth Moon" for the dualistic theory of creation which is the foundation of the Chinese world-conception). The rat, who lives in darkness and is by nature destructive, the peaceful ox, the hare disporting itself in the moonshine, the shy yet mischievous monkey, the obedient dog, guardian of the house at night, the slow pig with eyes bent earthwards, all partake of the *Yin*, or passive, essence, whereas on the other side of the Zodiac, symmetrically opposed, we find the terrible tiger, the mighty dragon, the swift horse, the sun-loving sheep, and the cock, or fowl, who announces the dawn—all essentially male, or *Yang* symbols. The inclusion of the snake among them is puzzling. He seems to have been made *vis-à-vis* of the pig across the celestial sphere because, according to the adage, "snakes and pigs are hereditary enemies." Natural history confirms this juxtaposition, since pigs eat serpents and are themselves immune from the poison of snake-bite.

In addition to the "Twelve Earthly Branches" the Chinese also have a lunar zodiac known as the *Lunar Zodiac*

Huang Tao, or "Yellow Road," consisting of twenty-eight constellations that were early mapped out and served to subdivide the path of the moon. "These stars," says de Saussure, "are very nicely paired across the celestial sphere so as to enable the full moon to be used as an index to the sun's position." [2]

Now, astronomy in China having as its main object weather prediction and the fixing of the seasons, is, notwithstanding, much confused with astrology and divination.[3]

The Sacred Number Five The Celestial Dome which covers the earth like an overturned rice bowl was accepted as the fifth cardinal point. The number five is a magical cypher that runs through the Chinese cosmogony —a semi-sacred figure, though how and why it came to be so is too long to explain here. Suffice it to say that this numeral is the mysterious connecting link, moral as well as mathematical, between the cardinal points and the seasons—

---

[2] The reader is referred to de Saussure's works for a detailed description of the part played by the lunar zodiac in Chinese civilisation, and the light it throws on early intercourse between Eastern and Western Asia, through its similarity with the Hindu "nakshatras" and the Arab "menazils."

[3] Science and superstition both influenced the making of the Chinese calendar. It is peculiar that some theories of the earlier Chinese astronomers, like the Hun T'ien system which considered that heaven is a complete sphere in whose centre earth reposes, were considerably nearer to the modern scientific conception than later theories. Dealing with this point, A. Forke in his *World-Conception of the Chinese* (p. 22) quotes Gaubil as saying that "Chêng Hsüan spoke about the rotation of the earth, but in a very confused way. In Greece, Aristarchus of Samos, about 300 B.C., was the first to teach the rotation of the earth round the sun on its own axis, while the firmament of the fixed stars remained motionless. The doctrine, overthrowing the cherished ideas about the importance of our earth and its inhabitants, could not overcome the common prejudices, and fell into oblivion. Copernicus did not know about this theory, and had to discover it again."

four, and one added to represent the centre of the universe and the middle of the year; the fivefold group of elements—wood, fire, earth, metal, and water; the colours—green, red, yellow, black, and white; the organs—spleen, heart, liver, lungs, and kidneys; the animals—sheep, fowl, ox, dog, and pig; the tastes—sour, sweet, bitter, acid, and salt, etc. Such divisions are peculiar not only in themselves but in the way they control one another and influence the daily life of the Chinese race. "They form," says Wells Williams, "a chain of causes acting and re-acting through and with each other,—causes whose explanation is peculiarly well fitted to serve the astrologers as a basis for their predictions."

These diviners, who worked out a most elaborate astrological compass, rely largely also on the lucky or unlucky influences of the stars, and especially those of the aforementioned 28 Constellations popularly known as the "Palaces of the Moon." Indeed, so great is their power over the weather and the affairs of mankind that the official almanac gave a careful list of them, with recommendations as to which dates are propitious or unpropitious for certain rites, duties, and pleasures. *The 28 Constellations*

The following is a typical, though abridged, example of the "Moon Palaces" poster with its lucky and unlucky days, such as used to be found in nearly every Chinese home where all important events of life were guided by it.

Constellation 1. *The Crocodile* 蛟. Unlucky day for buying land or building houses. Funerals bring further grief and the repair of graves means misfortune, though daughters may be safely married. If granaries are opened, the rice will either spoil or be destroyed by insects.

Constellation 2. *The Dragon* 龍. Composed of seven stars supposed to resemble this mythical animal. A day of general ill-luck. Anything undertaken is bound to fail. Calamity follows the soldier to battle, the bridegroom seeking a bride, the official taking up office, and whoever presumes to lay a foundation stone.

Constellation 3. *The Badger* 貉. Scarcely more fortunate. Children born on this day are deaf, dumb, or idiots. Business begun is doomed to fail. A kitchen-stove repaired brings fire to the house.

Constellation 4. *The Fox* 狐. Equally unlucky. Few dared shave, lest their heads be covered with boils.

Constellation 5. *The Dog* 狗. A propitious day for starting public works, building, digging canals, etc. Also for weddings.

Constellation 6. *The Wolf* 狼. Again unfortunate for all enterprises.

Constellation 7. *The Hare* 兔. Lucky for funerals, for weddings, and in a general sense.

Constellation 8. *The Porcupine* 貐. It is recommended to start silk culture under these stars. Happy is the man at whose birth they preside, for he will never lack servants, and his future is assured.

Constellation 9. *The Rat* 鼠. A day of thoroughly bad omen. Those who dig drains will be bitten by snakes and devoured by tigers. A couple who starts building a house will squabble all their lives, and their children will be undutiful.

Constellation 10. *The Leopard* 豹. A fortunate day in all respects, presaging domestic peace.[4]

Constellation 11. *The Griffon* 豸. Again a lucky occasion—especially to buy cattle. Mourners may not weep for the dead, as to do so is to experience sorrow upon sorrow.

Constellation 12. *The Bat* 蝠. A day of dissension. Brothers fight brothers like wolves and tigers. Disease also is rampant.

Constellation 13. *The Pheasant* 雉. In general, exceedingly propitious.

Constellation 14. *The Gibbon* 猿. An evil day for marriages, nevertheless he who builds a house will have a numerous posterity.

Constellation 15. *The Cock* 鷄. Again a prohibition against marrying on this day. Likewise, burials are unlucky.

Constellation 16. *The Crow* 烏. To build under these stars brings riches. To bury foretells abundant harvests to the family of the dead.

Constellation 17. *The Horse* 馬. Unlucky, especially to women, although this is the birth-star of Confucius.

Constellation 18. *The Earthworm* 蚓. Choose this day for a burial, and a successful official will be born in the family. Marry a daughter, and her son will become a high government functionary.

Constellation 19. *The Deer* 獐. Avoid all building under this star, lest law-suits result.

---

[4] These ten days are called "male" days, and the twelve following—"female" days.

Nor dare to bury the dead, lest ruin overtake the whole family and only one descendant be left alive.

Constellation 20. *The Monkey* 猴. Exile will be the portion of such as start a new house on this day. Equally, misfortune attends the sale of land or buildings.

Constellation 21. *The Serpent* 蛇. General discord and calamity attends all enterprises, marriages, funerals, and work of any kind.

Constellation 22. *The Stag* 鹿. Peaceful and happy the marriages contracted on this day, fortunate the buyers of land, and those who erect monumental arches.

Constellation 23. *The Sheep* 羊. Build no walls, lest the girl-children of the family die. But bury the dead, for this will draw honours on the descendants.

Constellation 24. *The Tapir* 犴. A lucky day for the increase of flocks and herds.

Constellation 25. *The Swallow* 燕. Generally unlucky. No man shall build a house, lest he risk having violent death in the family, suicide by hanging, and the drowning of children under three years of age.

Constellation 26. *The Ox* 牛. Again unlucky for every enterprise.

Constellation 27. *The Tiger* 虎. Heaven smiles on those who build under these stars. Happiness, long life, and treasures await the man who digs a canal, who opens a new gateway. His children will be numerous, rich, and attain to titles of nobility.

Constellation 28. *The Pig* 猪. Lucky in all ways — for newly-married couples especially.

Constructions, repairs, burials, all augur prosperity and riches.

This list is interesting for two reasons. First of all, it proves that, to Chinese belief, "nothing under the stars is beyond the reach of their hyperphysical control." Every event in social and official life—such varied things as commerce, shipbuilding, silk-culture, cattle-rearing, fuel-gathering, digging, draining, building, laying foundation-stones, literary competition, marrying, burying, travelling—all are within the scope of their influence. Secondly it gives a picture of Chinese mentality, showing the blessings that the people value, and the calamities they fear.

* * *

But, picturesque as they are, the Constellations are not a practical basis for fixing dates. Imagine, when asked one's age, having to reply by mentioning eight characters, or four groups of two cyclical signs, in order to give the year, month, day, and hour. Despite popular rhymes to aid the memory, most people are unable to do this off-hand. The result is that simple folk answer the question: "When were you born?" with the vague phrase: "In the year of the Rat, or of the Ox," etc., leaving the inquirer with an elaborate calculation to make, because it is not clear which Rat Year is meant, since in every sixty year cycle there are five of them (five times twelve equals sixty).

Because of these difficulties, the Government and the people early came to date years not by their cyclical signs, but by the reign-year of the

18   THE MOON YEAR

Emperor who happened to be on the throne. Since the advent of democracy, official correspondence is now inscribed "such and such a year of the Republic," the key-date being its formal establishment on January 1st, 1912.

The new style year begins on the same day as ours. But the moon year, fixed, as we have already mentioned, to start on different dates under different dynasties, has ever since 104 B.C. been reckoned to fall on the day of the first new moon after the sun enters Aquarius,—which makes it come not earlier than our January 21st nor later than our February 19th.[5]

The Chinese moons, except the first and twelfth which have special names, have come to be designated by numbers: 2nd, 3rd, 4th, etc., with the addition of *yin li*, "national style," or *yang li* "foreign style," the latter referring to the Republican calendar. This system of numbers is for practical purposes only. Poetically speaking, each moon has several flowery titles, some of which appear in our chapter headings.

* * *

"Joints and Breaths" Besides the division into months, the Chinese year is apportioned into twenty-four two-weekly periods of fifteen days each. These are continued on from year to year, irrespective of the lunar intercalations, are known as the "joints and breaths of the year," and correspond to the days

---

[5] Our ancestors did not always begin the year in January. Sometimes they started it in September, sometimes on the first of March. A memory of the latter system lingers in the fact that the last months of our year—"September" to "December," mean "7th" to "10th," whereas now they are, actually, the ninth to the twelfth.

on which the sun enters the first and fifteenth degrees of each zodiacal sign. The former are known as *chung ch'i* 中氣 or "principal terms," the latter as *chieh ch'i* 節氣 or "divisional terms." Most moons include two of these terms, but some will have only one, and some three. An intercalary moon, for example, has only one *chieh ch'i*.

The first of these "joints" falls between February 4th and 6th, but it is only possible to give their approximate dates, according to our calendar, because of the discrepancy between the 360 degrees of the Zodiac and the 365¼ days of the true solar term, which makes them vary a day either earlier or later, according to our reckoning. Solar periods added into a lunar calendar as they are, these "joints and breaths" still serve as a very accurate sub-division of the seasons. The farmers implicitly rely on them for sowing and harvesting. Indeed, they coincide so remarkably with atmospheric and climatic changes that many sensitive persons, both foreign and Chinese, suffer from headaches and heightened blood-pressure at every *chieh ch'i*. Chinese physicians, both in north and south China, recognise this fact so well that they prescribe special tonics for invalids and old people at these seasons. Even foreign doctors are beginning to admit this phenomenon.

The names of the *chung ch'i* and *chieh ch'i* have reference to the obvious changes in nature at the time they come round. Thus, the first is known as *Li Ch'un* 立春 or "Beginning of Spring" (about our February 5th). "It is an extremely good omen," says the peasants' almanac, "if the

Li Ch'un be clear and bright, for then the farmers will find ploughing easy."

The second term is *Yü Shui* 雨水 or "Rain Water" (about February 19th). After the *Yü Shui* there should be no more snow, but rain-showers may be expected.

About March 6th comes the *Ching Chê* 驚蟄 or "Awakening of Insects," when creation stirs after the winter sleep. The first thunder of the year, due on this day, is supposed to wake the hibernating dragon.

The *Ch'un Fên* 春分 or "Division of Spring" (about March 21st) marks the Spring Equinox. For the health of the country-side, rain should fall.

The *Ch'ing Ming* 清明 or "Pure Brightness" (about April 5th) is the true Spring Festival. If south winds blow, the harvest will be abundant.

The *Ku Yü* 穀雨 or "Corn Rain" (about April 20th) is the right time to sow wheat.

The *Li Hsia* 立夏 or "Beginning of Summer" (about May 6th) means the beginning of the hot weather.

The *Hsiao Man* 小滿 or "Ripening Grain" (about May 21st) indicates that the winter wheat, sown the previous autumn, is filling out its ears and may be harvested within the next few weeks.

The *Mang Chung* 芒種 or "Corn in Ear" (about June 6th) marks the limit of the grain-growing season and reminds the farmer that it is the last chance for sowing. If the rain tarries, prayers are said to bring it quickly, lest the ground become too dry. A thunder-storm on this day is considered lucky.

The *Hsia Chih* 夏至 or "Arrival of Summer" (about June 21st) is actually the Summer Solstice and the longest day of the year. According to truck-farmers, this is the time when garlic ripens and should be taken out of the ground. There is even a popular proverb which says: "At *Hsia Chih*, dig up the garlic."

The *Hsiao Shu* 小暑 or "Small Heat" begins about July 7th, and the *Ta Shu* 大暑 or "Great Heat" about July 23rd. The latter term falls in the midst of the three *fu* 伏 or ten day periods when the heat is greatest. Nevertheless, the farmers do not complain, however sultry it may be, as they believe in the old adage: "If it is not exceedingly hot in the three *fu*, then the five kinds of grain will not be of good quality."

The *Li Ch'iu* 立秋 or "Beginning of Autumn" comes about August 8th, but the heat does not, in fact, greatly diminish until the *Ch'u Shu* 處暑 or "Limit of Heat" (August 23rd), when summer is really over and the harvesting of the millet in North China begins. According to the peasants' proverb: "Should rain happen to fall at *Ch'u Shu*, then it will be difficult to retain the fruits of the earth."

The real dry weather, however, comes only with the *Pai Lu* 白露 or "White Dew" about September 8th, when the winter wheat to be gathered next fourth moon is sown.

The *Ch'iu Fên* 秋分 or Autumn Equinox falls about September 23rd. "If at *Ch'iu Fên* white clouds are abundant, then everywhere rejoicing will be heard, because of the prospect of a good harvest for the late crops. If, however, there should be thunder and lightning, then it is feared that in the coming winter rice will rise in price."

The *Han Lu* 寒露 or "Cold Dew" (about October 8th) sees the first leaves falling from the trees, and the *Shuang Chiang* 霜降 or "Hoar Frost" descends about October 23rd and is likely to bring the first film of ice.

The *Li Tung* 立冬 (November 7th) marks the beginning of winter.

The *Hsiao Hsüeh* 小雪 or "Small Snow" comes about November 22nd, and the *Ta Hsüeh* 大雪 or "Big Snow" about December 6th.

Finally, the *Tung Chih* 冬至 (December 22nd) is the Winter Solstice, when the Emperor used to sacrifice at the Temple of Heaven.

The *Hsiao Han* 小寒 or "Small Cold" falls about January 5th, and the last of the twenty-four "joints and breaths" is the *Ta Han* 大寒 "Great Cold," about January 20th. After this, the weather grows slowly warmer until the *Li Ch'un*, or "Beginning of Spring," comes round again.

* * *

Though these fortnightly periods are almost as old as the calendar itself, the Chinese did not use the week as a time division until some notion of it crept in with foreign influence, and the days, from first to seventh, are beginning to be known by their numbers, "first day," "second day," etc.

The old style day in China began, as in Egypt and with us, at midnight,[6] and was divided into two-hour "watches," each of which bears the name

---

[6] The Greeks and the Chaldeans started theirs at sunrise.

of one of the "Twelve Earthly Branches" with its corresponding animal.

We may compare Chinese time and Western time by the following table of hours:

Western Time.	Chinese Time.
11 p.m.—1 a.m.	Hour of the Rat
1—3 a.m.	Hour of the Ox
3—5 a.m.	Hour of the Tiger
5—7 a.m.	Hour of the Hare
7—9 a.m.	Hour of the Dragon
9—11 a.m.	Hour of the Snake
11 a.m.—1 p.m.	Hour of the Horse
1—3 p.m.	Hour of the Sheep
3—5 p.m.	Hour of the Monkey
5—7 p.m.	Hour of the Fowl
7—9 p.m.	Hour of the Dog
9—11 p.m.	Hour of the Pig

Nowadays, though watches and clocks are becoming common substitutes for the old-time sticks —long spirals made of clay and saw-dust well mixed, then lighted and marked at intervals to show when a piece corresponding to sixty minutes has been burned—and the clepsydras of various forms anciently employed, though Gregorian dates head official correspondence, and the picturesque names of the hour-periods are going out of use, giving place to translations of our numeric terms, the common people still follow the moon year because of its simplicity and its sentimental associations.

* * *

Gods of the Calendar

Furthermore, the old Chinese calendar whose roots strike deep into primitive nature worship has a marked religious significance. The popular Pantheon includes twelve Presidents of the Calendar cycle in charge of the "Twelve Earthly Branches," presiding over months and the watches of night and day. Another set of spiritual divinities is in charge of the "Ten Heavenly Stems." They are supposed to be the brothers of the first rulers of humanity, the "Celestial Emperor" and the "Earthly Emperor." A late Taoist invention, they have made their way into some Buddhist temples and are generally supposed to take turns presiding over the years from their distant star-throne.[7] The *Wu Fang,* or Five Directions —the Four Cardinal Points and the Centre—that correspond to the seasons, have their Patrons, known as the Five Rulers (*see* "The Tenth Moon"). In the Chinese mind all things sprang from the original Chaos, later divided into two polarities, the *Yang* and the *Yin,* already mentioned. These polarities combined and generated "five elemental forms"—the Five Seasons, Five Directions, the Emperors of the Five Regions, each with his own element, his own colour, and his own realm in nature. This is the conception, briefly stated, of how the creative forces of the universe work—"the hypothetical basis of Chinese medical and moral science, decorative art and architecture, even of the very structure of the language."

This, too, is the reason why the Chinese are so superlatively direction-conscious. Numerical sym-

---

[7] They are supposed to live in the five stars in the constellation of *Leo.* The Taoists have another group of cyclic divinities known as the *Ting Chia* spirits.

bolism and the points of the compass are, practically, instinctive in the Chinese race. "It is a matter of early training and habit through long centuries and, perhaps now, of hereditary aptitude . . . . After a night in an inn one traveller says to another: 'Did you rest well?' and his companion replies: 'Not too well. I was cold on the west side. I found this morning that the south-west corner of my quilt was not properly tucked in.' . . . . A blind man navigating a street full of obstacles turns to the east when told to do so . . . . and well-diggers, working at the bottom of a pit thirty feet deep, when the foreman at the top shouts: 'What are you doing? Can't you see you're going crooked. You'll have to dig a foot further to the south!' all immediately face about to the direction suggested."

The twenty-eight constellations of the Chinese lunar zodiac, which correspond to the days of the month, also have their patron spirits, a group of heroes killed in the wars between the Shang and the Chou dynasties, and canonised later by Taoism.

Moreover, Taoist divinities are resident rulers of various other star groups, each with their own individualities, their own powers. Prominent among them are the Spirit of the Northern Measure whose position marks the seasons and whose Governor is supposed to control Death, while his colleague of the Southern Measure is supposed to control Life. Associated with the latter are those curious star gods of Longevity, Luck, and Honours, whose triple images are found everywhere under the name of *Fu Lu Shou San Hsing* 福祿壽三星 (*see* "The Eighth Moon").

Special honour is paid to the Spirit of the Polar Star identified with *T'ai Chi* 太極 or *T'ai Yi* 太乙, the great Limit or Absolute, the Lord of the Five Rulers, the centre of the Chinese world because it never changes its position with relation to the rest of the universe, and the "inclusive union and the *Yang* and *Yin* Principles." Humanly speaking, this popular star-god was, according to Taoist legend, the eldest brother of the founder of the Chou dynasty—a good man whose virtues brought on him the cruel vengeance of the last tyrant of the preceding line.[8]

Finally, included in the elaborate system of Celestial Government, the Chinese have a special Ministry of Time under *T'ai Sui* 太歲 , the "Grand Marshal," spirit of the Planet Jupiter, with the aid of no less than 120 assistant officials, including patrons of light and darkness, spirits of the roads, etc. T'ai Sui himself is often called the "God of the Year," and is supposed to change his dwelling every year. The location of his new home is set by the almanac, and the lucky line for the twelve-month depends upon it. This is a superstitious relic of the old belief that the happiness of the Empire depended upon the movements of Jupiter, probably because this important planet, which takes such an unusually long time to revolve around the sun—twelve years—was studied with religious awe as far back as

---

[8] For Sun myths and Moon myths *see*, respectively, "The Second Moon," and "The Eighth Moon."

1000 B.C. Although the worship of the Master of Jupiter began very early, his regular cult was not established until the Ming or even the last Manchu dynasty, when it assumed great importance and sacrifice was made to him regularly at the *Li Ch'un,* the day that marks the beginning of spring One of the principal buildings in the Temple of Agriculture in the capital contains his shrine before which the last Emperors of China made their prayers.

Those ceremonies have vanished with the rest of the official cult. But T'ai Sui still enjoys private worship, rather through fear than love. He is a dangerous spirit, spreading disease and misfortune, and must be placated before new undertakings.

Although these Chinese time-divinities never held as important a place in their religious vision as did those of the Chaldeans and Egyptians, builders of temple-observatories ("zikkurats") and pyramids, nevertheless one of the stellar gods was considered the centre of the Chinese world. This, as we have said, is the Polar Star connected with the Spirit of T'ai Yi, the Great Unity or Absolute, and sometimes even with the Supreme Being. Emblem also of the Emperor who, on this impermanent earth, stood for permanency, the "Purple Planet" was believed to remain stable among the shifting constellations of the Heavens.

Yet not Jupiter, or T'ai Yi, nor any of the other stars that check her courses and mark the days of the year that still bears her name, receive the reverence accorded in China to the Moon. Deep sentiment for her, "The Shining Mirror," has existed since the earliest times. Partly from religious motives, and partly because of the deep conservatism which is the foundation of the Far-Oriental character, the Queen of Night remains the inspiration of the Chinese calendar. Where compromise was necessary to correct her, the Chinese compromised, admitting the principle of the two zodiacs. Thus, while the sun was allowed to regulate the seasons at the solstices, the moon retained this privilege regarding the equinoxes.

The result was a "book of the year" which, despite its apparent intricacies, has proved excellent for all practical purposes. In some respects it is more pliable than our own and, notwithstanding the acknowledged superiority of Western mathematics, better records the alternations of the seasons and their fractions. Indeed Painlevé, the well known statesman and scientist, asserts that, in his opinion, an adjustment between Chinese and Western methods of time-computation may give birth to the perfect calendar of the future.

CHAPTER II.

"THE HUNDRED GODS."

ONTINUITY is the most striking characteristic of Chinese life and thought. Nevertheless, the flowing stream of their civilisation— a civilisation morally antipodal and intellectually remote from our own, though not on that account to be judged inferior—has been interrupted in its course by the rocks and rapids of warfare, moral and material; retarded by the quiet waters of stagnation; deflected by whirlpools of religious fervour, and disturbed by currents of atheism and materialism. Even foreign conquerors and barbarian neighbours influenced a culture which always ended by absorbing them.

The dawn of history shows the Chinese as a group of clans living under a patriarchal system. They began to evolve from an earlier stage of hunting and fishing communities, ignorant even of the use of metals, when in the fourth or third millenium B.C. several mythical benefactors appeared, descendants of the first "Rulers of Heaven, Earth, and Men," each of whom is supposed to have reigned thousands of years. These benefactors include the First Perfect Emperors who taught the "black-haired people" how to "build nests," marry, kindle fire, till the soil,—all the rudiments of culture: Fu Hsi,

Shên Nung, Huang Ti, Shao Hao, Chuan Hsü,[1] later Yao and Shun, and the Great Yü, founder of the Hsia Dynasty (2205 B.C.).

Migration followed these crude beginnings of civilisation, and early colonists, whose race-cradle was the upper basin of the Yellow River, slowly fought their way to the Ocean. In the legendary age long before Christ—an age we may call the first period of Chinese social evolution—two pillars of thought, like the Pillars of Smoke and Flame that led the Israelites, guided the Chinese out of their wilderness, two parallel beliefs whose beginnings are untraceable,—the Ancestral Cult and Nature Worship.

Origins of Ancestor Worship

Herbert Spencer states positively that "Ancestor Worship is the root of all religions, and was probably coeval with the first belief in ghosts"; that "as soon as men were able to conceive the shadowy idea of an inner self or 'double,' so soon the propitiatory cult of spirits began." This second self took shape in the mind of the gross-feeding savage, alternately half-starved and gorged after a successful hunt, through vivid dreams of psychic phenomena,—dreams forcefully gripping his imagination, yet beyond his waking powers to explain.

Now the Chinese evolved the idea of a soul, or rather of several souls, for every living being, very early in their development, and a world of peculiar fancies grew out of this belief. At least

---

[1] These first Perfect Emperors are sometimes identified with the Five Rulers presiding over the Centre and the Four Cardinal Directions, although they are, in fact, of a different origin (*See* "The Chinese Calendar" and "The Tenth Moon").

one of every person's souls was supposed to have the power of leaving and returning to its body during sleep or unconsciousness. Hence the custom, still persisting, of "calling back the soul" when anyone faints or is delirious. Many a mother, watching over a sick child will try to do this. Standing in the doorway, she commands the baby spirit to return while the father, carrying the little one's clothes over his arm, replies from the threshing floor, playground of the village children: "I am coming! I am coming!" Both cry and response are unbelievably ghostly and pitiful.

While Western and Christian teaching maintains that the souls of the dead depart to another and a better world, the Chinese have always supposed them to hover about this earth. Through some mystical connection with the material body whose form they still retain, their ghosts keep in touch with mortal surroundings. Nor are human desires outgrown. Spirits still feel hunger, cold, and the need of affection without, alas! the power to obtain them.

Helpless, they depend on their descendants for help—for food-offerings and loving service. They do not ask much—scarcely more than to be thanked as founders and guardians of their homes. The greetings of their kindred are enough to make them happy and, though they require nourishment, the vapour of food suffices for them. But these simple rites they do exact, and to neglect the dead is not only cruel but dangerous for then, instead of watching over the welfare of the household and expressing their gratitude in blessings and prosperity, they become "hungry ghosts" with terrible powers of revenge. The Ancestral Cult, now developed into a ritual of dutiful reverence and affection, began in fear, and the "wish to please the ghosts of the departed was chiefly inspired, originally, by dread of their anger."

* * *

Ghosts as Nature Gods

The Chinese soon identified the earth-bound spirits of their dead with the rulers of natural forces. Their voices spoke in the thunder. They were enthroned upon the mountain peaks. They looked down from the stars. They pulled the tides to and fro. Their spirituality was expressed in the trees and the rain. Growth and decay, flood and drought, in fact "everything desirable and dreadful," was under their supervision.

Raised then to control the elements—in a word, deified as Nature Gods—it followed naturally that the spirits of powerful human personalities were supposed, having been leaders in this life, to take charge of the more important divisions of the universe. Thus the Primeval Ancestor Shang Ti—the "Original Man"—was made Supreme Lord, Ruler of Heaven.

Shang Ti

The powers and personality of Shang Ti are difficult to define since, in addition to being the mightiest spirit and the Royal Ancestor, he is also identified with *T'ien,* the visible Celestial Dome. Yet we have no real right to call him "God," because he is not thought of as the "Creator" or connected with the making of the world, which later Taoist legend declares was chiselled by P'an Ku out of the Original Chaos, from which he was himself born.

P'an Ku was pure myth, but Shang Ti appears to have been a mortal deified at some very remote epoch, later promoted to a tribal divinity and, finally, to the Supreme Overlordship. Confucius defines him as the "Governing Force of Nature and Supreme Arbiter of Human destinies." This dominating position he has held throughout the ages, and it is in his honour that the sacrifices at the Altar of Heaven—the most important of all Chinese sacrifices—were made. They plainly embodied the double tradition associated with Shang Ti. The Emperor alone could make them, since he alone was the proper person to worship the First Ancestor of the Chinese Race whose tablet was enthroned together with, yet taking precedence of, the Sovereign's own immediate ancestors.

In addition to the cult of Shang Ti, the oldest and most solemn of all, there are references in ancient Chinese texts to worship given by the Head of the State and his vassals to the "Hundred Ghosts" or "Hundred Gods," a collective term meaning all the ghosts or deified spirits, just as the expression "Hundred Families" is still used to denote the whole Chinese race. While, according to the oldest tradition, each soul should be ministered to by its direct male descendants, this was no longer possible when men travelled far from their native place and could not return, perhaps for many years, to their family graves. Provision also was necessary for offerings to spirits whose descendants had become extinct. Furthermore, as the gradual expansion of the Chinese race broadened the narrow bonds of kinship, made the community,—even if its members were gathered from different localities,—more important than the clan, and transferred allegiance from local hereditary chiefs to the Ruler of the State, it was admitted that those Ancestral Spirits whose noble deeds, in pioneer days, deserved general gratitude should receive posthumous reverence from the community in general, and in some public place of worship rather than in one private family shrine. These ancient traditions still survive in the masses said on periodical "all souls' days," and the offerings then made to "hungry ghosts." Both are part of a national ancestral cult in which all spirits share and especially such as, deserving well of the country, have special sacrifices and special shrines erected to their memory.

In very ancient days a few of the hundred ghosts were chosen to rule the mountains and the rivers particularly beloved by the people, and they received homage from the Head of the State till the last days of the Empire. Shrines to the Spirits of the Sacred Peaks, that still remain centres of pious pilgrimage, exist in Peking both within the Forbidden City and at the Temple of Agriculture. Likewise to the Spirits of the Four Rivers. This latter cult is infinitely old, indeed goes back to the days when the Chinese were fishermen as well as farmers (the character for "food" is derived from the character for "fish") and depended for their livelihood upon the rivers and the rain. Speaking in a broad sense, the antique River Spirits, created to be placated in times of flood, are responsible for one of the most popular festivals in China—the Dragon Boat Festival; for the choice of the Dragon as the emblem of Empire; for the incense still burned beside springs and wells and, finally, for the numerous sub-divinities ruling oceans and seas, lakes, streams and marshes —divinities later grouped into a "Ministry of Waters."

* * *

Another essential and early development of the Chinese world-conception was the dual notion of the *Yang* and the *Yin*, Masculine and Feminine Principles ruling the universe. They stand for the "yea" and the "nay," the active and the passive, the positive and the negative, the fructifier and the fructified. To one or the other all things belong, man and woman, the sun and moon, light and darkness, etc. Even among animals some are "masculine"—*Yang*, others "feminine" —*Yin* (*see* "The Chinese Calendar"). *Yang and Yin*

The Supreme Ruler and Lord of Heaven, Shang Ti, was naturally associated with the active *Yang* principle. For his Consort and spiritual complement it was inevitable that a race of agriculturists like the Chinese should invent a Mother-Goddess of Earth, personification of Feminine Fecundity. In fact, Hou T'u, the Earth Queen, appears in the earliest written records as one of the first deified spirits. Like the Father God of Heaven, Hou T'u was worshipped by the Emperor as vice-regent of the Supreme Powers and, therefore, the "only orthodox channel of spiritual communication with them." Later this conception of Sovereign Earth was sub-divided into minor spirits whose lesser dignity permitted the direct prayers of the people. Peasant-altars were with propriety raised to a number of Gods of Fruitfulness included in the pantheon—Gods of the Soil, of the Crops, even to a spirit of the Roads, now nearly forgotten, but doubtless of great importance when the Chinese settlers were expanding into unknown regions.

All these divinities were known and honoured before the VIIth century B.C. In the general course of religious evolution they became associated with gods having jurisdiction over special localities, such as the *Ch'eng Huangs,* or City Guardians, and the *T'u Tis,* or Lords of the Spot (*genii loci*). Next in antiquity are the "Household Gods" (*see* "The Twelfth Moon"), a changing spirit-group which generally included the Guardians of Gates and Doors, of the Central Hall, of the Well, of the Kitchen, the Mao Ku Ku, Patroness of Women, and the God and Goddess of the Bed Chamber.[2]

* * *

When the hundred ghosts evolved officially into gods with spiritual authority derived from his forefather the "First Ancestor," Shang Ti, and the Ruler assumed the position of High Pontiff as well as Head of the State, we reach the period of the true birth of the Chinese nation which was followed by the awakening of Chinese philosophical thought.

A gradual political evolution from the days when the Chinese were simply a group of clans

---

[2] *See* "The Twelfth Moon."

engaged in agriculture developed Elders who assisted their chieftains to rule, themselves later became headmen, led tribes far afield, and warred on barbarian neighbours. Conquered or compromised with, the latter infused new blood into the strangers, supplying them with wives, servants, subjects, and allies, and modifying imported customs and beliefs by the toleration they demanded for their own.

Early Dynasties At the details of this evolutionary stage, which must have been long and difficult, we can only guess, since no records are left, except a few stones, bronzes, and bones with puzzling inscriptions from the Hsia and the Shang (Yin), the first two Chinese dynasties (2205-1122 B.C.). Legend, so persistent as to be almost truth, stamps the Chou Wang, last sovereign of the Shangs, as an arch-villain albeit a man of remarkable ability. A mental and physical giant, a cruel tyrant and oppressor, he was finally destroyed by one of his vassals, with the help of the gods, after a bitter struggle. Yet, later, we have the curious anomaly of the canonisation of his adherents by the new dynasty whose enemies they had been, while the Chinese Nero was himself deified as "God of Joy" presiding over marriage and enthroned on the planet Venus (*see* "The Eighth Moon").

Chou Period Such peculiar and conflicting evidence seems to confirm the theory that the advent of the first strictly historical Chinese dynasty (the Chou Dynasty 1122-255 B.C.) was a compromise between a waning central power and a growing emancipation of clan chieftains from their overlord. Indeed, the Chou period when most of the origins and principles of Chinese civilisation became crystallised, and most of the laws, customs and institutions still in use to-day originated, is known as the Feudal Age of China. Split up into a series of warring kingdoms, constantly fighting one another or making aggressive campaigns against barbarian neighbours, China at that time was little more than an armed camp.

During this period of military domination when the Chous maintained but a shadowy power over the princes, and the princes often had scarcely more control over their armed vassals, the civilian population suffered terribly. War killed trade and ruined agriculture. Moreover, the burden of military equipment rested entirely on the farmers who were expected to furnish horses, chariots, foot soldiers and camp-followers. No wonder that a wave of despondency swept over the land—a despondency expressed in contemporary poetry and song. Greed and violence overrode ideals, and one by one the virtues that men live by went out like stars in a stormy sky.

"Age of awakening" Lao Tzŭ

But, as usual in history, the darkest hour was just before the dawn. Saviours of civilisation were to appear in a group of remarkable men such as the human race produces in its hours of greatest need. The first was Lao Tzŭ, called "the gray-haired child," who stressed the ancient Doctrine of the *Tao*, or Way. "Follow the True Path," he preached. "Remember, happiness is inaction, virtue is inaction, God is inaction. Then cast aside all that you have built up, all that you have accumulated, prejudices, conventions, knowledge, affections. Return to Nature. Merge into the Universe. Become one with the Infinite, and you shall be without limitation."

Another Chinese philosopher who approaches the purity of Lao Tzŭ's doctrine, and whose sayings are sometimes almost identifical with the words of Christ, is Mo Tzŭ. His motto was love and unselfishness, his teaching—non-resistance, much like that of Leo Tolstoi. At the opposite pole of human thought stand the Materialists headed by Yang Chu. Denying a better world after death, they advocated enjoyment of the good things to be found in this life.

There is no better proof of the depths of distress into which the Chinese had sunk mentally and morally than the variety of extreme philosophical remedies presented to them in this age. Between the Scylla of Lao Tzŭ's "Nihilism" and the Charybdis of Yang Chu's Epicureanism, they steered a difficult course until Confucius appeared and, like a shining beacon, guided them to a safe harbour.

* * *

Confucius, greatest of Chinese Sages, thoroughly realised and suffered deeply from the misfortunes of his age. But, instead of preaching to his countrymen about God and a future life, he said plainly: "This existence is fact, reality. It behooves men therefore to so live that right shall prevail and harmony exist."

Confucius

"Do unto others as you would they should do unto you," was the foundation of his doctrine. Thus the solace he sought was within everyone's reach, and the rules he laid down could be followed by all. The "ordinary man," knowing no better, must, however, abide by hard and fast regulations of conduct, while the "superior man," or the elect, finds sufficient guide in abstract moral principles, unquestioned and unquestionable.

Where most reformers formulate new laws, Confucius, avoiding innovations as he abominated extremes, turned to the past for inspiration and example. The heirlooms of antiquity were the corner-stones on which he built his teaching and his ideal of future perfection. Yet, while he loved old things and believed them good, he used discrimination in his choice of what to keep. To him and to his disciples we owe the preservation of what we may call China's Bible—her sacred records—a Herculean task whereby he earned the gratitude of his countrymen and of the world.

With unerring judgment Confucius discarded transient beliefs and frivolous or changing ideas, ripples on the surface of civilisation. Only the best thought and the best work of past ages was worthy to serve as a foundation for a new code. To control human action he went straight to the heart of race experience, intellectual as well as emotional; he turned straight back to the Ancestral Cult, mainspring of all virtues. But whereas, before his time, reverence for the dead was based on fear or the hope of favour, he made it an "imperative ethical principle."

While Lao Tzŭ and his followers stoically preached Nature's spontaneous inspiration as the supreme rule of life, and recommended mankind to try and achieve happiness through *Wu Wei*, or "Immobility," Confucius counselled action rather than inaction. Not emptiness, but fulness was his ideal,—fulness or completeness to be attained not through meditation or negation, but through the practice of the virtues of the "Happy Mean,"—benevolence, staunchness, uprightness, self-control and, above all, the principle of self-improvement and study leading to active achievement.

The example of the great deeds done by their great dead should be, according to Confucius, both a spur to further righteous impulse, and a debt to be scrupulously repaid. A man unable or unwilling to acknowledge and meet this obligation will never be able to regulate his own life. Without a sense of duty towards the Unseen, he will be like a ship that has no rudder. In human relationships he will be a failure, for the virtues that lie at the root of the ancestral cult are the theme of a mighty moral symphony enfolding the harmonious relations between parents and children, husbands and wives, sovereigns and subjects.

In an age of adjustment and expansion, the greatest characteristic of the work of Confucius was the suitability of his doctrine to those for whom it was intended and to whom it brought help when help was badly needed. It may be called utilitarian in the sense of taking the weal of man as a supreme goal. But this does not overshadow its constructive merits. It tightened the links of the family and of the community, discouraging individual selfish adventure. It preached that virtue and co-operation alone could save national life and unity from the dangers of a warring feudal age. By advocating loyalty and truth, and decrying all violence and disorder, Confucius saved China from the disintegrating fate of so many other Empires.

*   *   *

The "First Emperor"

The great moral revolution brought about by these teachings was in the IIIrd century B.C. followed by an equally violent political reaction. The movement towards a consolidation of all the hundreds of petty states scattered over the land known as the Middle Kingdom came to a climax when the mighty figure of Ch'in Shih Huang appeared. Born in a far northern principality, he was probably related to the barbarians whom his ancestors conquered. Nevertheless, he was a genius who ought to have earned the gratitude of posterity by three great innovations. He unified the country, suppressing the local war-lords and, under the title of "First Emperor," made himself Supreme Master. He established the principle, observed for two thousand years, that no official should hold office in his own province, and introduced a new and simplified code of writing without which Chinese civilisation could never have spread as it did after his day. But when the conservatives, retrenched behind the wisdom of Confucius, showed violent opposition to his reforms, Ch'in Shih Huang replied by burning most of the old classical books and burying some of the scholars alive. Fortunately, the loss to literature was repaired, from memory and stray copies of manuscripts hidden from the holocaust. The First Emperor, who determined to start history and culture over again in his own person, deliberately replaced the old ideals of scholarship with the study of the antique Nature Cult, also alchemy, and diverse kindred superstitious practices.

The opposition to scholars in general, and Confucius in particular, died out when Ch'in Shih Huang disappeared. Under the succeeding dynasty of the Hans (206 B.C.—A.D. 220), Chinese rulers whose brilliant age produced great generals, statesmen, and scholars, Confucianism was reinstated as the philosophy of the *literati*. This was, in fact, the period when the early and hazy Chinese philosophical beliefs came to be expressed in practically their present form. *Han Dynasty*

Let us glance back for a moment at the Chinese world-conception inherited from antiquity. In a general way, the whole universe was considered one animate undivided entity born from the Original Absolute, *T'ai Yi*, or Chaos, who engendered the Two Principles of the *Yang* and *Yin*, the Positive and Negative Elements. From these all Creation began. Their meeting or their separation, their harmonious co-operation or their balanced opposition form the *Tao*, or Way in a threefold sense, the Way of Heaven, the Way of Earth, and the Way of Mankind.

The idea of the *Tao*, with its suggestion of a guiding force, was infinitely older than Lao Tzŭ who preached it, or Confucius who adapted it,— was, indeed, much older than the "age of awakening" when most of the great philosophers appeared. Nevertheless, in one form or another, it was accepted by all of them as the fundamental creed of China.

The Confucian interpretation appealed most strongly to the Chinese people. Therefore it carried the day and became a rule of both official and private morality, seldom contested down to the last days of the Empire. With a vast vision, with a high ideal, theoretically perfect if not always practically attainable, Confucius presented the Imperial Government as the highest manifestation possible of the "Way of Mankind"—a machine destined to guide humanity along the right Road (*Tao*) because of its harmony with the Rule—or Road—of the Universe. In other words, it was part of the unified rule of Creation.

The principles of this high form of administration were laid down in the Confucian holy writings: the Five Classics and the Four Books with their moral precepts. The Perfect Example was the Supreme Ruler of the Chinese Race, heir of Shang Ti, Lord of Heaven. Directly from Heaven came his mandate to rule, but he might be deposed if his vices automatically deprived him of his right to act as intercessor between Heaven and his people, or his misconduct brought calamity upon his country.

Neo-Confucian School

Since the Chinese long imagined themselves the central pivot of civilisation, a point of view natural enough to a race surrounded for thousands of years by cultural inferiors, the sound yet flattering conception of Confucius took deep root. Indeed, his ideas received a new impetus under the Sung dynasty about A.D. 1000, when the Neo-Confucian School of Philosophers commented amply upon the Canon of the Master. It was then that not only the *literati*, long devoted to his teaching, but the people in general, in a renewed reverence for antiquity and his inculcation of the "Five Constant Virtues," in admiration of his principles of perpetual and universal morality, and in gratitude for his supreme judgment in giving his countrymen models of wisdom and virtue, took him to their hearts.

Even alien dynasties when they occupied the Chinese throne adopted his philosophy. The Manchus, for example, commented both officially and privately on his life and doctrine—commented so voluminously that they covered the whole range of Chinese thought. Their elaborate ceremonies were the supreme expression of his cult (*see* "The Imperial Ceremonies"), and their religion, though a combination of several beliefs, gave definite precedence to his teaching.

Lately it has become the fashion for an advanced and radical group of thinkers to deny the Master. Some Republican scholars, harking back to old currents of opposition, criticize his theories. On the other hand, Confucian societies are being formed to keep ancient ideals alive. He may indeed be criticized, but the Master is not yet dethroned. So deeply is his dogma rooted in the popular heart that it seems unlikely that the Sage of All Ages ever will be banished or replaced. In any event, this can not happen till the Sons of Han change beyond recognition.

* * *

Taoism

Having traced briefly the connection of Confucius with the primitive Spirit and Nature cult, and his interpretation of the *Tao* into not only a

personal but a State religion, let us see how the
first Taoist thinkers developed the same guiding
principle of the Way. Their thought-evolution
took a very different form. Instead of seeking
perfection through an attempt to infuse harmony
into practical life, the followers and disciples of
Lao Tzŭ sought freedom from the material. Their
ideal was to escape the body and its restraints, so
as to become free and all-powerful spirits. Early
in their career, we hear of men becoming *Tao
Shih*, "Wise Men of the Way," or hermits seeking
salvation in solitary retreats. Taoist treatises
recommend different methods of attaining sanctity
or immortality. Some bid men fast, others recommend deep breathing, others suggest elixirs,
talismans, and secret formulæ. The early saints
appear to have tried them all with varying
success. Many became healers, either pioneers in
the field of material medicine,—an art often linked
with the oldest roots of religion,—or spiritual
guides owing to their control over supernatural
beings. As soon as pupils and patients gathered
around these saintly physicians of the body and
the soul, we have a nucleus for the establishment
of a Church, such as was required for the spread
of Taoist doctrines.

When the Hans were on the throne this church
had already developed into a powerful organisation, with a regular hierarchy of priests headed
by the Chang T'ien Shih (*see* "The Fifth Moon")
and its own divinities, holy books, and rites.
Thus, from the beginning Taoism and Confucianism diverged. The former creed logically demanded churchmen, whereas the latter needed no

priestly intermediaries except the Sovereign as High Pontiff and his officials who became his delegates and, from the highest Grand Councillor to the humblest District Magistrate, his natural vicars in matters of religion as well as of administration. When ethics were not different from religion, and religion was not different from Government, official ceremonies, preceded by prayer and sacrifice, became a religion in themselves.

The Taoists, on the contrary, aimed to develop a popular and not a State church. While accepting many divinities of the original Nature cult, they re-christened them and arranged them in an orderly pantheon. Inventing a triple Paradise, they placed in charge a trinity of deities, collectively known as the *San Ch'ing,* or "Three Pure Ones," and gave precedence to the Original Creator P'an Ku. An important though later addition is the "Jade Emperor" (*see* "The First Moon"), while Lao Tzŭ himself figures in this august company as Lord of the Third Heaven, or "Guardian of the Golden Tower." The *Yang* and *Yin* principles are represented by the Tung Wang Kung and the Hsi Wang Mu (*see* "The Third Moon"), born from P'an Ku and his consort after the former had divided Chaos. The *Yang* and the *Yin* gave birth to the Heavenly Emperors, succeeded by the Earthly Emperors, ancestors of the first Chinese rulers, beginning with Fu Hsi.

In addition to these dominating figures, the Taoists developed an enormous number of lesser divinities. The old, old Chinese belief in a world populated by innumerable ghosts gave them an excuse to deify as freely as they pleased. The whim of an ascetic, the fantasy of an Emperor, the religious fervour of a Taoist pope, led to the canonisation of some favourite saint or distinguished Immortal. Local spirits were accepted as divinities to please local converts. Men of meritorious life were included in the Pantheon as examples to the community. Trees were believed haunted and holy. Even animals, rocks, and stones might be adored on one excuse or another. Whatever or whoever had power and was, therefore, supposedly, under the special control of invisible forces, was considered worthy of identification and worship. This theory admitted men like Confucius to a place on the Taoist Olympus, and even gods and saints from other rival religions, like Buddhism. Thus the Taoist Pantheon grew and grew till few men could remember the names of all the greater gods, not to speak of the lesser, and no mortal could find time to address every divinity in his prayers.

The priesthood, however, simplified the duties of the faithful by acting as mediators and exorcisers. They also, perhaps in order to make identification easier for their followers, gave human form to nearly all their gods. This even applied to deified tree and animal spirits. They were given parents and birthdays, wives and children, and elaborate biographies, adventures, and tactful descriptions of the methods by which they learned how to control the universe and attain godship.

Their images of bronze, wood, clay, or paper, while merely symbols of divinity to the elect, are

TAOIST ABBOT.

BUDDHIST PRIEST IN CEREMONIAL ROBES.

to the masses, whom the priests encourage in their faith, actual manifestations of the divine after an "induction ceremony." Idols leave the shop of an idol-maker complete to all appearance. But not until the priest has placed some tiny living creature, such as a baby-bird or a spider, inside the hole with the silver or cloth replicas of the vital organs, then seals this aperture and removes the paper scabs purposely left over the eyes, does the figure become "alive." The "size of an image or its material have little bearing on its effectiveness . . . . nor even the material of its vital organs strung together in a series and suspended from a hook. . . . The one and only thing that counts is whether or not the priest has introduced the spirit of the god into its body." Instances are on record of an image losing power or gaining it. In the latter case, when some battered figure develops magical faculties, a new shrine may be erected to propitiate the new spirit.

* * *

The beginnings of the Christian era, which saw Confucianism and Taoism develop into mighty doctrines and spread over the Land of Sinim, also saw the introduction of Buddhism. Strangely enough, Taoism paved the way for this new religion—Taoism with its theory of *Wu Wei*, or "Nothingness," subtly akin to Nirvana and its renunciation of worldly honours. Superficially, there was some similarity of doctrine and even of ideals. The principle of monastic life was common to both religions, and both recognised the beauty of sacrifice. But whereas Taoism sought

Buddhism

the salvation of the individual, whether in bodily or spiritual form, Buddhism preached a wider gospel of tenderness. In its highest essence, it was a faith founded on love and charity, practised not for any ulterior motive of salvation, but for their own sake. Besides teaching new respect for life, the duty of kindness to animals as well as to all fellow beings, it brought to the Chinese the notion of a world beyond this earth and a conception of sin. Sin the Buddha showed not as a crime against a particular individual, but against one's own soul, and capable of retarding the soul in its development, since all "states and conditions of being, all progress and all backsliding, are the sum of past actions. The sorrows of men were thus explained as evils of their own former acts in former re-births returned upon them. Life was expounded as one stage of a measureless journey whose way stretched back through all the night of the past, and forward through all the mystery of the future—out of eternities forgotten into eternities to be."

Yet, at the end of this long road lay the reward of a perfection beyond anything our gross earth-bound conceptions can grasp. For those whose understanding was more developed, this state of bliss was called Nirvana, a condition rather than a place. But for the simple people, it was called Paradise, the Paradise of Amida Buddha, Lord of Immeasurable Light (*see* "The Eleventh Moon"). On the subject of future punishment, the teaching was not less explicit. Hell awaited the wicked as surely as Paradise the good. But "hell was the penalty for supreme wickedness only, it was not eternal, and the demons themselves would at last be saved."

The human appeal of such a doctrine is obvious. This, together with the superior organisation of the new religion, enabled it not only to hold its own against the Taoist Church but, in a measure, to supplant it. At the present day, Buddhist monasteries have larger communities of monks, and Buddhist shrines are often more popular than Taoist temples.

* * *

The "Three Religions"

The rivalry between the two creeds led to many bitter struggles. Both among Chinese and foreign dynasties who occupied the Dragon Throne, we find monarchs who staunchly supported now one religion, now the other. Some Emperors went so far as to sell themselves as slaves to a certain monastery, on the understanding that the Court should pay a huge sum for their ransom. Others again destroyed the temples and disbanded the monks. Now and then ardent Confucianists took a hand at persecution when they thought either Taoism or Buddhism menaced the supremacy of their Master's doctrine. True, the latter did not concern itself with transcendental beliefs in China nor interfere with any man's private faith. But this breadth of view demands one condition for its calm indifference, and that is that no new creed shall tamper with the fundamental principles of Confucian ethics and the order of things based upon it. Hence the official decree that both Taoist and Buddhist monks must observe the ancestral cult, though this order strikes at the very root of monachism. How again can one reconcile the notion of Hell and Paradise with the idea of reward and punishment depending only on the Emperor, as God's Vice-regent on earth, who owned "everything under Heaven" and had power even to promote or degrade the gods? Surely, any attempt to diminish these prerogatives of the Son of Heaven was to interfere with the harmonious course of the Universe.

As a matter of fact, conservatism was seldom pushed to this extreme limit. Regulations which appeared during the long religious controversy were rarely enforced to the letter, and, indeed, seem chiefly intended to restrict the growth of monasticism and check the spread of unproductive and idle communities. They also established a principle of control over both churches, in case such control ever needed tightening up for the prevention of doctrines subversive to the Government under either religious slogan.

But, as far as the bulk of the people was concerned, the gods of all three sects seem to their simple minds just so many powerful spirits to be placated. Whether officially introduced by an Emperor, sponsored by the *Tao Shih*, or imported by the Buddhists, prayers and petitions to any and all of them may be useful and are, therefore, cheerfully given. One result of this popular eagerness to pray to an ever-increasing pantheon was a curious borrowing of gods from one sect by another. This explains why we find Confucius in a trinity with Lao Tzŭ and Buddha, and many saints peacefully invading the altars of their rivals.

It accounts also for the confused jumble of personalities found among "the Hundred Gods" (*see* "The First Moon," and each individual deity on the date of his particular festival). Whatever Buddha might think of the Ma Wang, God of Horses, their effigies appear side by side on the popular *chih ma* (luck-poster) used on the New Year altar. If the Niang Niangs are jealous of Kuan Yin, doctrinally speaking, they all nevertheless answer the prayers of anxious mothers. With spirit-essence everywhere, with ghosts made gods, and gods made ghosts, the people of China, in an attempt to discharge their duty to the immense concourse of the invisible, will invite priests of all their leading sects to the burial ceremonies of their dead, Taoist exorcists to cast out devils from their homes, and Buddhist monks to say masses for their souls. Are not all these righteous folk —and the teachings of Confucius also—but symbols of the *Tao*, sign-posts which attempt to point the Way—the Way of Truth that rules the Universe.[3]

---

[3] An interesting, important, and very ancient product of the Chinese nature cult is an elaborate geomantic science, popularly known as *Fêng Shui*, or "influence of Wind and Water," and other natural forces, on the well-being of mankind. Adepts of this science determine the position of towns and palaces, houses and graves, according to lucky or unlucky *Fêng Shui*, adding apparently useless towers and walls "to ward off evil influences," etc., etc. (*See* de Groot, *The Religious System of China*).

CHAPTER III.

THE IMPERIAL CEREMONIES.

HE official cult in China, an expression of the most antique beliefs of the race, consisted of sacrifices of three classes: Great, Medium, and Small, offered by the Emperor as intercessor for his subjects.

Though these sacrifices are no longer performed, they are of interest as the model on which the people themselves have copied their simple home ceremonies.

* * *

Great Sacrifices: (a) to Heaven The Great Sacrifices included the worship of Heaven, Earth, the Imperial Ancestors, the Gods of Land and Grain and (by special edict of the Empress Dowager Tz'ŭ Hsi) Confucius.

As the sacrifices to Heaven comprised the highest and most complete ritual, we shall describe these only.

A special Ministry, known as the Board of Rites (*Li Pu*), was in charge of the sacrifices, because any slight deviation from century-old precedent would impair their effect. The programme was written out and submitted to the Emperor several days beforehand. The Altar of

THE IMPERIAL CEREMONIES 55

Heaven and the temples adjoining it were prepared under the supervision of certain high officials, others were in charge of the whole bullocks, jade, silks, and food-offerings cooked in the "Kitchen of the Gods." The latter included beef, pork, mutton, venison and hare, fish, rice, sorghum and other grains, salt, jujubes, chestnuts, water-chestnuts, beetroots, celery, bamboo-shoots and different kinds of cakes. Each course was presented in dishes of strictly prescribed form made of blue china or bamboo. A rehearsal of the ceremony took place in the Board of Rites, whereupon the sacrificial jades and silks were displayed in the Palace for the inspection of the Emperor who sent them off with nine *k'o t'ous*. The text of the prayer to be read by him was also submitted. A new prayer was drafted for each great and medium sacrifice every year, and inscribed on a wooden tablet covered with paper of different colours: blue for Heaven, yellow for Earth, red for the Sun, white for other divinities. This prayer began with the date of the sacrifice, which was followed by the words: *"The reigning Son of Heaven, subject"* (followed the Emperor's own name, which was taboo even to his nearest relatives), when the monarch was addressing Heaven or Earth. In sacrifices to the Sun and to the Moon, the word *"subject"* was left out. When praying to his ancestors, the sovereign called himself *"respectful grandson"*; on all other occasions simply *"Emperor."* In the prayers to Heaven, Earth, to the Imperial Ancestors, and to the Spirits of Land and Grain (*Shê* and *Chi*) the text of the actual invocation began with the words: *"ventures to lift up the following prayer"*; for the Sun and Moon this term was replaced by the words: *"earnestly prays,"*—in other sacrifices simply by the sentence: *"offers sacrifice."* This gradation shows the distinctions the Chinese made in the hierarchy of their chief divinities. The Emperor signed the written text of the prayer with his own hand.

A three days' fast was obligatory before the sacrifice for the Emperor and all those who took part in it, princes and officials only, be it well understood. No foreigner has ever seen this grandest of Chinese ceremonies, and none of the native population were admitted. Persons in mourning and those who had any wound on their body were excluded from participating in the sacrifice. The rules of the preparatory fast (from which only men above sixty years of age were exempted) required that no wine, garlic or onions, should be touched. Neither was it permissible to visit sick people, to sweep cemeteries or attend funerals, to worship other spirits, to listen to music, to attend to lawsuits,—in short to do anything capable of distracting one's attention from the great solemnity. The fast was observed by the Sovereign for the first two days within the Palace, on the third, in a special "Hall of Abstinence," of which one was attached to all the greater altars. Both in the Palace and in the Hall of Abstinence (*Chai Kung*) a bronze Statue of Silence was put up, representing a man with one finger of his left hand on his lips, while in his right hand was placed a tablet inscribed with the rules of the fast.

The most solemn signs of reverence were shown

the Son of Heaven as he proceeded with a display of great magnificence along the "Road of the Gods" to the Altar of Shang Ti. He went out through the monumental porch of the "Meridian Gate," to the sound of the gong and bell hanging in its towers, between long rows of kneeling dignitaries. The streets of the city were cleared of all passers-by. Meanwhile, doors and windows were closed, and blue curtains were put up across all side lanes, so that nothing should intrude on the Intercessor during his progress.

Upon arriving at the Temple of Heaven, the Emperor visited the Altar and the sacred tablets in the temples: that of Shang Ti, Heaven, and those of the spirits sharing his worship, the Imperial Ancestors, the Sun, the Great Bear, or "Northern Measure," the Five Planets (identified with the Five First Perfect Emperors), the 28 Constellations of the Chinese lunar zodiac (*see* "The Chinese Calendar"), all the Stars of the sky, the Moon, the Masters of Clouds, Rain, Wind and Thunder,—in a word the whole "Ministry of Heaven." The Emperor paid obeisance to all these spirits; then inspected the bullocks, slaughtered and cleaned in strict accordance with the ritual, also the sacrificial viands, and retired for a last meditation in the Hall of Abstinence.

He was called by the officials in charge one hour and 45 minutes before sunrise and, after an ablution, proceeded to the tent where he changed into the antique ceremonial robes suitable to the occasion. Meanwhile, the tablets of the Supreme Lord and his acolytes were being transported with the utmost ceremony from their usual resting places to the Altar whose dome is the firmament itself, to be placed there on marble pedestals inside tents of blue silk, the bullocks on wooden tables beneath them, the viands before their shrines. Huge braziers and torches lit up the scene. Every attendant took the place reserved for him and knelt there on one of the three terraces. Then the Son of Heaven mounted the Altar preceded by two Masters of Ceremonies, one directing the proceedings, the other calling out loudly the next gesture the Sovereign was to make. Unless we bear in mind constantly the Chinese conception which made their ruler *the* link between humanity and the Upper Spheres, it is impossible for us to realise the reverential awe with which every movement of the Emperor must have been gazed upon at this moment. Will not the slightest flaw in the spring of a clock affect its working? To the Chinese, the Emperor was this spring as regarded their relations to the Universe.

* * *

The sacrifice to Heaven was the only so-called complete sacrifice, in so far as it consisted of nine "acts" (three times three—the Heavenly number) each accompanied by a special hymn:

1. *Meeting the Souls of the Emperors.*—The Son of Heaven approaches the middle terrace of the altar from the south to a yellow tent covering what was known as his "place of obeisance." To the sounds of the first hymn he is called to the upper terrace by the Master of Ceremonies with the invocation: "Light the incense and meet the

souls of the Emperors!"—meaning Shang Ti himself and the Imperial Ancestors. Then the Master of Ceremonies cries to the Emperor: "Go back to your place!"

2. *Offering of Jade and Silk.*—The souls of the Emperors having now supposedly entered the tablets, the Master of Ceremonies calls upon the Emperor to offer them the most precious produce of the land. The colour of the silk rolls differed according to the spiritual recipient: blue for Heaven, red for the Sun, white for the Moon, etc.

3. *First offering of the bullock.*—Same ceremonial as above.

4. *First libation,* accompanied by a dance performed by eight groups of dancers with halberds and shields, known as "the dance of military leadership."

After this the Emperor prostrated himself on the central stone of the highest terrace of the Altar of Heaven (considered to be the centre of the world) while a special official read the prayer prepared beforehand, invoking help and requesting the offerings to be accepted. Obeisance to the associate spirits followed, and the "dance of the blessings of civil administration" in which the dancers carried long flutes and feathers.

5. *Second offering of the bullock.*

6. *Second libation.*

7. *Removal of the offerings.*—In obedience to antique tradition, the Emperor received from an official some of the wine and meat sacrificed, and partook of them. The rest was divided among

the Palace officials later. This was known as sharing the "wine and meat of luck."

8. *Sending off the Emperors and Gods,* a repetition of the ceremony of "meeting."

9. The Emperor watched the burning of the bullocks in tiled ovens and of the other offerings in open iron braziers. Then the official in charge announced: "The ceremonies are finished."

The first and foremost among all Chinese sacrifices was the great sacrifice at the Temple of Heaven on the Winter Solstice, briefly outlined above. On this, the shortest day of the year, the *Yang,* or male principle, for which Shang Ti stood in Heaven, and the Emperor on earth, was, so to speak, re-born and increased in strength, like Osiris in ancient Egypt. It was then, therefore, that the Supreme Lord was especially implored for the "bounties of the six beneficent influences." Another great sacrifice took place in the same temple sometime during the *Li Ch'un,* beginning of spring, when plentiful crops were prayed for. At this ceremony the tablet of Shang Ti alone appeared. A special service beseeching rain was held in the same enclosure in the fourth moon at the beginning of summer. The Imperial cortège was on this occasion much simpler than usual; the Emperor and his assistants wore "plain clothes and rainhats" and no bullocks or wine were offered, prohibition to kill animals being one of the means of obtaining the eagerly hoped for showers. An impressive service was held at the Temple of Heaven on the accession of a new Emperor to the throne, when the tablet of his predecessor was enshrined there. Finally, events of particular note were sometimes reported to Heaven through the channel of medium sacrifices.

* * *

The Temple of Heaven ceremonies were typical of the rites of all other sacrifices, which differed only in the number of hymns, prostrations and libations; in the quantity and kind of offerings, the cut and colour of the sacrificial robes, the shape of the vessels used, etc.

Thus, the sacrifice to Earth consisted of eight parts and eight hymns, that to the Imperial Ancestors of six, to the Spirits of Land and Grain of seven; the same number was adopted for the Sun and for the Patroness of Silkworms. The sacrifices to the Moon, to Confucius, to the planet Jupiter, and in the Pantheon, consisted of six parts and six hymns. The Small Sacrifices consisted of only one part, and there was no singing and posturing.

* * *

(b) *to Earth*

The second of the Great Sacrifices was held in honour of Earth, the Consort of Heaven in the Chinese dualistic system (*see* "The Hundred Gods" and "The Second Moon"), in a temple-enclosure to the north of the capital, because the north corresponds to the *Yin* principle. Conversely, as the south belongs to the *Yang,* the Temple of Heaven was built in the southern city. Furthermore, in the Temple of Heaven nearly all the walls are curved, whereas in the Temple of Earth they are rectangular. Blue in the first, the roof-tiles in the second were yellow. The Earth-offerings did not rise to the sky in smoke, but were buried in the ground. Finally, the service to Earth took place on the morning of the Summer Solstice, whereas the Winter Solstice witnessed the complete worship of Heaven, because between these two terms the days grow shorter and the *Yin* element, for which Earth stands, comes into the ascendant. The Imperial Ancestors shared the worship of Earth as they did that of Heaven. Moreover, we have seen a "Ministry of the Skies" associated with the same worship. Tablets representing a "Terrestrial Ministry" likewise surrounded the spirit-shrine of the Mother-Goddess. These tablets stood for the Five Sacred Mountains (*see* "The Tenth Moon"), several other classical mountains, the hills protecting the Imperial graves, the Seas of the four cardinal directions (supposed to surround the earth), and the Four Sacred Rivers: the Yangtze, the Yellow River, the River Huai in Honan, and the River Ch'i in Shantung. The accession to the throne of a new Emperor and other remarkable events were reported to Earth with the same ceremonies as in the case of Heaven.

* * *

(c) to Imperial Ancestors   Third in importance were the solemn sacrifices offered at the end of the year and at the beginning of each of the four seasons to the Imperial Ancestors (of the reigning dynasty) in the palatial buildings known as the T'ai Miao, or "Supreme Temple" situated immediately to the south-east of the main entrance to the Forbidden

OFFICIAL ROBE FOR SACRIFICE AT
TEMPLE OF HEAVEN, PEKING.

THE SHÊ CHI T'AN.
(THE ALTAR WHERE THE EMPEROR SACRIFICED TO THE EARTH GODS).

City, and which, until quite recently, very few outsiders have ever entered. Great events were here reported to the departed Lords with special reverence, and the enthronement of the spirit-tablet of a monarch who had recently died was, in itself, one of the greatest functions at the Imperial Court. Associated with this worship we find the tablets of certain princes of the blood.

All these ceremonies were, so to speak, "privately" duplicated by the performance of similar rites in the "house-chapel" inside the Palace, the Fêng Hsien Tien, where obeisance was also made by the Imperial family to its forefathers on the 1st and 15th of every moon, and on diverse other occasions. All these sacrifices were supplemented by special offerings in the "Ancestral Gallery," the Shou Huang Tien behind the "Coal Hill" in Peking, and at the Imperial Mausolea. Members of the reigning house generally visited the latter at the time of the *Ch'ing Ming*, "the First Festival of the Dead" (*see* "The Twelfth Moon"), performed the customary *k'o t'ous* and lamentations, and strewed a little earth over the huge tumuli, symbolically "repairing" them as a peasant does with the tiny grave-mounds in his fields.

* * *

(d) *to Gods of Land & Grain*

The fourth of the Great Sacrifices took place in the Shê Chi T'an, an altar and group of buildings to the south-west of the Palace, directly opposite the T'ai Miao. Here the Gods ruling the land—(*Shê*), and those controlling the grain-crops (*Chi*)—the food-supply of the country—were worshipped. Every territorial division had its own *Shê* and *Chi* spirits, the greatest, naturally, corresponding to the whole Empire. To sacrifice at this altar was, therefore, one of the exclusive privileges of the sovereign, one of the symbols of his right to rule. Before the sacrifice, the altar was covered with earth of five colours, corresponding to the four cardinal points and the centre. A special stone tablet was erected in the middle, and was after the ceremony taken down and buried in the ground until the next occasion. Associated with the tablets of the Great Shê and the Great Chi were those of two prototypes of their divine productive forces, sages of the days of the Perfect Emperors, known as *Hou T'u* (Kou Lung), who taught the people how to distinguish the qualities of the soil, and *Hou Chi* (Chu Wang) who helped his father, Shên Nung, to spread the knowledge of agriculture.

These sacrifices took place in the spring and autumn, and their main object was to pray for weather favourable to the season.

* * *

(e) to Confucius
The last of the Great Sacrifices—and the only one retained by the Republic—is that offered to Confucius, in the second and in the eighth moons. It will be described in "The Second Moon."

* * *

Medium Sacrifices:
The divinities and their acolytes worshipped in the Great Sacrifices included all the main figures of the classical Chinese cult.

The importance of their underlying idea was, however, further stressed by the promotion of some secondary spirits to individual worship, embodied in the Medium Sacrifices.

First among these came the offerings made to the Sun on the morning of the spring equinox, and to the Moon on the evening of the autumn equinox, at two once magnificent altars to the east and to the west of the capital. The distinction between the *Yang* and *Yin* principles was marked here by the fact that the altar of the Sun was raised, whereas that of the Moon was hollowed out. *(a) to Sun and Moon*

The rulers and great men of former dynasties, whose tablets are gathered in impressive simplicity at the Li Tai Ti Wang Miao in Peking, known as the "Pantheon," or "Temple of Emperors and Kings," were honoured by medium sacrifices in the third and in the ninth moons. *(b) to "Emperors and Kings"*

A very important sacrifice was that to the Patron of Agriculture, Hsien Nung, identified with the Perfect Emperor Shên Nung, offered at a special altar in the south-west of the capital, in the third moon, at the beginning of the farming season. It was followed by the well known ceremony of "Imperial Ploughing" when the Sovereign, assisted by his highest dignitaries and a hundred old men selected from the peasants of the metropolitan district, personally made six furrows in the sacred field with a plough drawn by an ox. Thereupon, his attendants, in order of precedence, finished the field in the same manner. The grain sown was carefully tended and harvested, and served for the offerings at the Imperial sacrifices. This very antique ceremony combined the ideas of setting an example in their work to *(c) to the Patron of Agriculture* the nation, and of personally providing for the requirements of the ancestors of the Son of Heaven.

(d) to the Patroness of Silkworms — The sacrifice offered at the same time of the year by the Empress to the Patroness of Silkworms[1] at her altar (the Hsien Tsan T'an) in the northern part of the Sea Palaces was practically modelled on the sacrifices at the Temple of Agriculture and at the Shê Chi T'an, for the ceremonies, both in the Palace and at the shrine, were similar to those of the sacrifices performed by the Emperor. A picturesque note was added by certain details, such as the Empress and her attendants plucking a few leaves from the sacred mulberry-grove near the altar with little sickles, the handing of these leaves to ladies in charge of

---

[1] The Patroness of Silkworms is identified with the wife of the mythical Emperor Huang Ti. She is credited with having first discovered and taught the art of silk-weaving.

A quaint popular legend states, however, that she was, in remote antiquity, the daughter of a man who was kidnapped by neighbours. Despairing of his return, his wife promised the hand of her daughter to anyone who would rescue the captive. Thereupon, the horse on which the husband had ridden out and which had returned alone, grew restive, broke away and vanished—to reappear a few days later with his master, whom he had found, on his back. When told of his wife's promise, the old man grew angry, saying that it could apply only to humans, not to animals. But the horse would not have it that way: he refused all food, stamped and reared, insisting on the fulfilment of the agreement.—His owner finally killed him and hung his hide in the court to dry. No sooner, however, did his daughter pass that way, than the hide flew down, enveloped her and vanished into the sky. After ten days' search, the hide was found under a mulberry tree, but the girl had been turned into a silkworm and was busy spinning the threads of future garments. Great was the grief and the remorse of the parents, until one day they saw a vision of their child, riding the same charger, and heard her say that, because of her sacrifice on their behalf, she was now a Princess in Heaven.

In accordance with this legend, the patroness of silkworms is sometimes represented with a horse's hide thrown over her head and shoulders (whence her name: *Ma T'ou Niang*, or "Horse-Head Lady") and is, under this guise, specially respected in the province of Szechuan.

the precious insects, the inspection at later periods (mostly through delegates) of their growth, of the washing of the cocoons in the sacred moat existing for this purpose, finally of the making of the silk which was used on occasions of Imperial worship, on the same principle as the grain raised by His Majesty's hand at the Temple of Agriculture.

Prayers for rain and for assistance in cases of calamity caused by nature were the object of sacrifices to the "Rulers of the Skies" (the Spirits of Thunder, Clouds, Winds, and Rain) and to the "Terrestrial Rulers" (Spirits of Sacred Mountains, Seas, and Rivers) whom we have already seen reverenced at the Altars of Heaven and Earth. Their individual places of worship are in the enclosure of the Temple of Agriculture. *(e) to Spirits of Sky and Earth*

There also we find the Altar to the Planet Jupiter, *T'ai Sui*, "the Master of the Year" (see "The Chinese Calendar"), a mention of whose cult, performed in the 12th and in the 1st moons, concludes the list of Medium Sacrifices. *(f) to the Planet Jupiter*

* * *

"Small Sacrifices," to the number of thirty (in very few of which the Emperor personally took part) were offered: to the Patrons of Medicine (2nd and 11th moons), to the God of Fire (6th moon), to the God of Literature (2nd and 8th moons), to Kuan Ti (2nd, 5th, 8th moons), to renowned servants of the last and of preceding dynasties (2nd and 8th moons), to several dragons, spirits of rivers, lakes and springs (all in the 2nd and 8th moons), to the Spirit of the *Small Sacrifices*

T'ai Shan, to the City God (Ch'eng Huang) and to the Polar Star. The Emperor sacrificed to this star, his symbol, on his birthday, and all his subjects were supposed to worship the particular luminaries under which they were born on the same day, a service they repeated in the New Year Moon. (*See* "Star Festival"—The First Moon).

Although observed throughout the Empire by officials, the Emperor's birthday did not affect the everyday life of the people any more than the Republican anniversaries do nowadays. It was, therefore, not a national holiday in our sense of the word. The death-anniversaries of a late Emperor or Empress, when the prohibition of certain public rejoicings was more or less strictly enforced, meant far more to the man in the street.

* * *

Most of the spirits represented in the Imperial Sacrifices were worshipped also on a minor scale on the monarch's behalf by officials in special temples throughout the provinces, prefectures and districts of the realm. Even the "Ploughing Ceremony" was repeated by the chief local magistrates, who did not forget either the reverence due to the Sages and Heroes of the Empire in the localities they guarded.

But in none of these celebrations could the individual—whatever his rank—partake, either as officiating member or as part of a congregation. If admitted, all he could do was to make the ritual genuflections, perhaps burn some incense on his own behalf.

CHAPTER IV.
THE TWELFTH MOON, OR "BITTER MOON."

HE Chinese people have no Saturday half-holidays, no idle Sundays. They can not afford such luxuries. Life competition is too keen in a densely over-populated country, where ancestor-worship —that bridge between mortal homes and a world of spirits needing offerings and loving service—makes a duty of large families, where, without sons, there is the tragedy of the broken line.

But if the pathway of the seasons brings few days of rest to the toiling masses of China, there are, at least, six "Great Festivals" to break the monotony of everyday life:—three Festivals of the Living (*Jen Chieh,* from *jen*—man: the New Year, the "Dragon Boat Festival," and the Harvest Moon Festival; and three Festivals of the Dead, *Kuei Chieh,* from *kuei*—spirit: the *Ch'ing Ming,* the 15th of the 7th moon, and the 1st of the 10th moon). And the greatest of them all, the longest, the gayest, the happiest, the noisiest, is the New Year. Then every man lays aside his work for as long as he can afford leisure. Frugal fare gives place to feasting. Re-union takes the bitterness from habitual separation. Amusement, like a bright thread, colours the drab pattern of dull daily lives.

Indeed, it is difficult for us Westerners to grasp the full significance of the Chinese New Year. Our Christmas, our Easter, and whatever national holiday we celebrate, all taken together really mean less to us than *the* great festival of their calendar does to the hardworking Chinese. Socially, it signifies re-union. Morally, it represents the idea of resurrection, the re-birth of the year, since it includes the *Li Ch'un,* or beginning of Spring, and also of life in general—the period when the *Yang* and the *Yin,* the Male and Female principles, meet and fuse into the Harmony which rules the world; when Heaven mysteriously fructifies the soil for the crops which feed the nation; when the pulse of every living creature beats faster in response to budding forces. Materially, it stands for rejuvenation both in the home and in the market place. Personally and commercially, men turn over a new leaf, strive to pay off old debts in money and loyalty, and start with a clean sheet on which they hope to write better success and greater happiness.

* * *

As is but natural, much preparation is required for the prolonged festivities to come with the coming year. People begin to make ready in the twelfth moon, often called the "Bitter Moon" (though it has other names, such as the "Moon of Offerings"), because it includes the periods of the *Hsiao Han,* "Small Cold," and the *Ta Han,* "Great Cold," two of the "joints and breaths" of Nature into which Chinese astronomers divide the Moon Year. (*See* "The Chinese Calendar").

THE TWELFTH MOON, OR "BITTER MOON" 71

A preliminary feast of very ancient origin, the *La Pa* *La pa ch'ou*, is held on the eighth day of the *Ch'ou* twelfth moon. Early in the morning, as soon as it is light, the women begin preparing the thick porridge (*la pa erh*) which, as a national dish, corresponds to our mince pie or plum pudding. Many ingredients are needed: whole grains of several kinds (but never meal or flour), special *lao mi*—"old rice"—long stored in granaries where deadly white scorpions lurk, beans, nuts, fruits, and four varieties of sugar. When cooked, a steaming bowl is offered first to the ancestral tablets. Next, every member of the family gets a share and, finally, what is left is sent to relatives and friends, at whose homes custom decrees it must arrive before the stroke of noon.

Legend tells us that the first *la pa ch'ou* was made by a poor mother whose unfilial son drove her to beg food from the neighbours. One gave a handful of grains, another a tray of fruits, a third a bowl of beans. Hence the number of ingredients composing this "mess of pottage", still eaten by high and low with the underlying idea of brotherhood.

The Buddhists, who so adroitly pinned their own festivals to the older holidays of the Chinese calendar, adopted the *la pa ch'ou* as a "remembrance feast" for their beloved Kuan Yin (*see* "The Second Moon"). They eat it in commemoration of the day when, before leaving home to become a nun, she gathered grains and fruits for her last meal under her father's roof. Buddhist monks are adepts at making this porridge, especially those of the "Lama Temple" in

Peking, the Yung Ho Kung. Some is offered before every image of the Goddess, and some is sent out to parishioners who, in return, give alms.

At this same season, friends send to friends, and relations to relations, soup made of white cabbage which has been buried in the earth since the coming of the first frosts. In shallow pits, like graves, laid carefully in rows with a thin blanket of earth between each and a roof of thick dried mud above, these favourite vegetables not only keep fresh but sprout in the darkness. The tender white leaf-shoots make delicious eating and fore-tell, according to whether they be sweet or sour, good or evil fortune to those who receive them.

House Cleaning

The twentieth of the "Bitter Moon" is marked in the calendar as the "day for sweeping the ground." This means that a regular house-cleaning takes place in every home. The heaviest cupboards are moved and, for once in the year at least, the heaps of dust accumulated behind them are dispersed. "Sweep carefully, my daughters," counsel grand-dames, too old to sweep themselves, "sweep carefully, lest the specks of dirt you neglect to drive away fly into your eyes and blind you." For they believe the tiny particles have power to do that. About the same time, rich people re-lacquer their front gates, whitewash outer walls afresh and re-paper windows, while poorer folk scrub and patch as best they can. Such renewal of material things signifies a desire to scruff off the dying year, with its defects and failings. It is a universal desire. We find it in Mexico, where people used to destroy their old furniture and buy new, to be in harmony with a season re-born. We find it still among the peasants of Europe who scrub, and polish, and paint, at Easter time. The difference in the date of renewal is simply due to a difference of climate. Everywhere the underlying idea is the same.

Once house-cleaning is over, Chinese matrons begin laying in provisions for the holiday feasts. Peddlers come down the streets carrying their goods in baskets slung on a bamboo pole, running with short steps to keep the burden properly balanced. They call their wares with musical cries: "Here's garlic for the vinegar to season your *chu po po* [special New Year dumplings]! Here's pork to fill your patés!" Standing on door-steps, the women bargain shrilly, while the peddlers squat down in the dust before them, weighing out purchases on little copper scales.

* * *

Thus time flies by in busy preparations till the 23rd (or, in South China, the 24th), when high and low, rich and poor alike, sacrifice to Tsao Wang or Tsao Chün, the "Kitchen God," before he leaves for Heaven where he makes a report on the behaviour of each family during the past year. *The Kitchen God*

The rôle played by the Tsao Wang [literally "Prince" (*Wang*) of the "Oven" (*Tsao*)] is of exceptional importance in China. Few of the gods are older than he. None are more universally worshipped, for, originally identified with the inventor of fire—Heaven's greatest material gift to man,—he has grown to personify the Hearth, pivot of the Home. Since the Han dynasty—when Imperial worship was first offered to him by the great Wu Ti in 133 B.C.—he has also become a censor of household morals like Agni, the old Brahman Fire Divinity. Both are messengers between Gods and Men. Did the idea of their goings and comings originate in the flashings of lightning between earth and sky with which the gods sent the first fire to our world? Very probably. Johnston says, in *Lion and Dragon in Northern China*: "There is some reason to believe that the Hearth God was once regarded as an anonymous ancestor of the family, though, nowadays, this relationship is ignored . . . "

The Taoists, who wanted to draw such a popular god into their temples, invented myths of their own to explain Tsao Wang, giving him human form and diverse attributes. Then the Buddhists also adopted him. They had to, because no religion makes a universal appeal in China if it denies the antique nature gods and the cult of ancestors. They excused themselves, however, for taking the deity into their pantheon by saying that their Tsao Wang was not at all the person worshipped by the Taoists, but a "King of the Kinnaras" (a fabulous race of celestial beings) who became a Chinese priest under the T'ang dynasty, and was appointed at his death to preside over the vegetarian diet of the monks. This is, of course, a lame defence of what is evidently a self-interested accommodation to popular notions.

Though, nowadays, there is a special Fire God in the Chinese pantheon (*see* "The Sixth Moon"), the Tsao Wang still holds his popularity as Hearth Guardian and Heavenly Censor, who metes out to every member of the family the length of his days and his share of wealth. Not only on the great festivals, but regularly at every new and full moon, incense is burned before him. Tsao Wang's birthday is on the 3rd of the eighth moon, when the Cooks' Guild of Peking burn incense at his temple outside the Hata Mên, a very shabby abode for such an important person. But his shrine is in every Chinese kitchen—a little niche behind the cooking stove, blackened with smoke and often full of cockroaches which the people call "Tsao Wang's horses." Inside the flimsy reliquary made of bamboo, wood and paper, a gaudy picture of His Highness is pasted. Sometimes it shows him sitting with his horse tied beside him, sometimes as an old man with an old wife sharing his throne, sometimes as a youth standing with a "memory tablet" (*tsao pan*) in his hand, to note down what he shall say to the Lord of Heaven.

Such pictures are renewed each year. They sell for a copper. The destitute, unable to afford even this infinitesimal sum, put up a plain sheet of red paper in honour of the Kitchen God, with his name and titles written on it, and a sentence referring to him as the Ruler of their "miserable household." A popular street-verse excuses all deficiencies:

> O God of the Hearth!
> Here is a bowl of water, and three incense-sticks.
> This year, I am living very miserably.
> Next year, perhaps, you shall eat Manchurian sugar!

Not in the meanest hovel do men dare ignore the Protector of the Hearth, who subtly combines the functions of an essential home divinity, a friend, a mentor, and a spy, "enveloped in the pungent atmosphere of smoke and the family dinner."

We should, naturally, expect to find the women playing an important part in sending off the Oven Prince. But custom forbids. Since the Hearth and its presiding God represent the corner-stone of the home, it is fitting that his worship be entrusted to the head of the family, the Master of the Household. He, and he alone, is responsible for seemly behaviour in the kitchen, enforces order there, and cleanliness,—old hygienic rules concealed under the name of etiquette or reverence,—forbids his women-folk to comb their hair within sight of Tsao Wang's shrine, or wash their hands before it, or sharpen knives on the range he guards, or drop chicken-feathers into the fire. He, and he alone, may make the bows of reverence and set a match to the Tsao Wang's image when the god ascends on high.

At the propitious hour for departure, food-offerings are made to the Heavenly Messenger,—all sweet things: melons, cakes, candied fruits, and a special white sweetmeat made of sticky rice. With this, or with honey, his lips are smeared, so that he will not talk too freely in the other world, and what words he says will be sweet and flattering; all about the good deeds of the family, nothing about their shortcomings. Gossip has it that in some homes opium is substituted

THE TWELFTH MOON, OR "BITTER MOON" 77

for sugar to make him drowsy before he goes. Or else his portrait is dipped in wine. Thus he arrives at the "Pearly Throne" of the Jade Emperor, tipsy and good humoured, tolerant rather than critical. "Boys will be boys on earth, Your Majesty," says he, making his report in jovial mood with the mellow sympathy of a man of the world.

After his feast, Tsao Wang's portrait is carried out into the principal courtyard, sometimes in a miniature sedan-chair. Or, if the palanquin be omitted, the paper image is set up on an improvised altar, between candles and incense-torches to light the way. Bows are made before it, and prayers murmured. Then it is set alight, and Tsao Wang starts skyward in a chariot of fire, while straw is thrown into the flames for his horse to eat, and a cup of tea poured on the ground for him to drink, as even the beasts in China are expected to appreciate tea. Meanwhile, peas and beans are thrown on the kitchen roof, to imitate the sound of his departing footsteps and the clatter of his horse's hoofs. These vegetable offerings will bring luck to cattle and ensure plentiful fodder in the year to come.

The service ends with a salvo of fire-crackers in an apotheosis of noise. Noise is a national necessity in China, and crackers an essential part of every ceremony. While gods and men delight in this holiday artillery, devils fear it and shrink away. Therefore, the popping and crackling, the booming and sputtering that makes night hideous throughout the Chinese New Year festivities and condemns the sensitive foreigner to wakefulness, serves a triple purpose. It honours the Immortals, it disperses evil spirits, and it delights the natives, with their happy faculty of listening without hearing in the midst of a carnival of uproar and deafening sound.[1]

When the Heavenly Spy has gone, people breathe more freely. They have given him the best send-off they can afford, and it is a great relief to have a few days respite from supervision, to be able to comb your hair in the kitchen, or spit towards the oven, if you feel like it. As the proverb says: "Once the cat's away, the mice climb over the bamboo fence."

* * *

Presents Now the period called "Little New Year" begins. The streets are full of people. Some gather round the stalls of peddlers selling sesamum and pine branches, talismans against devils. Some are buying flowers in those practical paper hot-houses which Chinese florists have used for over two thousand years. Here every plant has

---

[1] The character for "fire-cracker" is the same as that for "whip," because in very ancient times, before crackers were known, thin bamboo rods were used instead. These, being dried, were lit at one end and brandished about like whips. As the fire reached each joint, a loud pop resulted and frightened away any devils who might be lurking in the neighbourhood. Modern crackers are little cartridges of coarse paper filled with powder and damped clay. The smaller sorts are strung on strings with a continuous fuse that explodes one after another. An interesting variety is the "twice sounding cracker" which consists of a double chamber with a double fuse. It explodes a second time high in the air, whence it has been hurled by the first fuse.

Catholic missionaries, pandering to this harmless if exasperating Chinese weakness for noise, sometimes let off fire-crackers at the most solemn moment of the mass. It is a real deprivation to the Pekingese when the police, in times of disorder and uncertainty, forbid the firing of crackers at New Year, lest they be mistaken for the rifle shots of mutinous troops and alarm the timid.

THE TWELFTH MOON, OR "BITTER MOON" 79

been carefully bedded in loam, and every bud wrapped in paper long ago, and this skilful forcing brings the dwarf fruit-trees and the peonies into full bloom at exactly the right time. Purchasers choose their plants in pairs, then watch them packed for delivery in big baskets warmed by a tiny charcoal brazier and covered with heavy quilts as a protection against the frosty air.

As all shops will soon be closed for several days, everyone is busy getting gifts for friends, for New Year in China means presents—as Christmas does with us. Happily, in the East, one is saved harrowing decisions about what to chose. Convention long ago limited the choice of gifts. No Chinese aunt can send her small nephew a set of improving books instead of the bicycle his heart desires; no wife can give her husband a box of bad cigars. Useless trifles, that prove a nuisance instead of a pleasure, may not be offered.

What then does custom sanction? For members of the family—silks, ornaments, and even jewels. For distant relatives and friends—growing flowers (never cut blooms), fine tea, rare fruits, and food, especially food. In China, food is always useful and appreciated. "There's never a morsel," says the proverb, "but can find its way to a mouth." Besides, food-gifts are symbolic. They show that the giver has more than the bare necessities of life, and subtly express a desire that this superfluity of good things may pass on to others. Perhaps, to our way of thinking, a roast duck is not a poetic present. But is it not much more useful than a plaster bust of Napoleon? And, though we hate to admit it, who honestly prefers a badly made boudoir-cap to a succulent dish of macaroni?

Live fowls and prepared dishes are equally popular New Year gifts among the Chinese. The well-to-do offer both—an assorted lot of eatables packed in two round lacquer gift-boxes. It is bad form to accept everything sent. A polite person chooses a few things, say the live duck, with the red "joy mark" painted on his snowy neck, a paté or two, and half a dozen of those little red oranges that keep juicy in the bitterest weather. He tips the bearer according to the value of what he takes, and returns the rest with a message that the gift is far too generous to be accepted in its entirety. The system has one great advantage: what is refused by one person may be passed on to another, and vice versa. Among the poorer people, a pair of new shoes is an appropriate and welcome gift.

* * *

Settling Day
As soon as the rush of customers is over, shopkeepers close their accounts for the year. Now high and low make frantic efforts to collect what is due to them, and settle their debts. Instead of paying bills monthly, as we do, the Chinese have three yearly terms for squaring liabilities: the New Year, the Dragon Boat Festival (5th of the fifth moon), and the Harvest Moon Festival (15th of the eighth moon). Between times people, especially the poor, live largely on credit, even for household necessities. Therefore, when a paying day comes round—and the greatest of these is New Year—there is an undignified and often pathetic scurry in both private and Govern-

THE TWELFTH MOON, OR "BITTER MOON" 81

ment circles to get hold of ready cash. The custom, stronger than law, requiring the individual (and the nation) to pay up if possible, sometimes gives rise to funny situations. For instance, a man may steal to satisfy his creditors and thus *save his reputation*. A hired muleteer or carter may work the animals in his charge until they are lame, just to get them back to his employer on time to settle up accounts. At a season when everyone is pressing and being pressed, curio dealers will sell at a sacrifice to meet their obligations. It is very much a case of

> "Big fleas have little fleas upon their backs
>   to bite 'em,
> And these again have other fleas, and so on
>   *ad infinitum.*"

Foreigners see the struggle in miniature among their servants who like the usual New Year gratuity of a full month, or half a month's wages, paid well in advance. We can never hope to fathom the combinations of those who are "robbing Peter to pay Paul." The Chinese themselves say: "When a single cup of water is needed in three or four places at once, life indeed becomes difficult." Even honest folk, not unwilling to pay, will rarely do so until dunned. For such, moral suasion is used. A creditor will stalk his prey through the streets or, if this is not sufficient "loss of face," camp on his doorstep. Less delicate methods coerce the poor—bad language and even blows.

When a man sees no hope of paying up, he hides till New Year's morning. Then he is safe till the next settlement day (the 5th of the 5th moon), at least theoretically. Practically speaking, if he has the luck to escape the vigilance of his creditors during the last watches of the old moon, he can still be pursued after sun-up in the new, provided the man to whom he owes money searches for him with a lantern. "The light, by a polite fiction, indicates that it is still dark, therefore still yesterday, so the debt may still be collected without violating the New Year, when money transactions are forbidden." Of course, a man chased by a person with an unpaid account in one hand and a lighted lamp in the other, when it is broad daylight, is disgraced in the eyes of his neighbours.

Happily, there is a place of refuge for poor debtors in most Chinese cities—the courtyard in front of the temple of the "City God." Here, all the comedy troupes in town are morally obliged to give free (or for a small tip from the local official) performances in honour of the patron deity. These theatricals begin about the 24th, and last till the end of the Bitter Moon. Then "the play's the thing." No creditor, spying his debtor in the crowd, dares to demand payment, lest he himself be set upon by an audience disturbed by his importunities. A similar exemption seems to have existed in ancient Greece during the Elusinian Mysteries, when no man could lay his hand upon another, and one who pursued for debt could be condemned to death.

* * *

Lucky Symbols Once the hurry-skurry of the "settlement" is over, people put up new "luck-bringing inscriptions." These are strips of red paper with "fortunate phrases" written upon them, each craftsman using those suitable to his particular calling. On a merchant's premises, we find: "Successful in all Affairs," or "Acquisition of Treasures," or "Great Prosperity." Some tradesmen go farther, and underline their suggestions to the gods by hanging samples of paper money above their gates, on the principle that like attracts like, and "money goes where money is." An inn-keeper records his wish that "guests may descend in clouds this new year," and farmers express a desire for good harvests.

In private houses, the fortunate phrases always concern wealth, longevity, the gift of sons, and official promotion—the Chinese ideals of life. Never, as with us, is a wish expressed for amusement, travel or adventure, because Chinese want continuity, not variety. Favourite mottoes are: "Accomplishment of wishes," a phrase duplicated on both sides of a door lintel; "Ten thousand generations and long duration," or simply "New Year, New Joy," the last two put up in pairs wherever fancy dictates. If circumstances permit, people like to have the inscription: "As you leave the door, may you meet happiness," opposite the house. A crafty man will even paste a poster on his neighbour's wall with "Prosperity to those facing me" written upon it, hoping the gods will believe this the neighbour's prayer on his behalf.

Luck-papers are always red, except in cases of recent mourning when blue is used, or in temples, where yellow is the correct colour for inscriptions.[2] The talismans may be large or small, simple or elaborate. Sometimes, they are only lacy papers, with cut-out characters, called the "Five Lucky Happinesses that Knock at the Door" —Felicity, Honours, Long Life, Joy, and Riches— fluttering above a poor man's threshold according to the old Sung dynasty custom. Sometimes, they are handsome wooden panels with a single large *fu*—the character for "luck"—beautifully written to order by a needy scholar. The popularity of this decorative character dates from the Ming dynasty when, as now, the Chinese loved conundrums. According to legend, people in those days painted a barefoot woman, with a lemon pressed to her bosom, on their gates. Useless to ask the reason, for therein lies the riddle destined to be guessed with grim humour by the reigning Emperor Hung Wu (A.D. 1368–1398), founder of the dynasty.

While wandering through the streets of his capital, Nanking, to enjoy the Feast of Lanterns, the Sovereign—himself a man of low birth—saw the poster, and suspected a skit on his Empress who had large feet like a peasant woman. It looked like *lèse-majesté* rooted in anti-dynastic feeling. It had to be suppressed. It was. To all families without the objectionable cartoon the angry and suspicious ruler ordered a paper given with the character *fu* written upon it. Then came

---

[2] In all countries red is the colour of joy and life, but the Chinese trace their adoption of scarlet as the festive, or "luck," colour to the peach-blossom. This flower has been considered by them since remote antiquity as a symbol of the sun, because it blossoms at the spring equinox and renews itself each year, following exactly the phases of the "Heavenly Light." Hence the superstition that peach-wood is efficacious against devils who walk in darkness, and that peach-colour preserves from evil influences.

the drastic order to his troops: "Kill the inmates of every house on which no *fu* is pasted."

The bitter lesson is still vaguely remembered. The "happy ideograph" still retains its popularity. Not only does it embody the idea of submission to the existing government, but it means and attracts good fortune. Under the Imperial regime, one of the most valued gifts to an official at New Year was a *fu* written by the Sovereign's hand. Under the Chinese Republic, the President honours his Ministers in the same way.

The command of custom in China to protect property with written "luck phrases" is far-reaching, and still very much alive. We find them on such dissimilar things as cupboards and junks, on the pack-saddle of the mule and the harness of the camel, on the shaft of the wheelbarrow and the bucket in the well. The farmer sticks them on his plough and his pig-sty, on his mill, on the headband and the crupper of his ass, on everything that is his. Then, to make assurance doubly sure, he adds additional luck-symbols—pictures of Ch'ing Kuei, "Annihilator of Devils" [for, in China, the danger of mischief-making devils is ever present], of Chia Kuan, the "House Official," of the trinity of Fu Shên, the "God of Luck" (*see* "The Eighth Moon"), Tsai Shên, the "God of Riches" (*see* "The First Moon"), and the "God of Longevity" (*see* "The Eighth Moon"), known as Lao Shou Hsing, of little boys and pomegranates which attract posterity because these fruits have many seeds, of peonies for riches and honours, of the Grass of Immortality (the *ling ch'ih ts'ao*), of the Cash Tree, emblem of the God of Wealth, to attract business prosperity, of the bat for luck, etc. The choice of symbols is often due to a pun on words, the name of the thing represented being identical in sound with that of the blessing desired. Thus *fu*, "bat," and *fu*, "luck," are pronounced alike. *Hsi*, for "magpie," corresponds to *hsi*, for "joy," *p'ing*, "vase," to *p'ing*, "smooth" (without obstacles), *an*, "saddle," to *an*, "peace." In the same way, the terms for "a deer" and "an honour" correspond; "a sacrificial cup" and "official titles;" "a silver ingot" and "prosperity." That is why these designs are often found in pictures made for New Year, on birthday gifts, and in all sorts of ornamental motifs, from carvings on furniture to embroideries and dress-trimmings.

Gate Gods Finally, in addition to all these protections against malignant spirits, new "gate gods" are put up on the double panels of the front door. Their brilliantly-coloured figures, pictured in full panoply of war, are guardians of the home *par excellence*, and descend from among the oldest of the Chinese deities. Taoist legend traces their origin to two brothers of remote antiquity who lived under a peach tree so large that five thousand men, with arms outstretched, could not encircle it. Protectors of mankind, they vanquished demons by throwing them to tigers, and it is in their memory that district magistrates used to put up peach-wood images over *yamên* doors, and paint tigers on official gates.[3] Later,

---

[3] The tiger—king of beasts in Eastern Asia—has grown to be regarded in China as the sworn enemy of evil spirits, especially such as might seek to harm the dead. For this reason we find tigers carved on tombs and monuments.

THE TWELFTH MOON, OR "BITTER MOON" 87

the peach-wood figures were replaced by tablets and these, in their turn, by pictures of human figures printed against a background of peach blossoms.

The original "Devil Slayers," nowadays, are confused in the popular mind with two generals of the great Emperor T'ai Tsung (A.D. 627-650). After his unlucky expedition to Korea, this sovereign, a prey to rage and mortification at his ill-success, fell sick, and night after night teasing imps surrounded his uneasy couch. The Court physicians were powerless to help. Then two of His Majesty's favourite generals begged audience and, falling upon their knees, said: "Sire! we have spent our lives killing your mortal enemies. We have piled corpses on corpses, as if they were ants. We do not fear evil ghosts. Let us watch outside your gate at night and vanquish them." "So be it," the Sovereign answered. Fully armed, the faithful servitors posted themselves on guard till the devils and nightmares disappeared, and the Son of Heaven recovered. "There is no need for you two to sacrifice yourselves any longer," said he, when he was well again, "you also require sleep. I therefore command the Court painters to do your portraits in full armour. These shall be pasted on the Palace gate, that We may never more be troubled by ghostly enemies." The idea of thus deceiving supernatural beings is typically Chinese; nevertheless, it succeeded. There was peace in the Palace, until the devils discovered an unguarded back-door entrance. Then, a third general, named Wei Chin, volunteered to keep watch here, and to this day we sometimes find him associated on protecting posters with the other two Guardians of Gates.

From the palace to the humblest homes, the custom spread—as so many customs did—of using pictures of the warriors to protect the house. It still persists, a curious and typical example of the continuity of Chinese superstitions. You can see them for yourselves in almost any Chinese town, these ancient warriors that have stood guard for thirteen centuries, and are likely to stand for thirteen more. Banished from the portals of foreign-style government offices, they still keep watch over the homes of the people who believe in them. Why not? Surely, even the tradition of loyalty and self-sacrifice should tend to preserve from harm.

* * *

After the home, and all that therein is, has been carefully protected "from goblins and ghaisties and things that gae 'boomp' in the nicht," one day in the Bitter Moon, the 29th, is set aside for visits to parents and relatives who do not live "under the roof." However busy people might be, this was considered an essential duty—especially in Peking. Even now, when family ties are loosening, more's the pity,—since the family system with its mutual obligations and responsibilities has long been the cement holding the country together, —the *Chu-hsi,* or day of "close visits," is still observed. Pupils are also expected to call upon their teachers and say "farewell to the year," since the respect due to them is only second to that due to father and mother. The rest of the day may be devoted to charity. Kind-hearted folk will go about the streets followed by a servant carrying money. If women are heard crying in a house, custom gives such the right to enter, inquire what is wrong, and assist those in debt or difficulty.

Now comes the last day of the Bitter Moon, the last day of the dying year. The women are prisoners of their domestic duties, preparing and superintending food to satisfy the appetites, both mystical and bodily, of gods and guests. On New Year's Day, no knife, chopper, or sharp instrument may be used, lest it cut luck, and, so far as possible, cooking is avoided on the holiday. All sorts of delicacies are prepared, therefore, in advance—dishes with poetic names like "Gold Cash" chicken, or "Fairy Chicken," "Kidney Flowers," and Lotus Root Balls. But the *pièce de résistance*, in North China at least, is the *chu po po*, or meat dumpling. Meat is an extravagance in most Chinese homes, but poor indeed is the kitchen in Peking where on New Year's Eve the sound of the chopper is not heard mincing the pork to fill the paste-patties.

Should any one of the dozen seasonings be forgotten, a little maid is hurried off to the provision shops where late customers are still served through a wicket window. Long after dusk, long after other stores are closed, these remain open with half an eye, so to speak. Street lighting in Chinese cities leaves much to be desired, and electricity in many inland towns is unknown, so the temples hang out special lanterns to guide late shoppers back to their houses through unpaved streets.

When the cooking is finished, water is drawn for household use during the next forty-eight hours. During this time wells are closed. Their guardian spirits—not the least important among the Household Gods—also need a holiday. It is discourteous to disturb them.

Then a final dusting is given to reception halls. "Such a pity, such a pity!" old servants grumble, "to-day we clean, to-morrow guests will make everything dirty again." True. The floors will be littered with melon seeds and food spilled on polished tables. The best and rarest curios, set out on blackwood stands, risk being knocked over by the excited children. Nevertheless, priceless vases are brought out of cupboards, and the finest pictures the family owns are carefully unrolled and hung for the occasion. Gift-flowers are placed where they will show to best advantage. Dwarf fruit-trees, all a-bloom, are set out in pairs. Lemon, orange, and "Heavenly Bamboo" trees a-flame with red berries, are arranged to strike a bold note of colour against sombre carved wood partitions. "Buddha's Fingers"—curious lemons shaped like a half-closed hand—are laid in handsome porcelain bowls on shining grains of uncooked rice.

After they have "redded up," or seen their servants do it, ladies bathe in their apartments, while the men go out to the public bath-houses, stopping afterwards at the barber's stall at the street corner to get a new year shave and a hair-cut or, among old fashioned gentlemen, to have their

THE TWELFTH MOON, OR "BITTER MOON"

queues dressed. Two curious customs prevail in South China just after dark. Young boys traverse the empty streets calling out: *"Mi-sow,"* or "I sell my folly," or "I sell my lazy habits, in order that next year I may be wiser." Gray, in his book *China,* says "this custom is very common in Canton, also in the provinces of Hunan and Hupei." A superstition, practised both in Kuangtung and Fukien, is called "learning by the mirror." Whoever seeks an omen for the New Year places a sieve upon an empty stove, and on this sieve a basin of water and a looking-glass. He, or she, then silently steals out and listens carefully to what the first passers-by are saying. Good words mean good success in the next twelve moons, bad words portend ill-luck. Some people call this custom "meeting the fortunate head."

The omens must be consulted before eight in the evening, as after that hour a little rest is usual. The watch-night is at hand—the longest night of the whole year, when no man may sleep. It is the vigil of the great feast.

Before the home ceremonies begin, pine branches and sesamum stalks are laid down in the outer courtyard like a carpet, the idea being that they will crackle under the ghostly footsteps of any lurking devil, and give warning. At least one devil is expected, Pi-hu-tze, who comes to steal the New Year cakes. He takes from the poor to give to the rich. "This is but just," the people say, "since all the rest of the year the rich give to the poor." Then the front door is closed and sealed with crossed strips of paper, not to be opened again till New Year's morning, lest luck leave the house. Should a knock be heard, no one will answer for fear of admitting a goblin. At the same time, cypress branches are burned to "warm up the year"—making a cheery bonfire with flying sparks like iridescent jewels.

* * *

Triple Rites

As the New Year approaches its birth, the very oldest roots of religion are made visible in the three essential ceremonies performed in every Chinese home: the worshipping of Heaven and Earth, the worshipping of the Household Gods, and the offerings to the Ancestral Tablets. All three are twice repeated, once before midnight as a farewell, once afterwards as a greeting. The rites are similar, though details differ, and each ceremony is performed by the head of the household in the name of the family.

Now when we Westerners speak of a family, we mean a man's wife and children. But the Chinese family is a much larger group. As marriages take place early, and sons bring their brides to live under the paternal roof, it may consist of several generations gathered together in the same compound in houses constructed by successive extensions to meet their requirements. In early times, it might constitute the population of a village, or even a small town.

Of necessity, a patriarchal organisation grew up to keep order in a home shared by so many persons. The power of the head of the house—always the oldest male or the descendant to whom, by reason of age or illness, he delegated his authority—was theoretically supreme. All must obey him. Furthermore, females must obey the males,

THE TWELFTH MOON, OR "BITTER MOON" 93

wives their husbands, sisters their brothers, and, in principle, younger members older members. Even children must observe among themselves the domestic law of seniority, a rule of precedence enforced gently but firmly,—less firmly nowadays than of old.

Nevertheless, the Master of the Household still remains high priest of the religion of the home, responsible to gods and ancestors for all his clan. He, and he only, may approach the altar-table prepared in honour of Heaven and Earth for the watch-night service, with food-offerings, candles and incense, the two channels essential in China for divine intercourse. In the southern provinces, where rice is the staple food, a wooden vessel filled with it occupies the centre of the altar, surrounded by bright flowers, branches of cedar, and ten pairs of chopsticks. In the north, the grain country, millet, or *kao liang* (sorghum), are used instead. The service called "presenting the New Year rice, or grain," is a simple thanksgiving for the Staff of Life vouchsafed abundantly in the old year. When repeated in the first watch of the new, it expresses the hope of similar favours to come. *Worship of Heaven and Earth*

Both ceremonies take place in the open courtyard under the stars. Big, round, silk lanterns on bamboo tripods throw a soft glow of light on the solemn scene, while, high above the pointed eaves, a special "Heavenly Lantern," hoisted on a mast tipped with fir branches, casts a glittering halo around the house. Thus the scene is set. The Master appears in his long silk gown and reverently approaches the altar-table gay with neat piles of apples and oranges, and pagodas made of sweetmeats. Red candles of carved wax are burning upon it, in high candlesticks with a spike in each socket upon which they are impaled. A torch of incense-sticks flames in an old bronze vessel with dragon handles. Thin streams of blue smoke shed a warm perfume towards the star-spangled sky. Fire-crackers boom heavily as the Master kneels on the silk cushion placed for him, and bows to Earth and Heaven, once, twice, three times, touching his forehead to the ground. The deep, conventional *k'o t'ous* (*see* "The First Moon"), the waxen gestures of his clasped hands, are traditional, sanctified by the centuries, and the whole ceremony is intensely solemn. One feels that it lays bare the soul of a people.

Household Gods — Next comes the turn of the Household Gods to receive a share of offerings and worship. Their cult is older, much older, than that of many gods enshrined in temples. Long before Buddhism and Taoism brought divinities to China, they were already familiars of the home.[4] Unchanging in the principle of their functions, their personalities vary at different times and in different places. But nowadays, in North China at least, there are Five Principal Household Deities: Tsao Wang, God of the Hearth, the T'u Ti, or God of the Locality (*see* "The Ninth Moon"), the Gods of

---

[4] The Household Gods are of later origin than the Nature Cult, but are already mentioned in the *Li Chi*—one of the Five Chinese Classical Books, dating in its present form from A.D. 200, but embodying much older traditions. The oldest are the Hearth and Door Gods. The original Chinese *lares* were: (1) The Gods of Doors (2) of the Hearth, (3) of the Middle Chamber, (4) of Gates and (5) of Roads. The last mentioned had a special cult, now linked up with that of the Ch'eng Huangs. (*See* "The Hundred Gods" and "The Ninth Moon").

THE TWELFTH MOON, OR "BITTER MOON"

Gates, the Gods of Doors, and the Mao Ku Ku, patroness of women, needlework and lavatories. To these are sometimes added: the God of the Central Hall, the Spirit of the Well, and the God and Goddess of the Bed, male and female spirits who protect the sleeping chamber and keep babies from rolling off the *k'ang*, or brick sleeping-platform. The last named receive special offerings of fruits and wine, sour ginger, and eggs dyed red, placed on a table beside the bed, and special prayers are said to them for peace in the inner chamber throughout the year.

At the same time, they share in the feast provided for the Household Gods in general, since they form part of that little group of Lares and Penates whose worship may never be neglected, whom later religions were powerless to displace and, even, in the case of Tsao Wang (as we have already remarked) forced to adopt.

Their New Year altar is spread much as for Heaven and Earth, and their service is practically identical. Again the candles, the incense, the firecrackers and the food-offerings appear. Again the bows are made, while paper-money in the form of ingots covered with gold or silver foil, and *chih ma*—in this case a full package of 100 sheets of paper (12×18 inches in size), roughly representing most of the Chinese deities,—are burned in their honour.[5]

---

[5] Paper money is commonly used in China to satisfy the greed of gods, to whom the popular mind attributes the same desire for wealth as among mortals. There are various kinds of mock currency. Ordinary sheets of white paper, stamped with coins, are used chiefly as offerings for the dead. The gold and silver ingots are a survival from very ancient times. Indeed, as far back as 200 B.C., rich people buried lumps

These *chih ma* ("horse papers"), printed with portraits of the gods, have a curious origin. In far-off times, the Chinese sacrificed live animals, and especially horses, to their deities. Later came wooden images, as "a sign for the thing signified," then stuffed cloth-figures and, finally, under the T'ang Emperor Ming Huang (A.D. 713-756), paper horses were used. The *chih ma* of to-day are a vague survival of statues of patron saints, who were deified heroes mounted on horseback, or with horses beside them. When rough coloured prints of the gods came to be used for household worship, the expression *chih ma* remained, though all signs of a horse had long since disappeared from most of these posters. Every Chinese city still has printing-presses where they are turned out in thousands, and at every festival they are bought and burned in honour of the gods.

Worship of Ancestors

After the Home Protectors have been served and satisfied on New Year's Eve, the Ancestors receive their share of reverence. How primitive Chinese religious beliefs were rooted in ancestor worship has been explained already (*see* "The Hundred Gods"). The cult still embodies all the traditions of the race, the uprightness of ages of law-abiding culture—above all, the intense Chinese feeling of gratitude to the dead to whom they believe the living owe all they have and are.

---

of metal in tombs. Despite criticism and remonstrance from men who felt it undignified thus to hoodwink the gods, paper imitations gradually supplanted the real treasure. To this day, paper money is believed to be the coin of the spirit realm, and is used on all occasions for dealings with ghosts and gods.

THE TWELFTH MOON, OR "BITTER MOON" 97

Thankfulness for benefits received is the keynote of this last home ceremony before the dawn of the New Year. It is a simple service but, trifling as the rites may seem, few will neglect their performance. There is no imperative rule about prayers. The whispered invocations are short and few. The food-offerings are selected out of the family cooking, and the Dead Souls are satisfied, like the gods, with the fumes of the dishes set before them, leaving the substances to their living descendants, for vapour feeds the spirit, though flesh needs flesh. The food-offerings are afterwards removed and eaten by the family in accordance with the ancient rites which commanded that "sacrificial offerings shall be reverently consumed by those present at the Sacrifice."

Again we find the Master of the Household charged with the duty of expressing, for one and all, what the family owes to past generations in this service of remembrance. This he does with deep bows and genuflections before the ancestral tablets in which the spirits of the dead are supposed to dwell.

The first ancestral tablet may have been originated by the wooden statue put up under the Chou dynasty in 350 B.C. by a grateful sovereign to the loyal subject who sacrificed himself for his lord (*see* "The Third Moon:"—The Cold Food Feast), though some say that it was a filial son under the Han dynasty who, having lost both his parents, set up their statues and served them as in life.

Nowadays, the ancestral tablet, though it varies a little in size and shape in different provinces, is a simple strip of wood set in a wooden pedestal, with the posthumous name of the deceased upon it in raised or gilded characters, and the dates of birth and death. There is an important red dot on the tablet and the process of "dotting,"—sometimes with red ink, and, sometimes, with blood from the comb of a white cock,—is an essential part of the funeral ceremony. The tablet thereby becomes "spiritualised," just as an idol, once its eyes are "opened," becomes a god. The practice of "tablet-dotting" is mostly confined to better-class families, and used to be accompanied by much ceremony. It is said to be a relic of the living sacrifices once made at the grave.[6] In rich homes we generally find a special Hall of Ancestors, usually a separate building, or else a room set apart for the purpose. Among the poor, a special shrine, a shelf, or even a cupboard will serve. Of course, even in a large hall it would not be possible to find room for the tablets of thirty or forty generations, which is not an unusual lineage for a Chinese family. But the number of mortuary tablets in a shrine does not usually exceed five or six, or go beyond the third generation, only the recently dead being thus represented. Remoter ancestors are inscribed, as far as preserved records allow, on scrolls hung in the ancestral hall, where distinguished members of the family also have special laudatory inscriptions. None but direct descendants can claim a share in ancestral worship. The eldest son inherits the tablets, with the duties of family high priest. Younger sons, founders of homes

---

[6] *See* de Groot, *The Religious System of China.*

of their own, having no tablets to worship, sometimes place, high up on the *west* wall of the main living-room, a black bag covered with a square of white paper, in which are placed the names of the family ancestors. Daughters share in the cult only temporarily as, when they marry, they necessarily become part of a new clan with different roots.

* * *

After the ancestral rites are completed, a meal is taken in common by all members of the family. Outsiders, even the most intimate friends, are debarred from this feast, known as the "making-up feast to say goodbye to the year." All quarrels are supposed to be forgotten by those who gather round the festal board, to enjoy the smoking *chu po po* and dainties that are eaten with hiccups and other evident signs of enjoyment, permissible at a Chinese feast.[7]

Certain curious local customs distinguish the New Year's Eve supper from all others. In some towns a saucepan with lighted charcoal is put under the table as a protection against fire, or a new clay stove with a ring of silver or copper coins round it. From these pieces of money omens of good or evil fortune are drawn. Again, people amuse themselves foretelling the weather with bits of lighted bamboo, picked up with chopsticks

---

[7] Without any outsiders, seldom less than twelve persons sit down at table on this occasion. In the "Sacred Commands," mention is made of a little family party where nine generations of one family joined in the feast. Another home is cited where seven hundred mouths were daily fed. No wonder, with so many conflicting characters and personalities, that a "making up meal" is a necessary and popular feature of Chinese life.

and thrown into a pile of offering-papers. The embers are divided into twelves heaps and, while they burn out, predictions are made.

A little rest, perhaps another meal, and the New Year is at hand with new duties and holiday pleasures.

CHAPTER V.
THE FIRST MOON, OR "HOLIDAY MOON."

AT midnight on the *San Shih Wang Shang* (the last night of the dying year) members of a family present New Year wishes to one another. Among old-fashioned people this is done with much ceremony. The master and mistress of the house seat themselves, rigid as Buddhas, on two stiff chairs in the reception hall where all those "living under the roof" appear and *k'o t'ou* before them.

The order of seniority is strictly observed even in minor matters of etiquette. Husbands and wives, being of the same generation, are not expected to bow to one another, except in the rare cases of notorious shrews,—called *ma chieh ti*, or "revile-the-street persons,"—who may be ordered to make obeisance to their lords in penitence for bad temper or insubordination. But children and grandchildren come forward in turn, the eldest first, the youngest last, and, of course, all the boys before the girls. Dressed in his, or her, brightest and best robes, each "descendant" kneels on the flat cushion placed ready, bows three times with forehead touching the ground, and murmurs: *"ying tang"* ("I ought"), which is equivalent to a renewed vow of obedience for the coming twelvemonth.

Secondary wives or concubines (generally described by the polite euphemism of *hsiao hsing* or "Minor Stars," though a Western-educated Chinese friend calls his "my first wife's assistants"), daughters-in-law, and family retainers, being, so to speak, "outside the blood" and merely grafted on the family tree, yield precedence to even the youngest of direct descendants. They also, however, must *k'o t'ou* before the "House-Father and House-Mother"; their congratulations are couched in courteous phrases which mean "taking leave of the year," and at the same time they express thanks for the money-gifts made to them the day before by the head of the family.

These *k'o t'ous* ("*k'o t'ou*"[1]—literally—"to knock the head") are going out of fashion. Indeed the precise and formal etiquette for which the Chinese were famous is relaxing generally. But the duties owed by children to parents, and by younger to elder brothers, still remain in force to

---

[1] To perform the classical Chinese genuflection, a person first kneels, then touches his forehead to the ground from three to nine times according to the rank of the one he is greeting. The Emperor always received "the three kneelings and the nine prostrations." Some of the gods were also entitled to them, though others were only given the three head-knockings. This custom, universal under the Empire, has, under the Republic, been replaced at official receptions by a deep bow, and even at private ceremonies it is becoming rare. There were formerly eight "gradations of obeisance." The lowest (called *kung shan*) consisted in joining one's own hands and raising them either to the breast or to the forehead. This salutation corresponded to our handshake. The second (called *tso yo*) was a low bow with hands joined together. The third (called *ta hsien*) meant bending one knee as if about to kneel. The fourth was the *kuei*, or actual kneeling. The fifth, sixth, seventh and eighth comprised the true *k'o t'ou*, made one, three, six, or nine times. Another form of salutation grafted on to Chinese etiquette by the Manchus, and also gradually falling into disuse, consisted of bending one knee and lowering the right hand to the floor.

some extent. The first-born are shown respect because they have "a birthright in ancestral worship, in the division of property, and in the direction of the home after the father's decease."

The apparent coldness, to our eyes, of Chinese New Year greetings is rooted in a social sentiment totally different from our own. The Oriental code forbids a display of affection between relations. Hence, even in intimacy, Chinese personal contacts are a fastidious and restrained business. After babyhood, parents seldom embrace or kiss their children. The most tender of husbands gives no hint in public of his feelings towards his wife. Even women rarely touch or caress one another but, as throughout the East, use the same reserved salutations as the men folk.

* * *

Triple Rites Repeated

During the hour of the Tiger, that is to say between three and five a.m. (*see* "The Chinese Calendar"), when no cock has yet crowed, cypress and pine branches are again spread in the courtyard, and the head of the house goes out to break the seals on the front gate so carefully closed the night before.[2] He knows that "there are seventy-two evil influences constantly besetting unguarded doors." Therefore, he murmurs some such lucky phrase as "May the New Year bring us riches." This little ceremony is called "opening the door of fortune."

The triple rites are then repeated in honour of Heaven and Earth, of the Ancestors, and of the

---

[2] Both these customs correspond, in a way, to the formal "closing" and "opening" of seals under the Empire in all the Government offices, before and after the New Year holidays.

returning Household Gods with Tsao Wang at their head. Once again the "Heaven and Earth tablet,"³ found in practically every home in China, is placed in the courtyard under the clear sky, and the thanks for past protection, offered early on New Year's eve, are repeated as petitions for further favours in the New Year.

The Ancestors too receive offerings and prayers again for continued blessing and protection. Their second service is held in the guest-hall where the pedigree scroll, carefully wrapped up and put away in ordinary times, is hung, and the tablets of the forefathers unto the fifth generation are brought with due ceremony and set in their proper places.

Tablets of ancestors beyond the fifth generation (after which the influence of forefathers seems to diminish in an increasing ratio) are either kept in an ancestral temple or reverently burned in those cases where it is not practical to preserve all of them. The record of distant ancestors is inscribed on the pedigree scroll, "often a beautiful work of art painted to represent a temple or grand family mansion. The names of past generations are inscribed in successive rows, so that the space devoted to each name looks like a spirit-tablet in miniature. . . . In some parts of China, it is customary to have family portraits painted for the purpose of preserving the 'shadow sem-

---

³ The tablet to Heaven and Earth is worshipped by most Chinese families on the first and fifteenth of every month as well as on all important occasions. "When a bride comes, this worship is part of the wedding ceremony; when a son is born, it is part of the general rejoicings; when a birthday is celebrated, it is part of the festivities, and when a dwelling is built, it is part of the house-warming."

THE FIRST MOON, OR "HOLIDAY MOON" 105

blances' (*ying hsiang*) of ancestors as sacred heirlooms in the family temples. Like the pedigree scrolls, such portraits are exposed to view on solemn occasions only. De Groot compares these family portraits with the *imagines maiorum* of the ancient Romans."

Throughout the two weeks that the "spirits" are supposed to remain in the house (usually till about the 16th of the holiday moon), they are served daily with food and drink, while hot towels are given them to wipe their shadowy faces as if they were honoured mortal guests. The banquet offered consists of five kinds of food, five cups of wine, and five cups of tea. Ten pairs of chopsticks are placed upon the altar for their use, also a calendar, so that the dead may follow the feasts and festivals to come. In ancient times there was actually an impersonator of the dead at the ancestral sacrifices.

Not to invite the ancestors to share in the New Year festivities would be a grave breach of propriety,—a disgrace to the family who, absorbed by their own pleasures, forgot their duties to the dead. Even distant relatives, when making New Year visits, are led to the domestic altar and do reverence to the ancestral tablets. The last baby of the family will be carried there in his mother's arms and taught to bow his tiny head, dressed in a holiday cap with little figures of the eighteen Buddhist saints in gold, silver, or copper.

Though in some provinces the household gods are not supposed to return to their own roof-tree till the fourth of the New Moon, in Peking they arrive early on the first. They too are given a second service. Hungry and thirsty after their long trip to Heaven, they eagerly enjoy the fumes of this spiritual feast, and the warm welcome puts them all in a jolly humour for the coming twelvemonth. As for the Tsao Wang whose good graces are so essential, he is politely invited back to his shrine in the kitchen where he finds a new picture of himself set up in a new reliquary. Thus officially he is installed again as care-taker of the home with much burning of incense—for, "as breath to men, so is incense to the gods,"—and with much firing of crackers.

* * *

The New Year sun rises on muted cities in China. All the ordinary sights and sounds are absent. No blue-hooded carts, no little grey donkeys ply for hire, no peddlers utter their sing-song cries. The streets are deserted as on a New England Sabbath. Door-steps are littered with the wrappers of spent crackers exploded in the night. Gates are closed, shops tightly shuttered. Only the occasional sound of drums and cymbals from behind barred windows breaks the stillness. A private religious ceremony perhaps? Not at all. Simply a group of shop-apprentices having a little jollification behind the counter because, unless sent on some errand, they are not supposed to go out. Indeed nobody is.

The first day, or at least the early part of it, is usually spent at home. Primarily an occasion of re-union, it is regarded as a purely family gathering,—"every family apart." People remain with their own kith and kin or, failing these, with their business associates. To end the year, or begin it, supping in a noisy restaurant, as we often do, is unthinkable to old-fashioned Chinese.

Then, if ever, they want to be at home. Chinese society is so organised that the individual divorced by ill luck or misbehaviour from his family receives scant consideration, and ostracism from the clan-gathering at New Year is a serious matter. Many a returned student faces this tragedy. He has grown away from his conservative home, feels ill at ease among the old folks, or they with him. Yet, without his relations he is a cypher in Chinese life—a detached figure against an unfocussed background. "The old-fashioned family organisation had many advantages which compensated the individual for his state of subjection. . . A society of mutual help, it was not less powerful to give aid than to enforce obedience. Every member had to do something to assist another in case of need. Each had a right to the protection of all. Even the bond between masters and servants was so defined that the latter, though necessarily treated as inferiors, were also regarded as members of the household—trusted familiars permitted to share in the pleasures of the family, to be present at most of its re-unions, and to be cared for and supported when they became too old to work."

Though the power of the Chinese family as a whole is no longer so strictly exerted over individual members, those whose parents are still alive usually make an effort to return to the *chia*, or ancestral home, at New Year, and share in the pleasures and the duties of the holidays.

Apart from the essential "triple rites," there are lesser household ceremonies to be performed. The well, for example, must be re-opened on the 2nd of the 1st moon with a special prayer to the Spirit that guards it. When the first bucketful of water is drawn up, candles are lit, incense burned and crackers fired as usual.

* * *

God of Wealth

On the 3rd, a still more important rite must be performed. This is a home service to Tsai Shên, the God of Wealth, whose favours are so dear to the Chinese heart. Like the Kitchen God, his picture is found in most houses and shops. It is yearly burned and replaced by a new one. On New Year's afternoon, poor children go from door to door crying: "We have brought you a new Tsai Shên! Here he is under his money tree whose fruits are gold, whose branches drip with coins! Here he is with his magic casket! Open the box and you will find inexhaustible treasure. Shake the tree and wealth will descend upon you! Who will buy?" Of course, few dare refuse. Piety and prudence alike counsel men to keep in the good graces of this powerful and well-beloved god. They give freely that they may receive freely, and his altar-table is lavishly spread with flesh and fowl and the live carp that he prefers. The fish-peddlers, carrying their splashing burden in wooden tubs, "squeeze" on the price, well knowing nobody, if he can possibly afford it, will deny anything to the Giver of Riches.

Great is his power over all classes, learned and unlearned, religious and irreligious,—sincere and

THE FIRST MOON, OR "HOLIDAY MOON"

earnest the homage paid to him. Yet the origin of Tsai Shên is misty and uncertain, and his personality doubtful and confused. Some identify him with the Spirit of the North (birthday 15th of the 3rd moon, *see* "Four Diamond Kings" in "The Fourth Moon"). Others believe him one of five brothers, Regional Gods of Riches. But the most popular legend traces his descent from a deified hermit of Mount Omei, called Chao Kung-ming, in life a person "of infinite resource and sagacity." He could ride a black tiger; hence the tiger so often shown beside him. He could hurl pearls that burst like bomb-shells. Nevertheless, he was downed by an old and universal form of witchcraft. His chief enemy "made a straw image of the magician, *k'o t'oued* before it for twenty days, then shot a peach-wood arrow through the heart and eyes with fatal results to the living prototype." Fortunately, death in China is not always the end of a career—often rather the beginning. In this case, the soul of the Pearl Hurler obtained release from the Underworld through some sympathetic petitioner, together with the spirits of other heroes killed in battle. Sainthood followed. Then, extraneous legends attached themselves to the ex-hermit. Virtues gathered round him like a halo, till, finally, he was canonised and invited into homes and temples as the "pop-eyed, bearded God of Wealth holding a silver ingot" that we see to-day.

In China all the "essential gods," among whom Tsai Shên is included by courtesy, are worshipped at home. Neither Buddhist nor Taoist temples have weekly services which the pious are expected to attend as good Christians go to church every Sunday. There is, in fact, no such thing as community worship, except for monks. But on holidays many people will *kuang miao* ("visit temples"), an expression which implies no actual religious obligation. Nevertheless, the Pekingese are careful on the third day of the New Year to make an excursion to the Money God's shrine near the Race Course, and publicly show him their allegiance. The crowd in front of his red gates is intense,—all kinds, all classes eager to register their vows or make their petitions. The "brushing-off-women" do a brisk trade as they flap new arrivals with their rag dusters, hoping to earn a few coppers for adding to silk robes as much dirt as they flick off. Every beggar from the city is here. The blind crouch by the roadside with reed baskets ready for the contributions of the charitable. The lame, the halt, the lepers, whine for pity. Many poor wretches, whose sores are not allowed to heal because they arouse compassion, have dragged themselves for miles to be present on this happy occasion.

People come and go incessantly through the narrow entrance, further congested by the peddlers' stalls that crowd right up to the gate. "Lend me your light, lend me your light!" (equivalent to our "move out of the way") is the polite phrase of patient visitors trying to get past those who stop to buy toys, or cakes, or paper fish attached to bamboo sticks.[4]

---

[4] *Yü*, the word for fish, is a homonym of *yü* "surplus", therefore the fish has become a symbol of wealth.

THE FIRST MOON, OR "HOLIDAY MOON"

Once within the courtyard, the pious are enveloped in a haze of smoke rising from huge braziers full of gold and silver paper ingots that flash into flame and crumble into ashes, and of incense-torches still burning.[5] Their flames mount dangerously near the roofs, and the heat from this fiery furnace forces men to shield their faces with their sleeves.

Still, worshippers eagerly press forward to the open shrine of the Tsai Shên. Oblivious of the discomfort, the choking smoke, the deafening noise as bell and drum are struck to attract the attention of the god, they kneel before his images, for he appears here in quintuple form because of his association with the Wu Shêng, or Forbidden Spirits (*see* "The Second Moon.") An aged old priest, who looks like a fish in an aquarium because he keeps his mouth perpetually half-open, stands ready to light candles or give out incense-sticks and bundles of paper money. He has charge also of the bamboo tube filled with numbered tally-sticks which devout persons consult after their prayers. To ascertain the divine will, the tube is shaken till a stick falls out. Its number is noted and interpreted from a book of omens. But if the answer is not clear or is unfavourable, one is allowed to ask again.

---

[5] Of course, this is only common incense. The better kinds are not sold at temple-fairs because most people could not afford them. So we find these bundles of incense-rods, each about as thick as an ordinary lead pencil, but somewhat longer, sold for a few coppers to be used as the "Messengers of Earnest Desire." See *In Ghostly Japan*, by Lafcadio Hearn, for the poetic legends inspired in old China by "Spirit-Recalling Incense" and other rare varieties—legends of the days when incense had not only religious, but also luxurious and ghostly uses.

Sometimes, in South China, a platform is erected before Tsai Shên's temple. From this is fired a "wooden cannon loaded with a small charge of gunpowder, and a ball made of rattan. The ball rises some forty or fifty feet into the air, and innumerable hands are stretched out to receive it. He who catches it as it falls is specially favoured by the God of Wealth for the rest of the year, and is presented in Tsai Shên's name with an ornament for his ancestral altar . . . Upwards of thirty balls are sometimes fired from the cannon, but the luckiest man is he who secures the first ball."

* * *

After the "third day" people begin to go out again, and the peddlers re-appear on the streets. It is better form, however, to remain indoors in the family till the fifth of the New Moon, which corresponds to our twelfth night. Then the pretty home altar with its pagodas of cakes and pyramids of fruits, its luck talismans and its votive red candles, is dismantled. All decorations are taken down and every house is swept and cleaned again because, according to the popular saying, "the holidays are broken."

Small shops, which cannot afford to cease their trading any longer, re-open in the early morning with a grand fanfare of fire-crackers and more libations of wine and food offerings to the Tsai Shên. Ceremoniously, merchants invite him to their premises, saying: "God of Wealth, God of Wealth, please come in! Please come in! God of Wealth, God of Wealth! please give riches,

THE FIRST MOON, OR "HOLIDAY MOON" 113

please give prosperity!" Sometimes they "leave their doors open for the entire day for fear he may go by while they are closed, and a fortune thus be lost. An image from his temple is often carried through the streets, under which circumstances the worship is redoubled."

* * *

Now every man who has a decent coat to his back, or can afford to hire one, pays calls upon relatives and friends, taking care when he leaves home for the first time in the New Year to choose a lucky spot for his first footstep. To slip or fall on going out would bring misfortune on his own house and also on those he is about to visit. Equally important is the first person he meets. A woman is unlucky, but a priest is worse—the latter superstition being known equally in some Western countries. *New Year Calls*

A few years ago, when the "three thousand rules of behaviour" were still in force, the burden of New Year calls on a man of high position and, consequently, wide social connections, was so great that it was physically impossible for him to make all his visits in person. To those of inferior rank a servant might carry his master's visiting cards—strips of red paper about eight inches long and three inches wide, inscribed with the name and, in addition, some such phrase as "your humble servant;" sometimes also with pictures and emblems of many sons, official promotion, longevity, etc.

But to his teacher, to his blood relations, and to his wife's parents—if they lived within reasonable distance—every gentleman had to go himself and present New Year wishes. In the villages, all but the oldest generation of males were expected to make a round of their neighbours, the representatives of every family entering the courtyard of every other family and prostrating themselves before the elders who remained at home to receive visitors. *K'o t'ous* were made according to the priority in the genealogical table. As those oldest in years did not necessarily belong to the senior generations, it often happened that a man of seventy posed as the nephew or even the grandson of a mere boy. Such an absurd situation was redeemed for the individual concerned by the knowledge that he had conformed to custom—the grand consolation and excuse for everything in China.

Visitors' bows are accompanied by an expression of good wishes in stereotyped phrases. "May joy go with you on your way," or "May you have wealth in the New Year" are permissible, even elegant. But simpler sentences like *"Pai Nien,"* "Greeting the Year," or *"Ch'ing An,"* "Wishing you peace," are more commonly used. Many good wishes in China concern Riches and Peace, perhaps because the first is rare and the second, politically speaking, even rarer.

It is very important that only "good words" be said at New Year time. Children are specially warned not to use words of ill-omen, like "demon," "death," "coffin," "lion," "tiger," "elephant," or "snake." Lest they forget, or mischievously disobey, nervous parents stick up a red paper somewhere on the wall with the inscription: "Children's words do not count;" or a placard over the front door reading: "Heaven and Earth, *Yang* and *Yin* (the Male and Female principles), all things without danger from unlucky words." Sometimes naughty little boys who refuse to curb their tongues and risk drawing poverty or ill-luck upon the household have their lips rubbed with paper money. Whatever they say is thus transformed into a promise of good fortune.

Grown up people, of course, may be relied upon to find an ambiguous phrase which will circumvent the use of an unlucky word, not only in speech but in writing. A wise man begins his first letter on New Year's day with the character for Wealth, Happiness, or Long Life, well knowing that carelessness on his part may mean bad luck for the whole twelvemonth. The many taboos specially applicable to the New Year make all Chinese walk warily. Their "luck flower" is so delicate the least error may cause it to droop and die. Therefore, no loss of money may be referred to. Also to quarrel or to scold servants is to court misfortune. To gamble directly after a bath is to invite bad luck, and to change the dishes after one has begun to eat means death to the housewife. Each home, moreover, has its own particular prohibitions; hence the proverb: "When entering a family, inquire what are its tabooed words"—so as to avoid them. And each day there are certain things one must not do. For instance, never thread a needle on the 7th or 8th of the first moon, and never start on a journey on the 7th, 17th, or 27th, nor return home on the 8th, 18th, or 28th.

* * *

It would be wrong, however, to give the impression that the Chinese spend their New Year holidays assailed by terrors and weighed down by duties. Much time is devoted to amusement, and their pleasures, though sometimes boring from our point of view, are none the less thoroughly enjoyed.

Because the Chinese as a rule work so hard, leisure is in itself an amusement, and people are quite content to spend a portion of their long vacation sitting around in their best clothes on stiff chairs doing nothing. More active excitement is provided by the three "national vices": feasting, gambling, and theatre-going.

Though some devout Buddhists observe a "beginning and end of the year fast," which is the most valuable fasts of the year and brings great merit, most laymen compromise with the strict rule by taking one vegetarian meal in the twenty-four hours.[6]

---

[6] On principle, fast days in China are numerous, but any layman who keeps them regularly is unusually religious. These fasts are of various kinds, including "unbroken fasts" which last from year's end to year's end and imply a strict vegetarian diet. Such are observed only by priests and a small minority of exceptionally pious people. "Short fasts" may last for a few years or a few months only. Then there are one-day fasts, such as the "three, six, nine fast" observed on the 3rd, 13th, 23rd, 6th, 16th, 26th and the 9th, 19th and 29th, of each month; also the "one, four, seven fast" on the 1st, 11th, 21st, 4th, 14th, 24th, 7th, 17th, and 27th. Both sets are kept in honour of a vow made to any one of the gods.

Occasional fasts may be made at any suitable time. Favourite commemorative fasts are the "Goddess of Mercy Fast," observed from the 1st to the 20th (sometimes only from the 18th to the 20th, and sometimes during the entire month) of the 2nd, 6th, and 9th moons; in the 2nd because the 19th is her birthday, in the 6th because on the 19th she attained sainthood, and in the 9th because on the 19th she entered the Pantheon (see "The Second Moon").

Even to abstain from meat and, consequently, from the meat-filled dumplings inseparably associated with New Year's day, is a real merit among the Chinese to whom the pleasures of the table mean so much.

To "eat good things" is the common expression used for a holiday or a festivity, and New Year's calls are one long succession of feasts. "During the first part of the first month," says the proverb, "no one has an empty mouth." Though casual acquaintances, whose visits are short, may partake only of the cakes, sweetmeats, fruits, melon-seeds, and tea, which are constantly being handed round, relations are always pressed to stay and dine, nor does the code of etiquette permit refusal.

To them a regular banquet is served with a long succession of delicious dishes, dozens of which are offered in relays by a rich host. The Chinese, in cooking, use many ingredients to which we are unaccustomed, such as preserved eggs, seaweed, "birds' nests," sharks' fins, fish brains,

---

The "Sun Fast" occurs twice yearly, 19th of the 2nd and 19th of the 11th moons, and is kept for the forgiveness of sin; "Buddha's Birthday Fast," 8th of the 4th moon, for almighty protection. "Ti Tsang's Fast" on his birthday, the 30th of the 7th moon, lasts for three days and is intended for protection against evil spirits, forgiveness of sins and mitigation of the pains of hell. "Seeing the Stars Fast" means eating nothing till evening when the stars appear. Finally, the "Parents' Fast," intended to help deceased parents in the other world, occurs on one's birthday and is considered an act of thankfulness to the givers of life. (See "Chinese Religion Seen Through the Proverbs," by C. H. Plopper).

A man observing a fast was known as *tsai li*—"within the law," and often used to wear a special badge with a corresponding inscription so as to be spared tempting offers of wine, dainties, etc. In secret societies this sentence *tsai li* was sometimes used as a password for identification of members.

water-chestnuts, bamboo-shoots, sea-slugs, pickled fir-cones, soya-bean curd, ducks' tongues, cocks' combs, lotus roots, etc., and leave out many of the things we consider essential, such as butter, milk, cheese and cream. Nevertheless, the Chinese cuisine is exquisite, and even foreign connoisseurs consider it the equal of the French. Despite different conventions used in preparing them, Chinese sauces and seasonings—the basis of all good cooking—are no less delicate in flavour, and if, as gourmets tell us, "the stomach loves surprises," Chinese cooks excel in creating dishes that not only delight but puzzle the palate. Though duck or chicken may be served several times in the course of a feast, it is so well disguised we must admit their best *chefs* are "creative artists in the highest sense."

All Chinese food is cut up in small pieces before being served, as the Chinese consider it barbarous to bring big joints to the table, and carve them like a butcher; chopsticks belong in the dining-room, knives in the kitchen. The meal concludes with steamed cakes stamped with red characters meaning "happiness," or "longevity," and a bowl of rice. Every grain of this must be eaten to show one's appreciation of the good cheer though, occasionally, the host himself will grant a reprieve saying "the rice need not be finished." Since politeness requires the guest to taste at least of each dish as course follows upon course, a good digestion is essential and a hearty appetite is taken as a compliment to the host. Like the Scotchman who had no pleasure in smoking, because he put too little tobacco in his pipe when

THE FIRST MOON, OR "HOLIDAY MOON" 119

he drew on his own pouch, and too much when he borrowed from a friend, the Chinese say when a man "eats his own, he must not eat to repletion, but when he feasts with another, he stuffs till the tears run."

All this choice and, often, very expensive food is accompanied by tiny cups of warm alcohol of which there are several varieties. The yellow *shao hsing* variety has been pronounced by foreign experts "not only the best of Chinese fermented liquors, but one that is comparable to Spanish and other European wines." The common "burnt wine," three times distilled, known as *san shao* and distorted by foreigners into "samshoo," is less intoxicating. These and other rarer varieties flow freely but decently, and are followed by rare teas, stronger than wine and served in the same thimble cups.

As friend drinks to friend, and riddles are propounded, or quotations capped, the losers must *kan pei*—empty their cups. But, as a rule, there is more wit, more fun, and more laughter than over-drinking, and his innate refinement (according to his own conventions, be it understood) prevents a Chinese gentleman from losing control of himself. Chinese etiquette requires that the "superior man in everything turn to scorn the madness of extremes."

At the same time, custom permits a guest to express his satisfaction after a good meal by a series of loud hiccups, nor is it considered a breach of manners to remove one's long outer gown, or retire from the table for a walk or a smoke, during the pauses before a new series of dishes appears. Of course, no ladies are present. Except in the most modernised Chinese homes, the latter feast by themselves in their own apartments.

After a banquet which may last several hours, towels wrung out in hot water are passed around. Host and guests then adjourn to another room for the favourite amusement of gambling in one form or another. Games of chance are in many parts of China forbidden by law. Yet, they are universally enjoyed. In fact, gambling is the Chinese national sport among high and low. Even peddlers carry dice with them, and children with only two coppers to spend will go double or quits with the stall-keeper to see whether they are to have two cakes or none. Among the better classes, various games are played, some with dice, others with cards about two inches long and half an inch wide. Kwazan, the Japanese philosopher and acute observer, says: "There is one thing indispensable to Chinese life, and that is gambling which has been elaborated into a great art and includes a thousand and one varieties of betting. . . . It is possible," he suggests, "that in China gambling is a transformation of that instinct to fight which, according to Carpentier, is one of the two instincts by which man is impelled. While Chinese children never play at fighting, there is not one of them who does not gamble."

The click of Mah Jong tiles is as characteristic a feature of the New Year holidays as the sound of fire-crackers. At almost every entertainment the guests end by sitting round a Mah Jong table under which they have carefully placed the clay

THE FIRST MOON, OR "HOLIDAY MOON"

image of a tiger, patron of gamblers. This beast of prey, reputed for his ferocity and, as we have seen, an acolyte of the God of Riches, is the terror of evil spirits; he is believed to frighten away the demon of ill luck with the rest. Perhaps that is the reason why he is called the "God of Gamesters," and is pictured in this capacity on a board or paper scroll, rampant and clutching a cash. Under the title of "His Excellency the Cash-Grasping Tiger" he appears on the street-signs that denote gambling establishments wherever these are allowed.

* * *

Though a man may, if it pleases him, begin his New Year calls and feasting before the fifth day of the first moon, no old-fashioned lady will go out to see her friends till after the holidays "have broken." To "pass the door" and enter the house of another sooner was considered unlucky, but the old custom is being relaxed even in conservative Peking.

When in force, however, the strict rule entailed no hardship on women accustomed to remain at home for weeks at a time. Life is largely habit. Their lives, shaped by social conventions which discouraged promiscuous gadding, were passed contentedly within the confines of their homes and gardens, till they themselves grew to be like charming formal gardens, full of old-fashioned virtues resembling old-fashioned flowers. Our age, so restless and disturbed, fails to understand how, for these women, the life of home—the still deep stream—sufficed. But it was so. Children filled their hearts and stifled their personal longings, and they were never lonely because, usually, several ladies lived together under one roof,—the first wife, the concubines and, perhaps, daughters or sisters-in-law. Sometimes, of course, they "ate vinegar,"—that is to say, quarrelled,—but on the whole the system worked well enough. They played games together, Mah Jong, or dominoes, or threw arrows into wide-mouthed bronze jars, such as we find in curio-shops without suspecting the use to which they were put. They gossiped, often spitefully, fed their tame gold fish, and passed hours at their embroidery, stitching into the symbols to decorate shoes or gowns their thoughts of happiness and their hope of sons.

Superstitions Certain family ceremonies are also in their hands. The "sprinkling of vinegar," for example, is always done by the lady of the house, the first wife. Obeying the old custom, she pours vinegar into a long-handled saucepan in which is placed a tiny clay stove filled with hot coals. Immediately smoke rises. Turning her face aside from the choking fumes, the bearer of this primitive censer carries it into every room, waves it in corners and under tables. Thus she hopes to purify the home and drive out lurking evil spirits.

Now evil spirits appear to be particularly mischievous around New Year time, as we have already mentioned in "The Twelfth Moon." Consequently, the services of exorcisers (*see* "The Fifth Moon") are in great demand at this season. Processions of priests in bright robes, walking from house to house to drive away demons, are a feature of the streets during the holidays.

Especially in homes where there has been ill luck or misfortune during the year, elaborate ceremonies are held—generally by a Taoist magician. "Attired in a red robe, blue stockings, and a black cap, the exorcist stands, with a sword made of peach-wood in his hands, before a temporary altar on which are burning tapers and incense-sticks. . . . Placing the sword upon the altar, he then prepares a mystic scroll. This is burned and the ashes put into a cup containing spring water. The exorcist then takes the sword in his right hand, raises the cup in his left . . . and utters the following prayer: 'Gods of Heaven and Earth! invest me with the heaving seal, in order that I may eject from this dwelling-house all kinds of evil spirits.' Having received the authority for which he prayed, he calls to the demons: 'As quick as lightning depart from this dwelling.' He then takes a bunch of willow which he dips into the cup and besprinkles the east, west, north, and south corners of the house. . . This done, he fills his mouth with the water of exorcism which he immediately ejects against the eastern wall. He then calls aloud: 'Kill the green evil spirits which come from unlucky stars, or let them be driven far away.' At each corner of the house, and in the centre, he repeats the ceremony, saying at the south corner: 'Kill the red fire spirits, etc.'; at the west corner: 'Kill the white evil spirits, etc.'; and in the centre: 'Kill the yellow devils, etc.' Meanwhile, attendants are ordered to beat gongs and drums. In the midst of the appalling din, the exorcist screams: 'Evil spirits of the East, get you back to the East, of the South back to the South, of the North back to the North, and those from the centre of the world I command to return thither! Let all evil spirits return to the points of the compass where they belong. Let them all immediately vanish!' Finally, he goes to the door of the house, making some mystical passes with his sword in the air, for the purpose of preventing their return. He then congratulates the inmates on the expulsion of their ghostly visitors, and receives his fee."

This ceremony is but one of many forms of sorcery. There are a thousand others, like the sprinkling of door-posts with the blood and feathers of a cock, overcoming a single demon by putting a padlock round its neck, etc. Most people seem indifferent to the means used to protect them from the *Kuei,* and some families will try as many methods as they can afford, both Taoist and Buddhist priests being invited to go through their mumbo-jumbo for warding-off misfortune. In addition to their incantations, the monks make a regular income selling charms, talismans and amulets of infinite variety prepared for every possible contingency, from a lucky rebirth to a protection against coughs and toothache.

The custom of "observing the year" is a form of amateur sorcery regularly practised in the "inner apartments." In almost every household there is sure to be some old retainer, perhaps a peasant woman, whose long service gives her the privilege of expressing her opinions to her mistress, and even of joining in the conversation when guests are present. Such a person would be well-versed in the science of omens, and

THE FIRST MOON, OR "HOLIDAY MOON" 125

believed capable of predicting happy or unhappy events. She would know, for example, that a raven flying over the house foretells calamity, that an owl only enters a room to bring trouble, and that magpies appearing before noon prophesy happiness, but later in the day warn people of the presence of devils.

She will remember that the first ten days of the New Year are the birthdays of animals and grains, the first of fowls, the second of dogs, the third of pigs, the fourth of ducks, the fifth of oxen, the sixth of horses, the seventh of mankind, the eighth of rice and certain cereals, the ninth of fruits and vegetables, the tenth of wheat and barley; and she will remind her mistress of the superstitions corresponding to each day. The pigs' festival requires an offering to the God of Wealth, perhaps because the pig is the oldest domestic animal in China and, though he does not "pay the rint," he provides the New Year's dinner. On the Ducks' Day public bath houses should re-open, and priests must be summoned to make the occasion auspicious. These worthies pluck a chicken, spray a mouthful of wine over the dead bird, bite the head off and, with the blood that flows, spray the door-posts of the establishment. This guarantees that no one will be drowned during the year in the sunken tubs of the bath-house, or asphyxiated in the thick steam. The 6th, birthday of horses, is a lucky day to visit relations, whereas on the 7th, birthday of mankind, it is wise to stay at home and eat red beans, seven for a man, fourteen for a woman, thus protecting one's self against sickness.

Memories of long-ago beliefs persist in the weather prophecies indulged in at New Year, and in the omens interpreted through little household happenings. For instance, if the wind blows from the south-east on the first day, the harvest will certainly be good. If heavy clouds veil the sky, food will be dear, and so on. All the old women are would-be astronomers, carefully observing the skies during this critical period and often right in their deductions, as old folks are apt to be. Thus they hold the faith of their mistresses, and their wise saws have become a popular inheritance, part of the everyday beliefs. Nobody doubts that "when the cat washes its face, guests will arrive," "when ants plunder their neighbours' nests, rain may be expected," "when the eyelids quiver, someone is confessing his sins," and "when the lamp-light flickers, wealth is hovering about the house."

As anything which breaks the home routine is a great event in the life of conservative Chinese ladies, New Year visits cause much excitement "behind the Orchid Door." If the house-mother's hair is still dressed in the elaborate old-fashioned way, a maid will spend several hours combing, perfuming, and arranging it, while Madame herself adds the flowers and jewelled ornaments with an expert hand. Make-up requires a long time too—first of all a generous application of flour instead of the finer face-powder, then bright patches of rouge on the cheeks. The result is effective, though, to our taste, crude. Finally, the new gown, essential to the New Year, is put on and admired by the entire household.

To foreigners, the fashion of a Chinese lady's dress appears unchanging. Perhaps this is because the general line, modestly concealing the figure, does indeed remain the same. But a close observer will notice many alterations in the last few years even, though the general silhouette, which differs so little in its straight slim lines from present Paris styles, persists. Pleated petticoats, once universal in China, are much less worn; indeed in the North they have almost disappeared. Jackets are shorter, sleeves tighter, and collars, held together with a jade brooch or pearl clasp, are lower. Instinctive good taste, however, keeps the Chinese beauty faithful to the severe straight neck-band from which her sleek head emerges like a dark flower from its sheath.

As the children accompany their mothers, they too are dressed in their best. For them red is used a great deal, as evil spirits hate the sight of it and keep at a respectful distance. The sedate little girls, with the precocious dignity of Infantas who have been taking part in some ceremony, are pictures in their little pyjama-suits embroidered with lucky designs,—perhaps the plum-blossom, symbol of beauty, the squirrel eating grapes (an old and favourite pattern meaning long life), the bamboo standing for longevity. Their little painted faces may be framed with gilded miniatures of the Eight Immortals sewn on a narrow head-band from which hang long black silk fringes. Small brothers, no less brilliant than themselves, will wear a short cloth jacket dyed to imitate a tiger skin, a "tiger bonnet" with upstanding ears, silk wadded trousers, "pussy shoes," and, round their necks, a "hundred families" cash-lock, welded from coins given by a hundred friends and, therefore, the best possible talisman to assure long life to a precious son. Even babes in arms are taken calling wrapped in a red satin cape and clutching in their tiny hands a few cash strung on a red string. These are the "coins that crush the years," and are offered to relatives in the hope that future years may not escape them, any more than the cash can fall from the knotted string. Should the latter break, it would mean that the child receiving the gift would shortly die. Therefore, a very strong string is always chosen.

Accompanied by their servants carrying gifts, the ladies start on their interminable round. In old Canton it was the custom to attach long sticks of sugar-cane to the sedan chairs in which they were borne from house to house, because sugar-cane was a particularly lucky gift. As a rule, however, the cut canes were not untied from the poles, "the will to give them away being taken for the deed."

In Peking, painted cakes and steamed bread worked into complicated designs replace the southern delicacy, while carts, motor-cars, or even rickshaws, are conveyances more commonly seen than chairs. Otherwise, the customs and entertainments are much alike. While the men are received in the outer apartments by the host and his sons and would not dream of calling upon the ladies of the house, the "inner apartments" are "at home" to their wives and children. Separate feasts are served, and both sexes enjoy apart the same games, the same innumerable cups of tea, the same sweetmeats, and gossipy stories. At the end of the day, hosts, like hostesses, are happy but weary after the long procession of "friends who come but never take their leave," of relatives, often hating each other in their hearts but paying each other compliments in sound traditional phrases,—weary too of passing well-tried jokes, or urging their guests to eat and drink too much out of sheer hospitality. Yet, next day perhaps, they must make their own rounds, over-eating and over-drinking in their neighbours' houses to show how they appreciate the cheer.

* * *

While New Year gaieties are in full swing, and in the midst of the coldest weather, the Chinese calendar announces the *Li Ch'un* or beginning of spring. This movable feast, falling on or about the Birthday of Grains, is observed with an ancient and symbolic ceremony which inaugurates the farmers' year. In China no peasant was supposed to plough his lands until certain State ceremonies had been performed. These sacrifices, indicative of the deep veneration for agriculture ingrained in the people, were performed in Peking by the Emperor in person, and in the provinces by the local magistrates.

Li Ch'un, Spring Ox

Originally connected with the worship of the Earth Gods, they included the slaughter of a live ox. A clay effigy later replaced the living animal and, until quite modern times—perhaps even now in remote communities—a paper ox or paper water-buffalo is used. Fukien province, one of the last to accept Chinese civilisation in its entirety, persisted until recently in butchering a real buffalo "whose carcass was divided among the various officials taking part in a 'meeting the spring procession,' and whose head was then always given to the Viceroy" (*Doolittle*). On the same occasion, a paper buffalo, painted in five colours, red, black, white, green, and yellow, respectively, represented the Five Elements of Nature, metal, wood, fire, water, and earth. According to one authority:[7]

"In Shanghai, and probably elsewhere, a real ox is sometimes used, led by a real child... Needless to say, no blood is shed nowadays, though it seems not unlikely that at one time a living child and a living ox were both offered up in sacrifice."

In the most usual processions which included musicians, dancers and singers, the Spring Ox was made of stiff paper and so was his driver.

When these images were being prepared for the approaching festival, a careful examination under official direction was made of the "newly issued almanac, and the effigies were dressed-up and decorated in accordance with the warnings of that publication. . . . Hence, the crowds of people who went out to watch the procession did so not only as a holiday diversion but also for the purpose of inspecting the colours and trappings of the effigies, thereby informing themselves of the agricultural prospects for the coming

---

[7] See *Shanghai Folk Lore* by the Rev. A. Box in *The Journal of the Royal Asiatic Society, North China Branch*, Vol. XXXIV, 1901-1902, and Vol. XXXVI, 1905.

year. . . . If they saw that the head of the ox was painted yellow, they knew great heat was foretold for the coming summer; if it was green, there would be much sickness in the spring; if red, there would be a drought; if black, there would be much rain; if white, there would be high winds and storms. The *Mêng Shan*, or 'spirit driver,' was also a silent prophet of the seasons. If he wore a hat, the year would be dry; if he wore no hat, there would be rain. Shoes similarly indicated very heavy rain; absence of shoes, drought; abundance of body clothing, great heat; lightness of clothing, cold weather . . . because the *Mêng Shan*, being a spirit, behaves in a precisely contrary manner to ordinary mankind, and his garments indicate exactly the opposite of what they would if he were a living man. Finally, a red belt worn by him indicates much sickness and many deaths; a white one, general good health."

The spring procession was halted thrice *en route* by a rider who thrice dismounted in front of the magistrate and announced promotion for that official, each time receiving a reward of cash.

Arrived at an open space to the east of the city, the local magistrate, his underlings and, sometimes, the bystanders also, went through the form of beating and prodding ox and driver by way of making them work as an example to the farmers. This had to be done with bamboos decorated with strips of coloured paper at the "exact hour when spring begins." Meanwhile, in front of their houses, people stuck a large piece of hollow bamboo in the ground, with chicken feathers in it, and this they still do, for we have ourselves seen it in Peking. As the feathers fly upward on the first breeze, supposed to blow at the moment when the ox is beaten, everyone knows spring has actually come.

After these ceremonies, the effigies are burned, while the crowd rushes forward to catch pieces of the charred paper as a luck-charm, and the officials go home, take off their winter fur-lined robes, and put on their spring costumes.

In some provinces, the spring ox was sacrificed on an altar erected in the fields to the God of Agriculture. This seems to prove that the underlying motive of the ceremony was to promote the fertility of the crops. "The dummy ox which figured at the Imperial rites at the Hsien Nung T'an in Peking and, by proxy, at the various prefectures, being broken up and distributed,—is in principle identical with the slaughtered *meriah* among the Khonds, or the lamb of the Israelites which was put on the door-posts, or the blood which was sprinkled over the worshippers of Thor, Odin, and Frey, at Upsala. The raw meat sacrificed to the local god of the soil was similarly divided amongst the worshippers in ancient China." (*The Dead Hand in China,* by Herbert Chatley. *Journal of the North China Branch of the Royal Asiatic Society,* Vol. XV, 1924).

Johnston, in *Lion and Dragon in Northern China,* says that "in Northumberland, England, it is, or used to be, a custom to hold rustic masquerades at the New Year, the players being clothing in the HIDES OF OXEN." See *Country Folk-Lore,* Vol. 4.

Why the beast that draws the plough should be beaten is nowhere clearly explained. It may have been, originally, an effort to stimulate the spring and promote the fertilisation of the soil, though the modern and popular meaning of the ceremony is to drive off the diseases rife at this season.

When the whips are made of willow twigs, as they often are, the vital energy of this tree, symbol of the Sun (*see* "The Third Moon"), is supposed to invigorate the ox for his task. Symbols of spring, the willow branches induce the ox to chase winter away, a hint confirmed by that old saying in the Book of Rites: "The clay ox at the *Li Ch'un* escorts the cold away."[8]

* * *

Feast of Lanterns

The official ending of the New Year holidays is the Feast of Lanterns, called the "Feast of the First Full Moon," and celebrated on the fifteenth of the first month.

The origin of this festival, when householders hang lanterns over their doors and put up fir-branches to attract prosperity and longevity, is supposed to date from the Han dynasty some two thousand years ago. It began as a ceremonial worship at the Temple of the First Cause (*T'ai Yi*, see "The Hundred Gods") from the 13th to the 16th of the New Moon, thus fittingly bringing the New Year holidays to a solemn close. The lantern displays that make of it one of the

---

[8] The willow being also a rain charm, and closely connected with water, points to a mystical connection between the fluid element and the ox. An ox or cow—the docile animals corresponding to the *Yin* element in the Chinese Zodiac—are the guardian spirits of the Summer Palace lake near Peking, of the Hun River at the Lu Ko Ch'iao, etc.

prettiest and most picturesque festivals in the
calendar were not a feature of the "First Moon
Feast" till eight hundred years later. And by
that time all religious significance was lost, except
for the home offerings to the Household Gods.

Custom evolved, as it so often does in China,
the older festival into an occasion of pure enjoyment, a social event celebrated in the Palace with
much splendour. Under the T'ang dynasty some
Emperors, not content with looking at the illuminations of their capital from high observation
towers like their predecessors, went out into the
streets incognito. So did their ladies. Amorous
adventures were frequent. The carnival spirit
held sway, and sovereigns and courtiers did not
return to duty till daybreak.

Several historic lantern festivals are described
in the old records. Among the most extravagant
is a fête ordered at Nanking by a certain Ming
Emperor. Ten thousand lamps were set afloat
on the lake, and the effect was so beautiful that
Buddha came down from Heaven to see it.
Ch'ien Lung gave a similar party in the gardens
of his summer palace near Peking, where every
pavilion—and there were several hundred—was
outlined with lanterns, every canal afloat with
little lighted boats, every tree hung with flaming
fruits, and the miles of marble balustrades decorated with tiny stars of light glowing in their
painted silk cups.

In the southern provinces, lantern-fairs used
to be held in public gardens. Parents of newborn children would come and buy a lamp, then
suspend it as a votive offering in a temple near

their homes. Those desirous of sons had their names and addresses painted on their purchases "which were forwarded to them at the end of the month, having been lit at the ever-burning lamps before some altar. The messenger who conveyed such a lantern was accompanied by singers, and he presented, together with the lantern, a lettuce in the centre of which was placed a burning candle with two onions at the base. A dinner was given on the occasion of this ceremony, and the lantern suspended in front of the ancestral altar. . . On the evening of the fifteenth it was customary. . . for members of a clan to dine together. After the feast, a large lantern which had been placed in front of the ancestral altar on New Year's day was sold to the highest bidder. . . . In some parts of Kuangtung province, a tree with many branches expressive of the hope that the clan might never lack representatives was placed in front of the altar in the common ancestral hall. Clan dinners were given in these halls from the first to the fifteenth day by those who had been successful in business during the past year, or to whom children had been born. Either on the seventh or the fifteenth of the month, dinners were given in each district to the poor by such of their neighbours as had male infants, or who had newly come to settle in the neighbourhood . . . and messengers were sent into the highways and by-ways to summon the guests by beating gongs."

These old customs are dying out. Still, many Chinese towns look like fairyland on the fifteenth evening. Where everyone carries a lighted lantern, perhaps two or three as in Foochow, the narrow streets are a dancing pattern of black and gold.

In the Yangtze valley the peasants put up beside the temples poles dressed with long strings of lanterns making tents of light. "Tiny lamps, like will-o'the-wisps, are set out on the ground, offerings to the spirits of those who have died before their time" and whose souls the King of Hell, Yen Wang (*see* "The Second Moon"), has consequently not gathered in. Unguided, they wander helpless over the earth, unable to find their way to the Halls of Judgment, and are apt to create mischief. That is why the country-folk put out lamps for them at cross-roads, near wells, marshes and rivers, or wherever they are likely to be. Even if they burn for only a few minutes, the tapers suffice to light wandering souls to judgment and re-incarnation.

"So far as Weihaiwei is concerned," says Johnston in *Lion and Dragon in Northern China,* "the Feast of Lanterns may be regarded as pre-eminently the holiday season for children. During several days before and after the fifteenth of the first month, bands of young village boys dress-up in strange garments and go about by day and night acting queer little plays, partly in dumb show, partly in speech, dance and song. Some of them wear the terrifying masks of wild beasts, a few assume the white beards of old men, and many are attired in girls' clothing. The children perform their parts with great vivacity, and go through their masquerades, dances, and chorus singing, in a manner that would do credit to the juvenile performers at a provincial English pantomime. They are, indeed, taught their parts and trained by their elders for some weeks before the festival. Every group of villages keeps a stock of masks, false beards, clothes, and other 'properties,' and there are always adults who take pleasure in teaching the little ones the songs and dances which they themselves learned as children in bygone days. In the daytime, the dressed-up children take a prominent part in processions to the local temples... At night, they carry large lighted lanterns and march amid music and song through the streets of their native village, or from one village to another, stopping occasionally in front of a prominent villager's house to act their little play or perform a lantern-dance."

In Peking, the lantern festival is less brilliant than of old. Gone are the days when the lantern-show in the capital included works of art worth a thousand ounces of silver,—some of lacquer, some of horn painted by well-known artists, and some fashioned of shells, carved, gilded and set with imitation jewels. On the big street leading to the Drum Tower there was, until a few years ago, a special display of lanterns carved from blocks of ice, trimmed with cypress branches and decorated inside with bright paper figures of men and mountains.

The shops of Lantern Street outside the Ch'ien Mên, however, still have a glittering exhibition which draws the happy holiday crowd and serves as an advertisement to attract purchasers. The varieties displayed are infinite,—all shapes, materials, decorations, sizes and prices, and all alight in the open shops. Wall-lanterns to put on either side of the front door are offered in pairs. Others are sold in sets, eight or sixteen, intended to be hung together and thus form a complete picture; "Guest-lanterns,"—large white silk moons decorated with the purchaser's name and lucky bats,—intended to light visitors across a courtyard to the reception hall, stand ready on bamboo tripods. Cheap paper lanterns cunningly made to copy living creatures hang from the ceiling; fantastic crabs with moving claws, dragon-flies with flapping wings, birds with swaying necks. Glass or gauze panels painted with historic scenes, mounted in carved wood frames, are displayed in great variety. Inside the shops there are many special lanterns for special purposes. Of such are those in the shape of little boys, intended for presents to childless families; "heavenly lanterns" to be hoisted on a high pole in the courtyard and decorated with fir branches; round toy-lanterns made to roll on the ground like a fire-ball, lanterns set on wheels, red paper lamps pricked with tiny pin-holes to form a lucky character, like "happiness" or "prosperity," "*tso ma têng*" or horse-racing lanterns which consist of two or more wire frames, one within the other, "arranged on the principle of the smoke-jack so that a current of air sets them revolving;" and, finally, cross-word-puzzle lanterns with riddles pasted on their sides intended to hang outside a scholar's home for the amusement of his literary friends.

* * *

ChengWu   As the gods in China have their part in every

holiday, pious people present new lanterns to the temples. There is a pretty show at the little shrine near the Tung An Mên in Peking. A tall mast with a cross-spar outlined in coloured lamps guides sightseers to this sanctuary where the outer walls are covered with pale transparencies painted with long-ago ladies and bold warriors, with mythical beasts and calm-faced Immortals brought to life by the lights cunningly placed behind them. Inside the crowded courtyard, against the very wall of the temple itself, more lanterns are fixed and some of them relate the strange story of Cheng Wu to whom the temple is dedicated.

Cheng Wu Ta Ti, to give him his full title, is one more instance of a mortal become a god. A prince and a swashbuckling soldier, he abandoned his rank, abjured his profession, and became a follower of the True Way. He started home again once, when his resolution weakened. But a timely meeting with an old woman, who was grinding a large iron rod into a needle and assured him that "anything can be accomplished through persistence," sent him back to the mountains and the anachorite's life.

This hero appears to have unconsciously followed the Scriptural command: "If thine eye offend thee, pluck it out." In his attempts to attain perfection, troubled by the frailties of his body, he declared: "If my vital organs offend me, I will do without them." Thereupon he cut himself open, removed the five vital organs, and lived thereafter heartless, lungless, and liverless, without apparent inconvenience. Hence his name of *Ching Tu Fo*, or "Buddha Without Vitals." But his heart, lungs, etc., turned into dragons and tigers which tormented the people. "Alas!" then exclaimed the much-tried Cheng Wu, "how can I become a god if portions of myself are causing misery?" He, therefore, retraced his steps, hunted down the savage beasts, and turned them all into musical instruments. As a reward, he was appointed Guardian of the North by the Supreme Lord of Heaven, with residence at the North Pole. His pictures show him as a serene old man, barefooted, with a long white beard, and sometimes surrounded by tamed tigers, dragons, serpents, and turtles. (For Cheng Wu's place among the Marriage Gods, *see* "The Eighth Moon").

The Republic still recognises the Feast of Lanterns officially by closing public offices, and by street decorations and displays of fireworks, called "the letting-off of flowers," in the public parks. Set pieces, in whose manufacture the Chinese, and especially the Cantonese, excel, are attached to a high scaffolding in a large round container; from this, seven or eight different fire-pictures will appear in succession, released by a spark burning through a set of strings. Tableau one a temple or pagoda complete in all its details with multiple roofs outlined in light, and even flaming bells hanging from the eaves, burns out. Its *débris* fall to the ground. Tableau two appears, an official in his chair; then tableau three, two warriors fighting, and so on, until the last and most elaborate scene fades into darkness amidst the applause of the delighted spectators.

THE FIRST MOON, OR "HOLIDAY MOON" 141

"Under the heavens," says the proverb, "there are no unending feasts," so after the last happy evening of the Feast of Lanterns, usually celebrated for three days, people settle down to another year of toil and grind. Now the New Year holidays are really over. Even the cake-shops take down the special signs which advertise the thin sweet wafers called *hsi yüan hsiao*, specially prepared for the first moon feasts, and, after an orgy of food, people go back to *pien fan*, or ordinary fare, "even the most inveterate rice bowls" — an expression used in China for a gourmand.

* * *

Star Festival

Yet, if the amusement season is over, there are still sacrificial duties to be done. The calendar commands that on the 18th (in some provinces on the 8th) men should "thank their lucky stars," because that day is known as the Star Festival. Now Chinese say that "the mind is a dark place, and men can not know what is coming to them from the sky." The planets and the constellations to them are not empty worlds, but the abode of sainted heroes who not only rule the skies but strongly influence the course of human destinies. Sacrifice is due to these spirits, long ago placed in charge of the heavenly beacons. Therefore, in the night, during the third watch, when all the stars are shining, an altar-table is spread in the courtyard, facing north. Upon it are placed two rough prints in colours, one representing the Star Gods, the other the Cyclical Signs so closely related to them, also a sealed envelope containing a chart of the Lucky and the Unlucky Stars. Some of these Star Gods, like the God of Literature, the God of Longevity, the Old Man of the Northern Constellation, are worshipped at other times, but all are included in this combined service.

The Gods of the Lunar Zodiac (*see* "The Chinese Calendar") are represented by the most "grotesque images, white, black, yellow, and red, ferocious gods with vindictive eyeballs popping out, and faces like lumps of putty," some of the male, some of the female gender.

The master of the house first worships the heavenly hosts as a group, then makes a special prayer to the star which presided over his own birth. This he does on his personal anniversary, but repeats in the first moon because the New Year in China is supposed to be the birthday of every man.

The food-offerings on the Star table are insignificant—only three or five bowls of rice-balls cooked in sugar and flour. But as the head of the house makes his bows, one hundred and eight little lamps disposed before the star tablets are lighted. These have special wicks or spills made of red and yellow paper, and the oil in them is perfumed. When they burn out, a matter of only a few minutes, each son of the house comes forward and makes obeisance to his own star by re-lighting three lamps. According to the brightness of the flame, he knows whether good or bad luck awaits him. Some people used to put lamps near the household well, the washing-stone, and here and there about their courtyards, giving the effect of tiny "guest-stars" come down to earth.

Women were forbidden to take part in this festival, and supposed to hide till it was over. Then they might join the family group and share in the *t'ang yüan*, or "round sweets," offered on the altar-table, while the junior members made their sacrifices to the God of the Hearth and the Door Gods, lighting lamps for them in the kitchen and in front of every door. Thus, as was customary and polite, they took this opportunity of remembering the familiar deities with whom the family has been on intimate terms throughout the year.

* * *

The very day after the Star Festival, the 19th of the first moon, is known as the "Rats' Flitting," or the "Rats' Wedding Day," when people are supposed to go to bed early, lest they disturb the rodents and the latter revenge themselves by being a pest in the house for the rest of the year. As a matter of fact, most people, tired out by the holidays, are quite ready to retire and leave the rats to their own devices.

Rats' Wedding Day

A curious legend is attached to this day. The popular Peking version runs as follows. Five hundred years ago a sleek rat lived in a cave somewhere in the northern hills. Partly for love of solitude, partly to escape the summer heat, this wise little animal made its home in the rocks where it somehow became changed into a woman. Now solitude may be suitable for rats, but is unendurable to women. So one day, when some charcoal-burners built a hut on the hillside near her cave, the rat lady went and knocked at their door. The man left at home to prepare a dumpling dinner for his companions was amazed to see a female in this lonely place. But when she offered to cook for him, man-like he allowed her to prepare supper while he smoked a peaceful pipe. Sitting by, he noticed a claw mark on the dough cakes. Then he looked closely at her hands, and discovered to his horror that her fingers had rat's claws. "This stranger," said he to himself, "must be a witch." So he seized a meat-chopper and lopped off one of the woman's hands, whereupon she instantly disappeared. Later, he and his companions, following the blood trail to the cave, found neither rat nor woman. "*Ai yah!*" they exclaimed, "she must have been an Immortal." And they went home and told their wives about it. Fearing revenge, the latter prepared food-offerings for the rats at once, and every year on this particular day their daughters, their granddaughters, and their great-grand-daughters, did likewise on the anniversary in remembrance of the strange lady. Thus they hope to encourage the mild and benevolent members of the species to protect them against the more destructive rats, and persuade the latter to forage on the neighbours' premises. The queer psychology, human and animal, underlying this legend, is peculiarly Chinese.

* * *

Gathering of the 100 Gods   The 19th day of the first moon has, however, a deeper significance in popular life. It is the only day in the Holiday Moon when newly married daughters may visit their parents, because it is primarily a "day of meeting." They take their example from the gods who gather together on this occasion and make a group-visit to the Yü Huang, or Jade Emperor, Taoist Lord of Heaven. This "Meeting, or Gathering, of the Hundred Gods," as it is called, is celebrated in the homes with food-offerings, bonfires of paper money to light the deities to the "lofty seat," and ladders made of strips of yellow paper to help them in their ascent to heaven.[9]

No altar could possibly give room to the statues of all the divinities, and at the home service held to *sung* them, or "see them off," as Chinese *sung* a mortal guest, a picture or a tablet will be put up with several rows of representative figures; in the upper row, for instance, Kuan Yin, Goddess of Mercy, the "Heavenly Lady," or T'ien Hsien Niang Niang, "Our Lady of Good Eyesight," and "Our Lady of Many Sons," and, below, Kuan Ti, "God of War," the God of Riches, and the God of Medicine which are, in fact, the most popular members of the Chinese Pantheon. In addition, a package of *chih ma* (*see* "The Twelfth Moon") dedicated to all the gods is also placed on the altar, to be burned at the end of the service which includes an offering to the ancestors, and to a symbolic inscription of the Heaven and Earth Spirits hung on one of the verandah posts.

The *chih ma of* the Hundred Gods, known as

---

[9] The term "100 Gods," as we know, stands for the whole Pantheon, which actually numbers thousands, in much the same way as the term "100 Families" stands for the whole Chinese race.

Po Fên, or *Wan Fo*, deserves special mention because the Chinese themselves consider it of great importance. It may be a single paper sheet with figures of a number of divinities printed on it, or comprise a varying quantity of sheets, each with a single portrait. The characteristic feature of the "heavenly sheet" is the curious and tolerant jumble of personalities, attributes, and dignities. Side by side, many-handed, double-headed, seated on thrones or riding mythical monsters, shrouded in fire, serenely passionless or wickedly wrathful, these rival images include some of the oldest divinities of the Chinese Nature cult—figures like Heaven and Earth, several Dragon Kings, the Household Gods, the Ch'eng Huangs, or City Gods, the fearsome Fire God, etc. Next may come Confucius and his Disciples, the Star Gods, the Gods of the Seasons and of the Cardinal Directions, the Gods of Riches, of Luck, of Long Life, of Medicine and Diseases. A numerous group of Deities of the Nature cult are humanised as Ministers of Thunder, of Lightning, and of Water. Nor is the historical Buddha absent in this comprehensive collection of divinities, or the important saints of his church. National heroes who have been deified—men like Kuan Ti, Yo Fei, and Lu Pan—are also present, so are the patrons, legendary or historical, of sundry professions, and heroes deified in comparatively recent times, like the XVIIth century graduate Chang, who, after passing his examinations in Peking, miraculously dreamed himself into a heavenly official.

* * *

Photo by the Asiatic Photo Publishing Co.

A SHRINE TO THE JADE EMPEROR.

ALTAR WITH SPIRIT-TABLET OF AN IMPERIAL PRINCE.

Many as they are, these gods are only, so to speak, a selection of the eight hundred who appear on the Yü Huang's register, in addition to a myriad saints. And the whole of this vast celestial multitude visits the Supreme Emperor once a year to report to this mighty Overlord of the Skies.

Jade Emperor

Despite his power and his popularity, the Yü Huang, "Jade" or "Pearly Emperor," to give him his most usual title, is a vague and misty figure of comparatively recent origin. Many legends cluster round the name of this Taoist Jupiter. While on earth, he appears to have been a very normal person, but unusually kind and generous to the poor. When his father died, this pious youth ascended his throne only to abdicate shortly afterwards, become a hermit and devote the rest of his life to acts of charity and good works. So much for legend.

Historically speaking, the "Yü Huang was born of a fraud and sprang ready-made from the brain of a Sung Emperor who pretended, in an hour of need (about A.D. 1000), to be in direct communication with a Supreme God of Heaven." (See *Myths and Legends of China*, by E. T. Chalmers Werner). Some authorities assert that the Yü Huang was originally identified with Indra, which would make him a Buddhist divinity incorporated into the Taoist Pantheon. Others admit he was invented outright by the Taoists who feared to be outrivalled by the Buddhists in popular favour. For centuries there was much bickering and sometimes bitter strife between the rival creeds; when the imported religion declared the trinity of the Buddha, the Law, and the Priesthood, the followers of Lao Tzŭ formulated a trinity of their own, the Tao, or Pure Way, the Classics, and their Holy Orders. The Yü Huang was set up as a rival to the Lord Buddha, and the three-fold Taoist Heaven over which he ruled imagined to outshine the Buddhist Paradise.

Simple folk consider the Jade Emperor a humanised impersonation of Shang Ti, or Heaven itself. As such, he has become exceedingly popular. His shrine is on many a mountain top, his image in many a village temple where, as so often in China, he is worshipped long after the roofs fall, in fact, until nothing remains of his sanctuary to charm the eye or conjure up visions of his supreme power or past poetry, except the crumbling figure of the All Highest himself. Still, on his birthday, the 9th of the first moon, some one will stick a lighted incense-torch in the cracked pottery incense-burner, and whenever the "Gathering of the Hundred Gods" is fêted, as it is at Po Yün Kuan outside Peking, crowds go to say a prayer to him.

* * *

Popular Amusements

The New Year is, as we have seen, primarily a family festival when men return to their homes to pay respect to their ancestors, to perform their religious duties, to greet the elder members of their clan, and enjoy a well-earned rest after a year's hard work. It is also the season of popular amusements.

In Peking and its environs, for example, innumerable fairs fill the people's idleness. There are flower-fairs at Hu Kuo Ssŭ, where narcissus, the New Year flower *par excellence,* peonies, magnolias, early double-peach blossoms, especially cultivated in Peking, are in flower, and "heavenly bamboos," orange trees, lemons and "Buddha's Fingers" are in fruit. There are "Devil Dances" at the various Lama Temples, fairs at the T'ieh T'a Ssŭ outside the Tung Chih Mên, at the San Kuan Miao near the Tung Ssŭ Pai-lou, at the Big Bell Temple, and many other shrines.

But one of the most important and the most popular fairs which may be taken as typical of them all, is that held in the Liu Lu Ch'ang, the century-old amusement quarter of Peking where emperors and peasants have always gone a-junketing.

The Liu Li Ch'ang bazaar is divided into two distinct parts. One is frequented mostly by the well-to-do classes, whose brass-trimmed rickshaws and modern limousines fill the narrow streets leading to a Temple of the Fire God. Here the courtyards are turned into a curio exhibition where the important antiquarians of the city display their jades, porcelains, and pearls. Their stalls crowd so close up to the shrine of the deity that the pious (among whom we find many merchants) have scarcely room to *k'o t'ou* before the red-faced Fire God, dressed for the occasion in a splendid embroidered robe.

Further down the same narrow street, an open square is the resort of the common people in blue cotton coats, the uniform of the land. This is a place of fun and jollity, reminiscent of the *foire de Neuilly* translated into the terms of Chinese amusements. The gay hues of patterned fruits set out under umbrella-tents catch the eye. A babellish clamour of voices arises from the open tea-shops and out-door restaurants, where gossips gather and *coups d'Etat* are planned. Peddlers cry their tempting wares. Little temporary stalls, arranged in lines, sell knickknacks of all kinds—artificial flowers for the hair, ribbons, garter-bands, women's ornaments, even cheap foreign goods like socks or mittens. Here stands a booth with iron-ware, yonder a book-stall with cheap popular novels, or "small speaks" as they are called. A Taoist priest, in slate-coloured robes and hair coiled on top of his head in the fashion of the Ming dynasty, sells amulets. His neighbour cries: "hot dumplings, steaming hot!" A whole lane is monopolised by the toy-sellers where tiny tots, clutching their New Year coppers, find choosing difficult. The tops and iron marbles, the foot-shuttlecocks with pink and blue feathers, the life-like soldiers made of crickets, the models of mud farm-houses, the miniature sets of furniture, the clay dolls and flannel dogs, appear year after year—ingenious trifles, made with the simplest materials and sold at a ridiculously low price, that never fail to delight Chinese youngsters who ask no novelties.

For the grown-up there are mat-sheds with peep-shows shocking to the fastidious, and photographers where Chinese flappers dressed in the extreme of the new fashion pose in company with young beaux in ill-fitting European clothes, clutching the steering-wheel of a cardboard motor-car, or sailing in a paper boat upon a paper sea.

Old-fashioned folk will always gather round the story-teller installed on a ricketty platform in a corner. An old man, with his queue composed of a handful of grey hairs, he stands in the midst of his circle of listeners beating two wooden sticks to mark his tale and call the attention of passers-by so that none may miss his entertainment. Beside him is a wooden table with a crooked leg, holding his fan, his tea-pot, and his squeaky three-stringed fiddle. A youthful assistant sitting behind him is ready to provide a musical accompaniment at thrilling moments.

For those who understand colloquial Chinese, there is much fun to be had in joining the crowd around this Oriental *improvisatore*. Good humour is essential since, before beginning the serious business of the afternoon, the teller of tales is likely to poke fun at the stranger in improvised verse.

"Here comes a stranger from far away, from far away"

he sing-songs,

"That stranger's nose is long and his nails are short,
Yet, perhaps, my friends, his pocket is full of cash,
For we have heard it said that foreigners are rich,
even if they have no manners."

Such personal allusions imply no lack of politeness. They are justifiable jokes to enliven the performance and may be made at the expense of any bystander, native or foreign. In fact, they correspond in spirit to the rabbit that our conjurers pull out of the fat gentleman's waistcoat.

After this jocular prelude, the story-teller deftly slips from the region of prosaic fact to the realms of fancy and romance. The tales he tells are drawn from history, or quasi-religious legends. With a wave of his fan, the old man calls up pictures of brilliant courts thronged with poets, philosophers and lovely ladies, of saints and soldiers, of favourites dramatically deposed. He weaves before our eyes the ever-recurring warp and woof of the tapestry of statecraft. From the dim past—and the dimmer the better, since a Chinese crowd will listen with more interest to the happenings of the centuries before Christ than those of to-day—he draws his heroes with subtle artistry and makes them live. Warriors like Chu Kuo-liang, the Mighty, Righteous and Crafty, and Yo Fei, the Chinese Bayard, strut the invisible stage; sovereigns like T'ai Tsung tread a stately measure, beauties like Yang Kuei Fei are graceful passing shadows, and saints like the Eight Immortals (*see* "The Fourth Moon") pursue their marvellous adventures; while gods like Erh Lang with his Heavenly Dog (*see* "The Second Moon") are summoned from the skies.

Because life, average life, is dull with the routine that kills romance, because the times are drab and governments the world over have forgotten the Roman adage that people need amusement as they need food, men crowd about the old story-teller who, with the modulations of his voice, can evoke the charm of the past. The work-bench is forgotten while Kuan Ti (*see* "The Fifth Moon") mounts his charger Red Hare; the plough is lighter for an hour spent in pursuit of the Heavenly Monkey.

So long as men seek illusion, the story-teller

THE FIRST MOON, OR "HOLIDAY MOON" 153

will remain a feature of Chinese life, invited to wealthy homes to amuse those of the "inner chamber," welcome in the village tea-shops on summer evenings where he is the peasants' living book of history, drawing his audience at fairs, and wringing hard-earned coppers from the pockets of the crowds by the same device our up-to-date magazines use for serial stories, demanding a new outlay for a new instalment.[10]

The herb-sellers and medical quacks appropriate a corner of the fair for themselves. Not far away are diviners with their cabalistic books ready for consultations. It is these soothsayers who choose lucky days for weddings, and who decide whether certain matches are permissible, for in China it is not one individual choosing another individual, it is one family choosing another with the sanction of these gentry from whom no secrets are hid. The best of them undoubtedly have second sight, such as is often the gift of men belonging to old races. They pierce the veil of the future partly by instinct and partly by an elaborate technique developed by years of study of ancient and bulky books. Among them are spiritualists, interpreters of *planchette* writing, and men of good

---

[10] The story-tellers belong to an organisation. Novices sometimes start their careers under an experienced teller of tales but, usually, in the old days, they were unsuccessful students. Their ideal will always be P'u Sung-ling (XVIIth century A.D.), most famous of Chinese novelists who, being unjustly rejected at an official examination, avenged himself by setting down, in the most beautiful style and purest language, weird stories of ghosts and goblins,—of fox fairies, brave knights and beautiful damsels,—in direct defiance of the conventions forbidding any scholar to defile his brush by touching on such unclassical subjects. An able selection from his *Liao Chai Chih Yi*, which all Chinese know, is contained in H. A. Giles' *Strange Stories from a Chinese Studio.*

education and standing who charge high prices for their prophecies. The naive brethren of the public square do simpler tricks and chiefly earn their fees consulting almanacs for the illiterate, and telling them the lucky days to start building their houses, to cut out their clothes, to sign contracts, sow their fields, bury relations, arrange parties, close rat-holes, and wash out water jars. Without the approval of the stars, as so interpreted, the average superstitious Chinese dares not move a finger in his daily life.

A street of side-shows may add interest to the fair. Here we find entertainers of various kinds. There may be a troop of girl-singers, *ch'ang ch'u ti*, indented to an *impressario* who hires them for family festivals, — women of notoriously poor moral character. Also girl-actresses, *nü hsi*, performing on an improvised stage,—women of somewhat better standing. Men-singers dressed as women sometimes act, while their daughters sing spicy songs to the accompaniment of castanets. But the acrobats draw the largest crowd. They have put up a small enclosure, rather like a circus, with a high pole in the centre and a narrow track where the "horse-girls" appear. The show consists of alternate turns by riders and acrobats. Most of the performers are little girls dressed in red silk. Small-footed women like bright poppies ride ponies quite as well as our own circus professionals. One girl, with her legs and tiny feet arrayed in leggings of black and white printed cotton stuff, balances a large earthenware jar and whirls it rapidly with her toes. Then she stops, while a smaller maiden

Photo by the Asiatic Photo Publishing Co.

WOMEN ACROBATS.

VILLAGE STORY-TELLER.

THE FIRST MOON, OR "HOLIDAY MOON" 155

climbs in and is bounced up and down on her bound feet. Hampered as they are, it is astonishing what these women can do. The next item on the programme is a trapeze act on the high centre pole. No net is required by police regulations, and the sight is dangerously thrilling. The pole itself sways, and the ropes of the trapeze look far from new. Nevertheless, the dare-devil stunts pass off safely, while a clown provides the comic element by a really marvellous trick of climbing up and sliding down by two ropes round his feet, keeping up a running fire of patter with the audience meanwhile.

When the bats come zigzagging through the square, the side-shows close, the peddlers pack their wares, and the crowds disperse. Weary babies clutching their toys nod drowsily on their mothers' knees, undisturbed by the shouts of the rickshaws making their way through the muddled traffic. Women go home pleasantly tired, to talk over the excitements of the day and show their purchases to the rest of their household. But the young men, loath to curtail the holiday, resort to fashionable restaurants for the jolly fellowship of feasting, or spend the evening at a theatre.

The drama is enormously popular with all classes in China, and high and low delight to escape into the world of make-believe created for them by the "Disciples of the Pear Garden." Though the conventions of the Chinese stage are strange to us—although not unlike those of our own in Shakespeare's day—they have the sanction of centuries and, being thoroughly understood, are entirely satisfactory to the Chinese, trained as they are to let imagination elaborate suggested stage properties. There is very little originality in the plots of the pieces presented. Many of the plays are really tableaux of gesturing and posturing, in which the literary side is quite subservient to the pictorial. But the supreme beauty and delicacy of the acting is always a delight.

From this artistic enjoyment nothing disturbs the audience, neither the clashing of cymbals and clangour of drums which irritate our ears in the military dramas, nor the discomfort and bad ventilation of most of the play-houses. The Chinese theatre fan sits blissfully serene in the midst of a continuous hubbub, undisturbed by food-hawkers, by the hot towels deftly tossed across the theatre, the constant coming and going of attendants and the loud discussions in the audience. Our own theatre etiquette requires us to sit silent even if we find a play boring, and we feel bound to give attention to a great and admired actor towards whom fidgetting and chattering are signs of disrespect. It is not so in China. A favourite player is a popular idol and gains hearty applause. Nevertheless, his hearers will talk through many of his speeches, stretch their stiff legs, call for tea, go out for little strolls in the middle of the play, and come back noisily.

The world of illusion he creates is evidently strong enough to withstand such material interruptions. Indeed, it is so strong that when the play is over the rude transition to reality, it is often expressly softened by parallel couplets written upon the front of the stage. These scrolls remind the dispersing audience that "Life is not a performance on the stage," and that "When the players make their exit, the tragic and the comic, the parting and the re-union, must instantly become a vanishing dream."

CHAPTER VI.

THE SECOND MOON, OR "THE BUDDING MOON."

N compliment to the Lord of Light whose festival falls on the second of the second moon, "the plants on the mountain are changed into jade". This is the classical way of saying that every little leaf and bud makes ready to greet the sun. Willow-tips turn green, lilacs prepare for blooming, and in every orchard fruit-blossoms are eager to burst their sheaths. Yet, if the season be late in North China, no flowers dare open except those sheltered in the paper hot-houses, for fear of bitter winds and sudden snow-flurries.

Sun Worship   Peking beggars have a proverb: "Now is the time to exchange wadded robes for new yellow cotton coats", meaning garments of sunshine. Theoretically, the days should be growing warm. Practically, the wretched tatterdemalions often shiver on street-corners as they repeat their cynical dictum.

The Sun God himself says in an address to the people: "My speed and strength are according to my own pleasure. No one can urge me forward, no one can stay my progress . . . The

THE SECOND MOON, OR "BUDDING MOON" 159

dwellings of all men I, the Great Male Luminary, visit with my light, and when I come forth the whole heaven is tinged with my brightness . . . You, however, the people, address me with reverence and respect . . . . for were I, in displeasure, to cease my shining, you would all die of starvation, inasmuch as the earth would no longer bring forth fruits". Copies of such addresses are sometimes distributed by rich men in expiation of their sins, or as a thank-offering for recovery from illness.

The Sun Festival in China, as elsewhere, is a relic of prehistoric days when the planets were first worshipped by primitive man. Mythologically, the sun has never had the same influence on Chinese imagination as the moon, and solar myths are comparatively few. Nevertheless, sun-worship still has a direct bearing on daily life. "The *fêng shui*[1] cult is a humble survival of the original priestly lore . . . and has served to keep the Chinese people as conscious of the movement of the heavenly bodies as the ancients and, therefore, naturally as keenly interested in the progress of the calendar and in the importance of the points of the compass."

While to the Emperor, as Sovereign Intercessor between Earth and Heaven, formerly belonged the right to perform the solemn State ceremonies at the Altar of the Sun outside the capital, humbler folk observed the festival in a humbler way. "Sun-cakes," not unlike our griddle-cakes, only thinner and varying in size, were prepared

---

[1] *See* "The Hundred Gods."

and skewered together with a bamboo decorated with a cock made of dough, coloured red.²

In most homes five saucers filled with these cakes, together with the usual red candles, paper money and incense, were placed on a table in the courtyard facing the sun sometime before noon. The short and simple out-door service began with *k'o t'ous* to the sun and ended with a bonfire of paper money in an iron brazier. An unusual feature of the simple ceremony was the part taken in it by the women, whose presence is so often forbidden at the offering-table. They were admitted because of the old superstition that after death people will have to drink the water they wasted while on earth. As the Lord of Day sees all things and knows exactly who has offended, housekeepers pray he will be lenient to them when the judgment day comes.

Various amusements were organised in the Palace to mark the holiday in olden times. The Court assembled to witness cock-fights, such as are still popular among the Malays.³ Meanwhile, the ladies withdrew to make merry tumult among themselves, swinging on swings decorated with

---

[2] In China the sun is supposed to be inhabited by a cock. This Oriental Chanticleer attends the rising and setting of his Lord, and his crowing is supposed, as in Europe, to chase darkness and evil influences away. But in the East this bird of good omen not only symbolises the dawn but also the New Year and, in general, all auspicious beginnings. The festival cock is red because, as we know, red is the colour of joy, light and vitality. A white cock is often carried in a basket on top of a coffin. His crowing is supposed to enable the soul to follow the body to the grave.

[3] Under the Emperor Ming Huang (A.D. 712-756) a special enclosure was set apart in the Imperial City for cock-breeding, and the children of five hundred Imperial guardsmen had charge of the birds and trained them to fight.

THE SECOND MOON, OR "BUDDING MOON"   161

bright silk streamers. It must have been a pretty sight to see Ming Huang's three thousand concubines,—"half angels" as he called them,—floating to and fro like brilliant flowers on a vine, clad in their costly robes of gauze, girdled like queens, their beautifully dressed hair decked with mock flowers, wonderful combs, pins, and ornaments of pearls and jade.

Less addicted to refined pleasures than the hyper-civilised Ming Huang, one of his predecessors, Chung Tsung, the fourth T'ang Emperor, commanded for the season of spring sports a tug-of-war with seven Ministers of State and their sons-in-law on one side, and five Ministers of State and four generals on the other. History does not relate which team won, but it appears that two Ministers of State were knocked over. Being old men, they rose with difficulty. Their struggles to get up reminded the Sovereign of beetles on their backs and caused him to laugh heartily.

Nowadays people go for amusement to the tiny toad-coloured temple of T'ai Yüan Kung outside of Hata Mên in the Chinese city of Peking. Here, in the single small sanctuary the Sun, Master of the Stars, is represented by a gilded wooden statue. Visitors burn incense before the image, then adjourn to the small covered courtyard where tea is served free at the expense of the neighbourhood by volunteers with yellow cloth aprons—men desirous of earning merit or fulfilling a vow of humility. More than likely, a wandering story-teller will be present to amuse the crowd with old sun-myths appropriate to the occasion. "Hear!" says he in a shivery, shaky sing-song voice, "hear of the beginning of the world, when neither form nor force was manifest and the earth was a shapeless mass that floated like a jellyfish on water, when dim gods appeared and disappeared. Hear how P'an Ku came forth and regulated the Cosmos, separating earth from heaven, how he forgot to set the sun and moon in their proper courses and they retired into the Han Sea, wherefore the people dwelt in darkness.

Then the 'Terrestrial Emperor' sent an officer called 'Terrestrial Time' with orders that the Heavenly Lanterns should come forth, take their places in the sky and divide night from day. They refused. So once again P'an Ku was called. At the divine direction of Buddha he wrote the character for 'sun' in his left hand and for 'moon' in his right. He went to the Han Sea, stretched forth his left hand, and called the sun, stretched forth his right hand, and called the moon, at the same time repeating a charm devoutly seven times. Whereupon the Lord of Day and the Lord of Night ascended on high and there was no more darkness."

The curious feature of this myth is its typical jumble of religious ideas so common in China. That Buddha should be giving orders to P'an Ku who was a purely Taoist conception, and a late one, shows how easily popular imagination overrides time and warring creeds.

* * *

Earth Gods

Having reverently saluted the Sun and kept holiday in his name, the Chinese Sovereigns and the Chinese people next turned to rites even more important in an agricultural country—the worship of the Spirit of Earth and the Gods of the Soil at the season when the year passes out of the dark tunnel of winter into the brightness of spring.

Since time immemorial men have attempted to propitiate the forces which regulate their food supplies. Even before the beginnings of settled society this duty was recognised. But as civilsation advanced and governments became stable, obvious reasons of State policy divided essential sacrifices into official and popular ceremonies.

Sovereign Earth, Consort of Heaven, conceived in this sense as the female portion of the Universe, naturally received homage from the Chinese Ruler, entitled Son of Heaven and Lord of Earth. Accepting the Throne, he also accepted the responsibility of official mediator between gods and men, and with it the right to direct and modify the cult of those generative forces which weld together, in one mighty instinct, the highest and lowliest of terrestrial creatures. Thus we find a succession of lonely figures holding the supreme power, bound close to the throbbing heart of the universe, canonising their predecessors like Shên Nung (2838-2697 B.C.), inventor of the plough and often called the "Spirit Farmer" or "First Patron of Agriculture," or even, as in the case of Liu Pang, founder of the Han dynasty, constituting themselves divine patrons of the crops.

Once the State worship of Earth as a whole was completed with exceeding ceremony (*see* "Imperial Sacrifices"), and an example of industry set by a Ruler who put his own hand to the plough, the local Rulers of the Soil,—popular substitutes for Sovereign Earth reduced to a specific locality, and sub-divisions, so to speak, of the grand Principle of Fecundity,—received sacrifice. The act is now performed out of doors by the farmers some time in early spring—the season of sowing (though the actual date varies according to provinces), and is repeated again in the second autumn moon as a thanksgiving for "value received." Either each family erects its own altar topped by a flat stone in its own fields, pours libations and invokes aid and protection, or a communal altar is set up where all the inhabitants of a village gather, including the women, who are very devout in this worship.

There is a whole group of divinities included in this composite worship of fruitfulness. The oldest and most important is *Hou T'u*, known as the Queenly, or the Earth Bosom, a very ancient and very hazy divinity, sometimes a male, sometimes a female, sometimes considered an incarnation of the ancient sages who invented agriculture or separated land from waters by canals.

Next in importance come *T'u Kung* and *T'u Mu*, God and Goddess of the Soil, who must not be offended by digging in the ground on inauspicious days, nor neglected when the plough turns the first sods for a new crop. This couple are likewise patrons of wrestlers who pray to them for the gentle art of "falling soft."

There is also a special God of Harvests and a host of lesser farm-gods, including the God of Green Sprouts, the Corn Spirit, the God of the

Five Grains who receives a five-fold offering, and the God of Locusts, also, for some unknown reason, patron of beggars. The last named is none other than the Terrible Marshal Liu (*Liu Mên Chiang Chün*), variously identified with a T'ang or Yüan general, though occasionally with a Sung graduate, and specially revered in Chihli and Shantung. South of the Yangtze farmers pray to him under another name at little shrines that very often contain only simple uncarved stones. Where no sacrifices are neglected, people selfishly hope that the Locust King will direct his plague to some locality more careless about his rites. "The Chinese of Weihaiwei say that, in spite of the devastation that locusts can work among crops, they are really not so much to be dreaded as many other insects who *have no king* and are, therefore, under no one's control and subject to no law. If monarchical government could be established among all the harmful grubs and flies, the happiness of labouring mankind would be materially augmented."

The Spirit of the Green Frog is also worthy of mention. This merry little creature loves wine, feasting, and theatricals. The wine presented to it is supposed to cure deafness, and when plays are held in its honour it hops upon the programme, indicating with a foot its favourite pieces. Throughout the Yangtze valley the Spirit of the Green Frog has become a kind of secondary water divinity. Though we have not been able to trace the deeper meaning or the obscure origin of the frog-symbol in China, it is curious to note that in ancient Egypt a lunar Divinity—the child-god Tot,

Frog Worship

son of Isis and Osiris (creators corresponding to the Chinese *Yang* and *Yin*), who controlled the rejuvenating floods of the Nile,—is depicted with a frog. We may imagine, therefore, how this amphibious creature became a patron of tillers of the soil, dependent as they are on the rise and fall of the waters to enrich their fields. The unusual popularity of frogs in those Chinese provinces where rivers or lakes abound would also seem to confirm our theory.[4]

In Central China frogs often have their own temples, and in some places they are worshipped by local officials on assuming office. In Chekiang, where the cult is still wide-spread, a frog-spirit may chose to visit private homes, in which case it must be brought back to the temple with music and drums. Such domiciliary visits are fortunately rare, however, as frogs are allowed great freedom in their sanctuaries. They may hop upon the altar, or invade the living rooms of the priests, and wherever they go they are fed by devout Buddhists.

In Chinese symbolism the frog also stands for riches, and a very popular luck-emblem is a boy playing with a frog and a string of cash. According to one legend, the boy is the Immortal Liu Hsi pictured with a bandolier of alternating eggs and coins. Liu Hsi, by origin a Pekingese and Minister of State under the King of Yen, once received a visit from the Immortal Han Chung-li (*see* "The Eight Immortals," Fourth Moon).

---

[4] The *Chan*, or Three-Legged Toad, is associated with the Moon, the Moist Planet. *See* "The Eighth Moon."

THE SECOND MOON, OR "BUDDING MOON" 167

As he sat in the reception hall, the visitor began piling one egg upon another with a coin between every two eggs. "That's a dangerous trick!" Liu exclaimed. "Not more ticklish than being Minister to your Prince," Han Chung-li replied. Liu understood the hint and, going forthwith to his Sovereign, accused him of being a usurper—which he was.

As a result of his frankness, Liu was dismissed from the State service. He then became an ascetic and occupied himself searching for the Pill of Immortality. Centuries later—in 1662—he appeared again, this time as a servant in the family of a merchant named Pei. His employers only realised they had an Immortal in their midst when he took the son of his master to see the Feast of Lanterns in Foochow, disappearing with him right out of a crowd and journeying over provinces in a miraculously short time. Finally, one day he fished from the well a frog which he tied with a coloured cord and placed upon his shoulder, jumping for joy the while and exclaiming: "I have been looking for this frog for years, and at last I have found him." When the neighbours rushed in, attracted by the shouts, Liu publicly thanked the Pei family for their hospitality, rose into the air above the gaping people and disappeared, frog and all. Because of the coins he carries, and his association with the miraculous frog, Liu is invoked for success in commercial enterprises and his image is pasted on the double flaps of a door.

Closely allied to the Farm Gods are the *San Kuan,* three primordial sovereigns called Lords *The San Kuan* of the Three Worlds, who are supposed to have a kind of ghostly superintendence over earth, sky, and water.[5]

This Taoist trinity, dating from the Vth century A.D., consists of *T'ien Kuan,* Heaven, *Ti Kuan,* Earth, and *Shui Kuan,* Water, and they represent the sources of happiness, forgiveness of sins, and delivery from evil, respectively. Their worship has passed through several phases, now neglected and again popular, especially among the peasants who pray to them to avert misfortunes.

The cult is most developed in Fukien province where the San Kuan are worshipped separately on their various birthdays,—T'ien Kuan on the fifteenth of the first spring moon, Ti Kuan on the fifteenth of the first autumn moon, and Shui Kuan on the fifteenth of the first winter moon. In addition to averting calamity and curing disease, the "Three Beginnings," as they are often called, record the good and bad deeds of mortals, wherefore no capital punishment was inflicted on their birthdays.

* * *

The Wu Shêng

Most curious of all the farm gods are the *Wu Shêng* or "Five Seers," worshipped to avoid disease in the farmers' poultry-yards and pigsties. They are sometimes identified with the Spirits of foxes, badgers, weasels, hedgehogs, etc. Although they were venerated under the Mings, the Manchu dynasty officially forbade honour to the Wu Shêng, who were classed among the

---

[5] Not to be confused with the *San Huang,* or Three Emperors. See "The Fifth Moon."

"Corrupt gods." On one count the Wu Shêng certainly deserved the Imperial prohibition. They appeared in dreams to girls and women, assuming strange forms and arousing evil thoughts. And they claimed as brides those who already had husbands. Thus their worship is about the nearest attempt at the deification of sensuality found in China, where the name of religion has never shielded or countenanced licentious rites or orgies. Doubtless, it is the Chinese reverence for the family which has practically purged their religions and their folk-lore of all indelicate relationships. Their pantheon has no counterpart of Venus or of Eros. No descriptions of the amours of their deities are recorded, and illicit love or passion, whether among gods or men, is never featured as in Hindoo tales and Greek myths.

In the case of the Wu Shêng, the exception proves the rule. Nevertheless, even the Throne was powerless to uproot these very old and well-beloved deities, though time after time protests were made against them, the last not so many years ago when a high official of Chekiang province denounced the Five Seers as workers of evil and authors of a terrible plague. To him the people answered: "What the gods have sent, the gods can take away. More than ever must we propitiate those who can work us harm." Here again we find reflected the idea that the relation of gods and men is one of commercial reciprocity. It is therefore logical to assume that when immunity from disaster depends on worship and sacrifice, these must be given.

Though the Wu Shêng are worshipped more elaborately in the south, possibly because it was further from Imperial control, their little shrines that look so much more like dog-kennels than temples to the gods, are common enough in the villages around Peking. Among the rural population the vaguer spirits and the older ritual prevail with little change, though in times of acute persecution the latter has been diverted, or even transformed. Sometimes the Wu Shêng temples were dedicated to other gods, notably to the God of War, and again popular piety disguised the banned idols by deliberately confusing them with Tsai Shên, God of Wealth (*see* "The First Moon"), multiplied into a five-fold personality for the occasion.

* * *

The Lung T'ai T'ou About the same time that the farm gods are worshipped, when spring stirs in the ground, comes the day—the second of the second moon—called the *Lung T'ai T'ou*, or "The Dragon Raises his Head," after his long sleep. His Majesty began to hibernate in the ninth moon at the beginning of winter, and then supposedly took the form of a tiny creature. But when the first warm day brings the first thunderstorms, he flies up on the lightning to the sky. The renewed vitality he typifies is shared by the insect world. Hence the popular name for the day is the "Awakening of Excited, or Torpid, Insects."

Passing through the villages of North China at this season, we notice that every farmer has placed either a block of ice on the heap of fertiliser stored in his yard, or else a paper pennant on a reed flag-staff. Both fetishes are meant to propitiate those tiny creatures that stir in sympathy with their lord, the mighty Dragon, and thus to avoid insect plagues throughout the coming year. Various other old and odd customs are connected with the occasion. In Shantung people get up before sunrise and cook a kind of dumpling which, as it rises, is supposed to assist Nature in her work of stirring the sluggish or dormant vitality of the animal and vegetable kingdoms. Household superstition requires water jars to be washed out and *k'angs* fumigated on the day the insects awake. Women are warned not to do any needlework, lest they should inadvertently prick the dragon, with the result that a boil or sore might develop on the part of the body corresponding to the spot wounded on the monster. A feature of the holiday is a good midday dinner, when dumplings are eaten and strips of dough with meat fillings. The former have a vague resemblance to ears, and the latter to whiskers, though a good deal of imagination is needed to see the similarity. Still, it is polite to inquire of Chinese friends on this occasion: "Have you eaten the dragon's ears?" or: "Have you savoured the dragon's whiskers?"[6] Everyone will understand what you mean.

In the evening children are supposed to wash their faces in a soup made from the shrub *Lycium chinense*, whose remarkable properties assure that they may never be ill and never grow old. "This reminds us," says Johnston, "of the old

---

[6] In modern days the latter have come to be identified with asparagus.

English belief that young people will preserve their beauty indefinitely by going into the fields on the first of May and washing their faces in May dew."

* * *

Erh Lang

On the 3rd of the second moon Pekingese dog-lovers visit the little yellow-roofed T'ang dynasty shrine on Hata Mên Street dedicated to Erh Lang, nephew of the Heavenly King, and his dog which "howls towards the sky."

Erh Lang gained fame through his pursuit of the Heavenly Monkey. His adventures are set forth in great detail in the Hsi Yü Chi, a record of a journey to the Western Paradise to procure the Buddhist scriptures for the Emperor of China and, in reality, a species of Chinese "Pilgrim's Progress."[7] The whole text is an elaborate allegory, and the victory of Erh Lang over the mischievous ape represents the ultimate triumph of the spiritual over the purely material. Sun Hou-tze, to give the monkey his popular name, was created by the Heavenly King "to skip and gambol to the highest peaks of the mountains, jump about in the waters, eat the fruit of the trees and be a companion of the gibbon and the crane." But this "ornament of all the mountains" was not content to keep his appointed place in the scheme of things. Ambitious, he became king of his kind; curious, he discovered the secret of immortality; mischievous, he overturned his Master's throne.

---

[7] For a detailed description of the adventures of the Heavenly Monkey the reader is referred to *Myths and Legends of China*, by E. T. Chalmers Werner, Chap. XIV.

Then Yü Huang (*see* "The First Moon") in great indignation called upon Erh Lang to pursue the knave who had eaten the peaches of everlasting life in the garden of heaven, and the pills of immortality compounded by the gods. "Hand him over to me," said the Heavenly King, "I will distill him in my furnace of the Eight Trigrams and extract from his composition the elements which render him immortal." This, however, was easier said than done, for the monkey, when hard-pressed, could change himself into such dissimilar things as a worm, a Taoist priest, and a Buddhist temple. Finally Erh Lang, by an extra metamorphosis of his own, outwitted his quarry, and the moral of this delightful tale of legendary lore is virtue triumphant.

The Heavenly Dog appears to have assisted his Master in the chase, and shares his altar where owners of sick pets come to burn incense, collect the ashes and administer them as a medicine to the ailing one. If the latter recovers,—as sometimes happens after the best medicines,—his devoted master or mistress buys a toy puppy with real fur and a red flannel tongue, and places this as an *ex voto* before Erh Lang, patron of dogs.

* * *

Taoist and Buddhist Hells In the second moon (on the first) there also falls the anniversary of King Ch'ing Kuang, who is in charge of the First of the Ten Taoist Hells,— he who sets the span of human life and orders the re-birth of men as women and women as men, so as to give all those whose merits equal their crimes an equal opportunity.

Chinese mythology did not develop the idea of an Elysium or a Tartarus. The conception of such places, or states, was imported by Buddhism and then adopted by Taoism. Both religions, whose habit of borrowing from one another has often made it difficult to distinguish whence certain popular beliefs originated, divided their Inferno into a number of Courts, and pictured the tortures applied to sinners on painted scrolls or in such groups of life-sized clay models as we find in those temples designated by foreigners as "Temples of Horrors." Two singularly complete and awe-inspiring examples are the Tung Yüeh Miao and the Shih Pa Yü (outside the Ch'i Hua Mên near Peking).[3] Here we see how well the artists seconded the priests in vividly depicting the torments of the nether world. Wu Tao-tze, the great painter of the T'ang dynasty, is said to have depicted such scenes so realistically that the mere sight of his work terrified the butchers and fishmongers of the capital into abandoning "trades against which the anathema of Buddhism was hurled, and drove them to seek a livelihood in other directions."

Although educated Chinese look upon these pictured hells with good-natured contempt, and the more enlightened monks know that they are not in accordance with the true doctrine, they persist as a warning to ignorant folk who need to see the results of sin with their own eyes.

The Buddhists count an enormous number of

---

[3] For a detailed description of the Taoist Hells, see H. A. Giles, *Strange Stories from a Chinese Studio*, Appendix.

hells, including 128 hot hells under the earth, eight cold hells, eight dark hells, and 84,000 miscellaneous hells situated on the edge of the universe. As, according to this faith, there is no such thing as everlasting punishment, these hells really constitute a kind of Purgatory. Sinners are re-born in a place of suffering merited by their evil deeds, and they pass on from stage to stage in the great cycle of transmigrations, at the decision of King Yama, the Supreme Lord of the Underworlds. Although this dark Sovereign has been one of the pitiless judges for ages past and must remain so for ages yet to come, neither he "nor any other inhabitant of that dismal region is regarded as eternally damned," but will sooner or later reach the end of the path which leads to salvation.

The Taoists, when they adopted the Buddhist conception of the Inferno, reduced and simplified it to Ten Hells organised much like a Chinese *yamên*, with a King in the place of the presiding magistrate and a staff of recorders, lictors, representing the *yamên* "runners" or police, and executioners whose robes are trimmed with tiger skins, emblematic of their fierce hearts.

The entrance to these hells, euphemistically called the "Land of the Yellow Fountain," is supposed to be situated at Fêng Tu in Szechuan, through a ten-storied building that stands against a mountain with grim rocks and splintered crags. The gates of this gloomy, menacing place are sealed, when at night the cries of the damned pierce the iron-coloured cliffs. One man, and only one, had the temerity to enter—a certain governor of the reign of Wan Li (A.D. 1573-1620). Torch in hand he passed through that dread door and descended to the subterranean city beneath. Here he was met by Kuan Ti (*see* "The Fifth Moon") and passed on from one section of hell to another. It is from his descriptions, scrupulously noted, that mortals learned the detailed arrangements of the hells and how they were divided into ten courts, subdivided into sixteen wards each, and ruled over by ten kings.

The First Court is really an ante-chamber where men's good and bad deeds are inscribed in the Book of Life and Death. Fortunate are those whose balance sheet is to their credit, or nearly, for they may return to life on earth; likewise those whose sins being few may be forgiven if followed by quick and sincere repentance. Wicked souls are dragged by devils to a lofty tower—the Evil Mirror Platform—whence they gaze into a large mirror suspended in mid-heaven and see their past.

The Second Hell, the Great Cold Hell of Thieves and Murderers, is supposed to be situated at the bottom of the ocean. Its torments are as varied as they are horrible. King Ch'u Chiang (birthday 1st of the 3rd moon) is in charge of this vast realm with its enormous lake of ice where sinners are pushed to and fro till their bodies are flattened out.

The Third Hell of Black Ropes is under the jurisdiction of King Sung Ti (birthday 8th of the 2nd moon). Here unfilial sons, disobedient slaves, rebellious soldiers, and disloyal Ministers suffer.

THE SECOND MOON, OR "BUDDING MOON" 177

The Fourth Hell, reigned over by King Wu Kuan (birthday 18th of the 2nd moon), is named the Great Hell of the Lake of Blood. To its terrible tortures are condemned, among others, those who cheat in commercial dealings, do not pay their rent, sell sham medicines, counterfeit coins, and allow their animals to be a nuisance to other people.

The Fifth Hell—the Inferno of Great Lamentation—is the special province of King Yen Lo, or Yama (birthday 8th of the 1st moon) himself, the personification of an old Indian god. Strangely enough this same Yama, supposed among the original Aryans to have been "the first man who died," was once regarded as a benign deity and reigned in a heaven where "there was neither heat nor cold, old age or death, and in which passion and desire were stilled." Two dogs with four eyes a-piece and enormous nostrils guarded this paradise, and the departed were advised to hurry past these fierce beasts, the only objects of terror connected in the early myths with the King of the Dead. *Yen Lo Wang*

Later, by one of those divine metamorphoses which we find among Oriental gods, Yama or Yen Lo appears as the terrible judge of the deceased, with an explanatory legend attached to his name. He has been a mortal king engaged in bloody wars. His crimes merit punishment. Therefore, he is re-born as Ruler of Hell. His eighteen generals, including his own son, and his 80,000 soldiers descend with him to Avernus and serve under him as assistant judges and jailers. But Yama himself does not escape his share of personal punishment for his crimes committed on earth, and three times in every twenty-four hours boiling copper is poured down his mouth.[9]

Some say that not Yen Lo, but his son Pao is in charge of the Fifth Court of Hell. The confusion arises because, at one time, the youth was degraded from the rulership of the First Hell for being too merciful.

The Chinese idea of gross physical torture is replaced in the Fifth Court by the more subtle one of moral suffering. Here sinners are taken to the celebrated "See One's Home Terrace," curved like a bow and enclosed by a wall of sharp swords, whence they look back on their earthly dwellings. The cruellest sting of pain is the memory of past pleasures. The proud king remembers the pomps of his court, irrevocably lost; the wise but wicked man, his libraries and instruments of research; the lover of art, his pictures, jades and bronzes; he who delighted in pleasure, his sumptuous feasts, his dishes prepared with such skill, his choice wines; the miser will remember with a pang his hoard of ill-gotten gold; the robber, his stolen wealth; the cruel, their deeds of blood and violence. They will remember all this and loathe themselves and their sins. They will rage and fume to think that they have lost the bliss of Heaven for the dross of earth, for a few pieces of tinkling metal, for vain honours, for bodily comforts. Dante in the *Divine Comedy* expresses the same sum of

---

[9] For further details on Yama, *see* the pamphlet on *Some Divine Metamorphoses*, by Baron Staël-Holstein, and *Handbook of Chinese Buddhism*, by Eitel.

human misery when he makes guilty souls look back "on their desolate homes and broken household gods."

The Sixth Hell under King Pien Ch'eng (birthday 8th of the 3rd moon) is reserved for those who blaspheme Heaven, Earth, and the North Star.

The Seventh, ruled by the majestic god of T'ai Shan (birthday 27th of the 3rd moon), the powerful spirit of the Sacred Mountain (*see* "The Tenth Moon"), is the Hell of Pounded Flesh. To this Court will be condemned those who use human brains to make medicine, or bones from grave-yards to glaze pottery, or who stir up others to quarrel and fight.

The Eighth, or Great Hot Hell, ruled by King Tu Shih (birthday 7th of the 1st moon) is for undutiful sons and all who have in general shown ingratitude or disrespect to their elders. Punishment is also inflicted here upon women who have hung out clothes to dry upon the house-tops. The Chinese think this displeasing to departed spirits because such obstacles interfere with their flight through the air.

The Ninth Hell, under the jurisdiction of King P'ing Têng (birthday 8th of the 4th moon), is a vast circular place enclosed by an iron net where executed murderers and incendiaries, and those who have committed one of the "ten great crimes" are judged, and tormented pending the re-birth of all their victims.

Finally, in the Tenth Hell whose sceptre is swayed by King Chuan Lun (birthday 17th of the 7th moon), known as the "king turning the wheel of the law," there are six bridges, corresponding to the six forms of re-birth, across which all souls must pass as they come from the other hells. Those who are to be re-born are handed over to the Spirit of Winds who conducts them to the Tower of Forgetfulness. Here they receive an intoxicating drink to make them forget their previous existences. The idea is similar to the Greek conception of the Waters of Lethe.

* * *

Kuan Yin, Goddess of Mercy

Fortunately the pains of all these hells, terrible as they are, can be mitigated. The Kings of the Courts, like many mundane officials, may be bribed. With incense and offerings on their anniversaries, penitents throng to their mercy-seats, begging them to punish lightly. Forgiveness of sins, remission of suffering, are attributes which belong in a measure to all the gods. But two lovely figures of compassion have sacrificed their own Buddhahood to intercede for sinners, Ti Tsang Wang (*see* "The Seventh Moon"), the personification of the idea of salvation, and Kuan Yin, Goddess of Mercy, who holds out forgiveness to all lost souls.

Kuan Yin occupies a unique place in the hearts of the Chinese people, a figure empedestalled that stands for tenderness and pity. Her very name, "She Who Looketh Down and Hears the Cries of the World," proves how she comforts in sorrow and protects in danger. Like Mary, guiding spirit of Rome, Kuan Yin is the Divine Intercessor of the Buddhist faith, and her resemblance to the

Holy Virgin is so striking that Abbé Huc, with old-fashioned intolerance, suggests that the Devil, through the sweet personality of the Pusa,[10] was once more imitating the true religion in order to delude souls.

To her worshippers, Kuan Yin is to day much nearer and dearer than the Buddha himself. She is more than a goddess. She is an instinct of the heart. "The men love her, the children adore her, and the women chant her prayers." To whatever god a temple may be dedicated, there is nearly always a chapel for Kuan Yin within its precincts. But her image is also found in many homes, and in many hearts she sits enshrined. "Other gods are feared, she is loved. Other deities have black scornful faces, her countenance is radiant as gold and gentle as the moon-beam." Because she draws near to the people, the people draw near to her, praying the old, old prayer: "Great Mercy, Great Mercy, oh! Thou Take-Away-Fear-Pusa! Save from Terror, save from suffering through thy tender woman's heart and mighty Buddha's strength!"

Images of Kuan Yin, often very beautiful and appealing, represent her in her various rôles. Now we see her as the Thousand-Armed Goddess whose extended hands indicate her love for all mankind. Again she appears as a figure very similar to the Madonnas of European cathedrals, with the same gracious droop of the head under

---

[10] The word Bodhisattva, or *Pusa*, denotes the last stage of perfection before reaching Buddhahood. Many *Pusas* are hindered in their progress toward Buddhahood because they are so active in saving others they do not have time for the contemplation needed to pass on to the final state.

the high mitre-like hood, the exquisite fall of the robes and the fine modelling of the tapering fingers. Not unfrequently she bears a child in her arms. This last representation is traced to Hariti, an Indian goddess and a bad *Yaksha* fairy who spread contagious diseases, and notably smallpox, until the Buddha converted her by imprisoning her youngest child under his alms-bowl. Thus she saw the error of evil and selfishness and became a convert to truth and mercy. Hariti, the repentant and holy Hariti, has now become merged into Kuan Yin, emblem of the eternal feminine, though in no sense the Goddess of carnal love, like Venus. Orientals distinguish very sharply between passion, or the fulfilment of desire, and affection which includes the joys and cares of the home—what wives, if they be fortunate, may count on. Holiest of all, to them, is mother-love that the Kuan Yin protects and typifies.

Among the many metamorphoses the well-beloved Goddess assumes to help mankind are the "White Robed," or "Luminous" Kuan Yin, who gives souls to little children,—a relic of Manichæism, the "Religion of Light,"—and the "Kuan Yin of the Southern Seas" standing on a rock amidst dashing waves, to rescue shipwrecked mortals from the tempestuous ocean. One of her titles, "Vessel of Mercy", or "Bank of Salvation," indicates her power also to save sinners struggling in the darker sea of misery and sin, and convey them to the shore of true knowledge.

Two of her favourite emblems suggest her gift of rain to thirsty lands: the vase of heavenly dew (*kan lu*), sometimes called the vase of immortality (*ch'ing p'ing*) or "pure vessel," that she carries in one hand, and the willow-branch, itself a rain charm, that she holds in the other. As "in Buddhist literature religious truth is often poetically referred to as the reviving rain that descends upon the parched earth . . . it seems natural and appropriate that the divine Pusa, who brings succour to the distressed and sheds upon them the dew of immortal bliss, or the 'wine of sweet dew,' should carry in her hand a magic willow with which she charms down from Heaven the Rain of The Good Law." Finally, in addition to all her other powers to help mankind, Kuan Yin takes charge of those who travel the lonely roads, protecting them from brigands and wild beasts.

Though it would be impossible within the limits of a single chapter to describe in detail the cult of Kuan Yin, we may sum up the position of this important Pusa in the Buddhist Church by stressing once again the fact that her pity, her mercy, and her love for suffering mankind, bring all those who are weary and heavy-laden to her altars. "None call in vain upon her sacred name, for in the Lotus of the Good Law it is written: 'If any tiny creature in pain or trouble addresses a prayer to her, then will the Pusa immediately hearken and bring deliverance. If any tiny creature clings for support to the potent name of Kuan Yin, he may be thrown into a raging furnace and the flames will leave him unscathed; he may be in peril from sharp swords, but the steel will break in pieces; he may be in danger of death by drowning, but the blessed Pusa will come to his rescue and set him in a place of shallow waters'."

Nowadays the average Chinese considers Kuan Yin a female deity, counterpart of the Taoist Mother of Heaven (*see* "The Fourth Moon," also "The Third Moon"—Hsi Wang Mu) and, like her, an adaptation or personification of much older cults in honour of the Mother principle. But this was not always so. For many centuries Kuan Yin seemed to hesitate whether to be born into the Chinese pantheon as a man or a woman though, actually, the category of Buddhist divinities to which the Pusas belong are exempt from sexual distinction and may take any shape. According to Indian Buddhism, Kuan Yin was a male, Avalokita, Lord of Love and Compassion,—the St. John of the Buddhist Church, favourite disciple of Buddha, born from the Master's tears of pity for the suffering world, and wearing, as a symbol of exceptional affection, a little head of Buddha in his hair.[11]

In China until the T'ang dynasty Kuan Yin was also pictured as a man,—and is still so represented in some Buddhist temples, with the robes and lightly-bearded face of an Indian saint. But about the twelfth century the god became a goddess—a metamorphosis which vastly increased the popularity of Kuan Yin who then became an idealisation of womanhood and a link between "the merely human and the unapproachably divine."

---

[11] In Lamaism, the attributes which distinguish Avalokita are the stagskin thrown over his left shoulder and the rosary of roses in his hand.

THE SECOND MOON, OR "BUDDING MOON" 185

While the *Sutras* give the scriptural version of the earthly and heavenly activities of the Beloved Pusa, of her refusal to become a Buddha and enter the peace of Nirvana, of her decision to remain within the wheel of change, subject to re-incarnations, birth and death, in order to save those in need, there is a more romantic legend of the life of Kuan Yin written down by a monk named P'u-ming under the Sung dynasty.

This beautiful story, transcribed by the pious scholar who was rewarded for it by a vision of the Blessed Pusa herself, embodies the true Chinese conception of Kuan Yin in language simple and direct enough for public reading. We have heard it in a village temple—an old and faded shrine which demands much from the eye of faith. Pines and ginkgos shade the single sanctuary, and bats whirl round the crumbling eaves with their little ghostly whisperings and soundless wings. A few visiting monks are assembled to listen, and such laymen as desire to hear gather in the shadows where white feathery moths circle about them like fragments of mist.

An old priest reminds his hearers that to-day —the 19th of the second moon—is the anniversary of the blessed Kuan Yin's birth,[12] and begs them hear in decorous silence the precious words in which her life is recorded. Then, folding his

---

[12] This is the date most popularly observed as Kuan Yin's birthday. As a matter of fact, the goddess has no less than three anniversaries, explained by popular tradition as commemorations of her three incarnations. Very devout people observe all of them—the 19th of the second, the 19th of the sixth, and the 19th of the ninth moon. In many homes incense is also burned to her on the first and fifteenth of every month, when her rosary is recited with much fervour by the faithful.

hands, finger-tips to finger-tips, he begins his recitation.

Long ago there was a king, named P'o Chia, of the Golden Heavenly Dynasty, whose throne was shared by a beloved Queen. Nevertheless, the royal couple was unhappy because no son had been born to them. The king felt this precious gift was denied him by the gods because he had obtained his crown by cruelty and bloodshed. "Let us go," said he to his queen, "on a pilgrimage to the Celestial Ruler of the Hua Shan, confess our sins and ask that we be blessed with children." Accordingly, they went and prayed, and their prayer was answered with a gift of three children but, alas! all three were girls.

As the maidens grew to womanhood, the king determined to marry them to men of his choice so that, as he had no son, a worthy successor to the throne might be chosen from among their husbands. The two elder daughters obediently fulfilled their parents' wishes, but Miao Shan, the youngest, gave herself up to prayer and meditation refusing all attempts to force her into marriage. After many futile efforts to persuade her, the king in anger drove her from the palace. Miao Shan, after wandering in the forest with the moon for companion and the winds for friends, took refuge in the Nunnery of the White Bird. Here, by royal command, the abbess did her best to discourage the king's daughter from joining the sisterhood. When arguments proved unavailing, she was set to menial tasks. These she performed joyfully, with generous self-sacrifice, and the spirits of heaven came to help her in her irksome duties. The Sea Dragon dug a well for her near the kitchen, a tiger collected fire-wood for her, birds picked the vegetables,—all so that she might give herself up without disturbance to the pursuit of perfection.

The king, more and more enraged as he realised the strength of his daughter's determination, ordered his troops to surround the nunnery and burn it to the ground. Miao Shan then took a bamboo hairpin from her hair, pricked the roof of her mouth with it and spat the blood towards heaven, when immediately clouds gathered and heavy rains fell. Beside himself with fury, her father ordered the princess to be beheaded, but on the day of the execution the sword broke in two as it descended on her neck, and a great darkness fell upon the earth, while a tiger dispersed the executioners and carried off the fainting Miao Shan.

When she recovered consciousness, the Princess found herself in a strange place where there were "neither mountains nor trees, nor vegetation, no sun, moon, nor stars, no habitation, no sound, no cackling of a fowl or barking of a dog." A handsome young man, "dressed in blue and shining with a brilliant light," met her and explained that he was the envoy of the Kings of Hell who desired to congratulate her on her virtue and begged to hear her recite her prayers. "Let me lead you," pleaded the handsome youth, "to the Underworld which is not far from this desolate region." Miao Shan consented on one condition: that the suffering souls be released from their torments to hear her pray. But while the Princess Nun made her petitions, Hell was suddenly transformed into a Paradise of Joy, and the instruments of torture into lovely lotus flowers. Then Yama, the God of the Inferno, stood aghast at this interference with law and order. "If such indiscriminate mercy be allowed," said he, "We are wholly undone. Our authority will be jeopardised and the whole constitution of the Inferno will be destroyed." He thereupon ordered her soul back to earth, where it reluctantly re-entered her body.

Again a wanderer in pursuit of perfection, the Holy One met the Buddha of the West who suggested she establish herself on the island of P'u T'o (south of modern Shanghai) there to work out her salvation in a monastery. "Fear not the fatigue of this long travel," said the Master. "See, I have brought with me a magic peach of a kind not to be found in any earthly orchard. Eat of it and you will never feel hunger or thirst. Old age and death will have no power over you." Then, calling upon the Guardian of the Island, he bade him transform himself into a tiger and carry Miao Shan on his back to her journey's end.

Nine years the nun remained in this retreat at P'u T'o until she attained her aim. She was then crowned "Queen of the Three Thousand Pusas" and of all beings who have skin and blood. The same Guardian God who carried her to the island sent out invitations for the ceremony, and the list of those present included the Dragon King of the Western Sea, the Gods of the Five Sacred Mountains, the Emperor-Saints to the number of one hundred and twenty, Thirty Six

Officials of the Ministry of Time, the Wind, Rain, and Thunder-Gods, the Three Causes, the Five Saints, and the Ten Kings of Hells. "Miao Shan took her seat on the lotus throne, and the assembled gods (among whom we notice various Taoist divinities) proclaimed her Sovereign of Heaven and Earth. Moreover, they decided that it was not meet she should remain alone on P'u T'o, so they begged her to choose a worthy young man and a virtuous damsel to serve her in the temple." Again her old friend the Local Guardian God was requisitioned to help her, and he found a young orphan-priest from a mountain hermitage, and Lung Nü, third daughter of the Dragon King. Both, having proved their sincerity, were accepted as Kuan Yin's attendants, lived as brother and sister by her side, and still appear in effigy in most of her temples.

With his daughter the Dragon King sent to the newly made Goddess the gift of a luminous pearl so that she might recite her prayers by its light during the dark hours, and one night, while devoutly on her knees, she had a revelation of what was passing in her old home.

Now, one of the Buddhas had petitioned the Yü Huang, Taoist Lord of Heaven, to punish the King her father for the sins he had committed in persecuting his saintly daughter. So Yü Huang obligingly called the God of Epidemics and told him to afflict the body of the wicked monarch with ulcers of a kind that could not be cured except through medicines given by Miao Shan herself.

As soon as the filial and forgiving daughter realised this, she disguised herself as a priest-doctor, was admitted to the king's bed-chamber and, to the horror of the assembled courtiers, declared that the only remedy for his disease was an ointment compounded from the hand and eye of a living person. "Alas!" groaned the royal patient, "who would ever give his hand or his eye? Where can I possibly procure this remedy?"

"Your Majesty," the disguised priest replied, "should send your Ministers, who must observe the Buddhist rules of abstinence, to P'u T'o. There they will be given what is required."

When the envoys reached the Sacred Island, Miao Shan bade them take a knife and cut off her left hand and gouge out her left eye, and carry them back upon a golden platter to the sick king. The miraculous ointment prepared from the sacrifice cured the sinful monarch, but on the left side only. "If you wish to be made entirely whole," said the priest-physician, "send your envoys again for the right eye and the right hand." So a second time Miao Shan gave of her own body. Meanwhile, the queen recognised by a scar the hand of her own daughter, and overcome by gratitude and emotion exclaimed: "Who else but his child would endure such suffering to save a father's life?"

When the miracle was accomplished, the king and queen with the grandees of their court set forth to visit P'u T'o, and after many adventures they reached the monastery and went to thank their holy daughter and worship her. Entering the temple to burn incense, they saw Miao Shan seated upon the altar, her eyes torn out and her wrists dripping blood with the sweet odour of incense. The king recognised his daughter, the queen fell swooning at her feet. As Miao Shan spoke words of comfort, the king sincerely repented, and prayed that his child be made whole. Instantly the miracle was accomplished. She regained her perfect beauty, descended from the altar and became re-united with her parents for whom she interceded with the gods so successfully that the king, who abdicated his mortal throne, was forgiven, and raised to the dignity of "Virtuous Conquering Pusa," and his queen received the title of "Pusa of Ten Thousand Virtues."

These "patents of nobility" were conferred through a special messenger bearing a "divine decree" from the Jade Emperor, and in the same document the hermit-disciple had bestowed upon him the name of "Golden Youth," by which he is commonly known, and Lung Nü that of "Jade Maiden," while Miao Shan herself was made "The Very Merciful and Compassionate Pusa, Saviour of the Afflicted, Miraculous and Always Helpful Protectress of Mortals, Sovereign of the Southern Seas and P'u T'o Isle."

Buddhist imagination chose an island for the principal shrine of Kuan Yin for several reasons. *P'u T'o*

A sea-girt isle seems an appropriate "place of doctrine" for the Goddess symbolically connected with the sea of life, and the actual choice of P'u T'o was doubtless made because it best agrees with the legends.

In the Xth century A.D., a Japanese Buddhist monk, named Egaku, obtained possession of a beautiful image of Kuan Yin at the Wu T'ai Shan (*see* "The Fourth Moon"), and desired to take the statue back with him to his native land. Suddenly, in the Ch'u San archipelago where P'u T'o is situated, his junk was arrested in its progress by a miraculous growth of water lilies which covered the whole surface of the sea. Arising at dawn, the holy man found his ship moving slowly and yet more slowly, like a person falling asleep, and all around him the sea was like one vast meadow of perfumed blossoms. Thick as snow, their ivory cups floated on the water, thick as the coils of innumerable serpents the long stems twined about the rudder till neither sail nor oar could move the ship. Then the monk, kneeling on the deck, prayed to the Goddess: "If it be thy pleasure to go with me to my country, open thou the way. But if thy desire is otherwise, I am willing to go with thee wherever thou wilt go." "Thereupon a soft wind, like the swish of a passing garment, breathed on the surface of the sea, the ivory chalices of the lilies closed, and clear water flowed over them, forming an open pathway to the island, and his boat came to land before the Tidal Echo Cave where he built a shrine for the image—a shrine famous for centuries as the 'Temple of the Kuan Yin Who Would not Go Away.'" Twice again, according to legend, Kuan Yin performed the miracle of the lilies, once when certain "Predatory Dwarfs" (a name the Chinese used to apply to the Japanese) came bearing tribute to the Sung Emperor and on their return from Hangchow, then the capital, landed at P'u T'o and carried off some precious relics. They, too, found their ship enmeshed in the tendrils of water-lilies, a prisoner of beauty.

THE SECOND MOON, OR "BUDDING MOON" 193

The same fate overtook a haughty Chinese official sent to the island-shrine by a Sung Emperor for purposes of worship. His irreverence towards the holy places was punished, and he and his held fast in the same net of flowers. But when he sincerely repented and begged Kuan Yin's forgiveness, a white ox emerged from the sea, ate the lilies, and permitted him to go on his way. Transformed into a white rock, the ox may still be seen, and to this day the sea is called the "Sea of Water-lilies."

Legends innumerable connect Kuan Yin with P'u T'o and the founding of its various temples. Now she herself arrives there floating upon a flower; now appears in a cave dressed in rich robes and jewels to save a storm-tossed Imperial barge; again, she hears the cries of distress of Ministers of State afflicted with four days of intense darkness on a return voyage from Korea, and illuminates the sea with a brilliant light. She has been seen in A.D. 1424 in the Tidal Echo Cave, dressed all in white, attended by the Dragon King and his Daughter, and gliding along a crescent-shaped arch of light from hill-top to hill-top.

Indeed, the spirit of miracles has ever rested on P'u T'o, that lovely rocky island four miles long and very narrow, with its tiny coves, hazy and heat-ridden, and its jagged promontories where waves break and foam flickers like marble shavings against the azure of the sea.

A Chinese Mount Athos, the government was, until quite recently, entirely in the hands of the monks who were independent of civil jurisdiction.

In this holy place no creature of the female sex, not even a hen, is permitted to sojourn from sunset to sunrise, though women may take part in the great pilgrimages of the second moon, provided they return at nightfall to the ships that brought them.

The arrival of the pilgrim ships is but one more proof that the Chinese, so often accused of irreligion, have faith in Kuan Yin. Thousands of the faithful come yearly, risking the storms that beat against this rocky coast, often landing with difficulty at the little pier. The number increases or decreases according to the condition of the merchant and agricultural classes. The more prosperous the season, the larger the number of pilgrims, but it is rare that there are not some from every province. Even distant Mongolia provides its quota of devout Lamas and laymen who flock to hear the birthday masses of the Goddess. Strange that the shrines of P'u T'o should form a spiritual link with the Lamaist Buddhist sects of Thibet and the far north; but the Dalai Lama himself is supposed to be an incarnation of Avalokita. Strange too that this island without women should honour a woman!

There are more than a hundred monasteries, big and small—exclusive of solitary hermitages—on this sacred island, with more than a thousand monks to serve them. None of the present buildings are very old. In fact, the only one of historic interest is the battered Pagoda of the Prince Imperial dating from the XIVth century. Fires and pirate-plunderings destroyed the temples of the first founders,—temples supposed to date from about A.D. 900. Even in their heyday, their grandeur and historical associations never compared with the shrines of Wu T'ai or Omei Shan. But the monastic centre of P'u T'o is unsurpassed for natural beauty. Apart from the three great temples,—the Southern Monastery, or Monastery of Universal Salvation, the Rain-Producing Temple, and the Wisdom's Salvation Temple,—visitors find, if they follow the stone-flagged path that leads to the highest hill, walks branching off in all directions to smaller shrines or holy grottoes. Yonder is a crag with a weather-worn inscription: "Even the stupid stones bow their heads"; beyond, a solitary rain-defaced image with little ferns around its feet. Not far away will be an old temple-garden where scarlet trumpet-flowers explode with tiny humming birds, where the leaves of the Maidenhair trees are frescoed by the sunshine on old walls. In neighbouring valleys stand the grave and dignified tombs of saintly abbots, set in meadows of the fragrant "Little White Flower" (*gardenia florida*) for which the island has been famous for a thousand years.

From the highest point, the central rock called "Kuan Yin's Pulpit," there is a view over the sea with its archipelago of islands, like opalescent sea-shells floating on the water, as well as all the main temples whose yellow roofs, indicative of Imperial distinction, glow in the noonday sun. Standing here the pilgrim feels that "all these separate shrines of P'u T'o are but chapels of one vast cathedral. And it is not only the monks who chant the praise of Kuan Yin in their great pavilions. It is not only from jars of bronze and stone that perfumed clouds rise daily to the throne of the Compassionate Pusa. From the sea-waves also comes the sound of the mighty anthem. The rain that patters on the temple roofs is the Rain of the Good Law that is poured from the unfailing vial of the Goddess. The winds murmur *Sutras* in the sacred caves,—and in the spirit-haunted woods the wild birds join in the adoration, while the 'Little White Flower' sends up to heaven, from millions of censers not made by the hand of man, the sweet fragrance of inexhaustible incense."

* * *

Maritime Goddesses   Though Kuan Yin, the Best Beloved, is the supreme and favourite patroness of those who go down to the sea in ships, and has to a certain extent supplanted all other maritime goddesses in popular favour, there are certain local patronesses of sailors such as *T'ien Fei, Ma Tsŭ P'o* and *Ma Chu* (the two last-named being probably regional deifications of the same figure) who are worshipped by most of the sea-faring families of the southern coast-ports. Ma Chu has many elaborate temples erected to her and numbers among her devotees a large percentage of those teeming millions who, in China, have their homes upon the water,—the folk who are born, grow to maturity, live and die on junks, or even on the narrow confines of sampans.[13]

Legend describes Ma Chu as the daughter of a Fukienese sailor born under the Sung dynasty

---

[13] Around Hongkong alone there are about 70,000 people who live on the craft that ply in the harbour.

and father of two sons. One day, while sewing, Ma Chu fell asleep and dreamed she saw her father and her brothers on two junks in the midst of a terrific typhoon. Like a filial daughter, she immediately seized her father's junk in her mouth and caught the junk of her brothers in her hands. She was dragging them both safely to shore when she heard her mother call and, like an obedient girl, instinctively opened her mouth to answer. A few days later came the news that the vessel belonging to her father had foundered with all hands—and one more maiden regretted not to have kept her mouth shut at the right time.

Nevertheless, because of her dream and in spite of her failure to save, Ma Chu remains a goddess of mariners. When sailors set out from port on a fair day, clear blue to the end of the world, they take with them ashes gathered from the incense-burners in her temples, carrying these precious relics in a small red bag hung round their necks, or fastened somewhere on their junk. When violent storms come, the sailors kneel in the bow with the incense in their hands and call on Ma Chu for deliverance, and if they safely reach port they make a thank-offering to their patroness. Sea-farers will tell you that the Goddess sometimes assumes the shape of a fire-ball travelling up and down the mast, a superstition not unlike our own St. Elmo's fire on the bow-sprits of sailing ships. Or, again, she appears to those in peril on the sea in a vision with her two assistants beside her, "Favourable Wind-Ear" and "Thousand-Mile Eye," whose very names indicate how useful they must be on the ocean.

Yang Ssŭ Lao Yeh, sometimes called General Yang Ssŭ, is another favourite god of sailors, because "he controls the surface of the waters by his influence with the Dragon King." [14] Usually represented as a child of about ten years, with a white face and white robes, he holds an axe in his hand, upraised to threaten the dragon.

Ma Tsŭ P'o, possibly, as we have said, a local variation of Ma Chu, is the special sea-goddess of Fukien province. Many miracles are ascribed to her. Indeed, her power as a protectress is considered so great that, whenever a new junk is launched, a feast is spread for her and theatricals organised to put her in good humour. Just as the ancient Greeks carried a statue to their tutelary god on every trireme, so the modern Chinese sailor has an image of this Lady of the Waters crowned with a heavenly crown, in a shrine on the left side of his ship, the left being the place of honour in the east. Morning and evening incense is burned to her. At dawn, when the faint wind from the west wrinkles the sea, and again at sunset, when there is not "breeze enough to move three hairs," the naked figure of a young fisherman may be seen standing erect in the prow of his boat, making his simple prayer for good luck through the day and safety through the night.

T'ien Fei, who has power on the seas in addition to many other qualifications, is the sailors' goddess who has suffered most from the increasing

---

[14] His birthday is appropriately celebrated on the sixth of the sixth month, the Dragon Month.

THE SECOND MOON, OR "BUDDING MOON" 199

popularity of Kuan Yin. In fact, T'ien Fei, or the "Heavenly Princess," is in some respects the Taoist counterpart of Kuan Yin. Both are worshipped as saviours of mankind. Both save from the peril of the ocean, both are protectors of mothers and givers of sons. Yet, these divinities "eye one another with no unfriendly feelings" as we may gather from the fact that the Taoist Queen of Heaven has a shrine on Kuan Yin's own sacred soil of P'u T'o.[15]

The origin of T'ien Fei is shrouded in mystery. She is supposed by some to have been born in the reign of T'ai Tsung, the Sung Emperor (in A.D. 980), the daughter of an official, and to have died at the early age of twenty-six after having given proof of her miraculous powers by walking on the waters near her village dressed in a red robe. Later she was given the title of "Righteous and Heavenly Princess" by Kublai Khan for service rendered to the State. It so happened that an important transport of rice was on its way by sea from the south to Peking. Now this transport often took place under difficult and dangerous conditions, with sudden storms, adverse winds, and cruel currents, to contend against,[16] and once a vessel laden with this valuable food cargo was being blown on the rocks. Then the

---

[15] Note that this is not the only Taoist shrine on the Sacred Island. The Buddha's Peak Monastery, situated on one of the heights, contains a hall consecrated to the worship of the Supreme God of Taoism, admitted here because he is regarded throughout China as the principal presiding deity of every mountain summit. Thus, there seems nothing inappropriate to the Chinese mind in enthroning him on Buddha's Peak.

[16] It was owing to these dangers and difficulties that the Grand Canal was built so that there need be no delay and no danger of shortage of rice in the capital.

sailors prayed to T'ien Fei who delivered them safely out of their dangers.

Most Chinese divinities have several attributions, and from that day to this T'ien Fei, Giver of Sons, has become a sailors' goddess. In many ships a picture of her is pasted in the cabin, and paper talismans blessed in her temples are kept on board. When disaster threatens, the first poster is burned. If this does not succeed in calming the elements, the second goes up in smoke and, finally, as a last resort—the third. Sometimes, in answer to devout prayer, the Heavenly Princess sets a red guiding lantern afloat on the face of the waters, or even appears herself in the skies, dividing the wind with her sword.

Some ship-captains, especially in Fukien province where the cult of the sea divinities is most developed, always keep a wooden wand beside her picture. When two mighty fish-dragons play in the seas, as they sometimes do, they spout up water against one another "till the sun in the sky is obscured and the seas are shrouded in profound darkness." These monsters approach when a storm is near and, therefore, as soon as they are sighted, it is customary to burn paper or wool to prevent them drawing the ship into the depths. Often the Master of the Wand will burn incense before the fetish in the cabin, first taking the magic baton and swinging it three times in a circle over the tempestuous sea. T'ien Fei will then command the dragons to draw in their tails and disappear.

* * *

THE SECOND MOON, OR "BUDDING MOON"

By a curious coincidence, honour is paid during the second moon to China's two greatest Sages, Lao Tzŭ and Confucius. We have already attempted to explain in "The Hundred Gods" the relative importance of the rôles played in Chinese life and thought by the religion founded by the former and the ethical system developed by the latter. Nevertheless, we must again stress here the influence that these two men have had over more than a quarter of the human race—an influence that still affects hundreds of millions of people.

Though a definite date, the 15th, is set as the anniversary of Lao Tzŭ's birth, popularly but not authoritatively fixed at 604 B.C., his historical personality is only vaguely established. How he lived, where he died, no man can truly say. His message appears to have fallen at first upon stony ground and legend says he decided to leave a land which did not want to hear his teaching. Mounting a black bull he rode towards the west and is supposed to have been last seen at a distant frontier pass. Here the "Old Child," as he was affectionately called because he was born with white hair, spent one night in a guard-house bastion of the Great Wall. The rough soldier in command of this lonely outpost appreciated the noble philosophy of his guest and was ready to enter into Lao Tzŭ's mind—a temple of intellectual loveliness. Together the two men of such opposite types spent the whole night in converse, and in those wild surroundings by the dim light of a pith-wick in an earthen bowl filled with bean oil, the Master, so friendless and forlorn, dictated

Lao Tzŭ

the words afterwards recorded in the Tao Têh Ching, one of the most remarkable books in the world though it contains only four thousand characters.

When all was written down, and the sun rising beyond the mountain peaks, the philosopher bade his friend farewell, mounted again upon his black bull and disappeared forever from the eyes of men.

It is highly improbable that Lao Tzŭ actually thought of founding a religion when he dictated the Tao Têh Ching, and it is certain that he would be horrified at the degeneration of his noble principles in modern times. For to-day Taoism has become a faith of charms and exorcisms presided over by priests who have for the most part degraded the Teacher's pure doctrines into a spiritual hocus-pocus to suit their own ends. This was perhaps inevitable. The truth which lies at the core of every faith has been clouded by media more or less fit for its transmission, and the followers of Lao Tzŭ, unable to live in the rarefied atmosphere in which their Teacher delighted, soon came down to lower levels bringing his doctrine with them. Thus a religion whose ideal was freedom from all form developed into a mystic system of individual salvation, grossly overlaid with superstitious practices at which the Founder would wonder as much as we do. Forgetting lofty ethics, Taoist leaders in the first centuries of our era abandoned the Abstract Way of Perfection to seek the Elixir of Life and the Philosopher's Stone in wild woodland and mountain retreats by methods which remind us of the Indian fakirs. They personified the visible forces of nature and the old animistic gods in mythical rulers, spirits, and deified heroes,—and created an immense pantheon of gods. Lao Tzŭ himself has been included among them, and this very attempt to set him up on a pedestal has actually dragged him down from his misty heights.

When we think of Lao Tzŭ to-day we no longer see him as the "Harbinger of Eternal Truth," but as a genial old gentleman with broad-brimmed hat and flowing beard astride a buffalo, as the painters present him, or else, officially as it were, in the temples as one of the supreme divinities of the Taoist Olympus.

* * *

If the figure of Lao Tzŭ is misty as a fairy-tale, that of Confucius is sharp and clear. A native of Shantung, he was born in 551 B.C., son of a valiant father who preferred the sword to the scholar's brush and gained fame in his fights against the barbarians—"arrogant knaves of the borders" of his province. Confucius himself inherited no military tastes or talents. He early took to teaching, drifted into politics, and died at the age of seventy-three a disappointed man, unhappy in his private life, for he had to divorce his wife who was a shrew, and was pre-deceased by his only son. Like Lao Tzŭ he felt "the message which he was divinely appointed to deliver had not been favourably received." His own words best trace the stages of his career:

Confucius

"At fifteen my mind was bent on learning.
At thirty I stood firm.
At forty I had no doubts.
At fifty I knew the will of God.
At sixty I could trust my ears.
At seventy I could follow my heart's desires without transgression."

As he judged himself and as men judged him during his lifetime, Confucius was a tragic person wandering sadly from State to State, attempting to reform Princes who would not be reformed, giving advice that only a few would take. But "outcast as he was in life, the value of his common-sense teachings was soon recognised," and the "uncrowned king," as he is often called, is now "as firmly fixed upon his throne as at any period during the twenty-three centuries which have elapsed since his death . . . . At the present moment, there is a Confucian temple in every prefecture, district and market-town throughout China, where twice every year, in spring and autumn, memorial services are conducted by the local officials." Since the first shrine was built in his honour before Christ, rulers have commanded that sacrifice be made to him. The last sign of official reverence was the Imperial Edict issued by the Empress Dowager Tz'ŭ Hsi so late as 1906, wherein Her Majesty ordered that the Medium Sacrifices until that date offered to the Sage be changed to Great Sacrifices at which the Emperor should personally attend. The true purpose of this Edict was to enforce and stress respect for the founder of a moral code that had held China together for scores of generations, at a time when this great stateswoman admitted and recognised the necessity of introducing a certain degree of modern learning.

Her dynasty has collapsed and times and governments have changed with bewildering rapidity. Yet the sacrifices at the Confucian Temple in Peking are still made as ceremoniously as in the days when the Son of Heaven himself or his nominee, a Prince of the First Class, offered tribute to the Sage of All the Ages. The Chinese Republican Government which lightly abandoned the services to Heaven, the Sun, the Moon and, in fact all "great sacrifices," still considers the spring and autumn ceremonies in honour of Confucius so important that, even in 1926, when there was no recognised government in Peking, the usual prayers and libations were offered to the Master in the name of a non-existent President of the Republic. This is indeed one of life's little ironies, for Confucius in his lifetime, like Wally, "dearly loved a lord," was a great stickler for form, etiquette, ceremony and caste, socially speaking, and would have agreed most thoroughly with Michael Arlen's modern hero who considered "democracy a drain-pipe through which the world must crawl for its health, but did not believe the health of the world would ever be good."

No foreigner has ever been present at the national rites held twice a year in Peking to honour the Sage. They take place at three o'clock in the morning and even the Chinese public is not admitted. But it is possible to attend, by invitation only, a rehearsal of the ceremony on the preceding afternoon.

Fortunate visitors whose "permits" are in order gather at the grand old sanctuary dedicated to the Master and find places reserved for them on the marble balustrade outside the great hall where they can see the whole service, part of which is held indoors and part on the terrace itself. The officiants, gowned and hatted in the ancient costume of the Master's day—in dark blue robes with light blue facings, belted at the waist, and wearing curious caps like a college mortar-board—slowly mount the three flights of marble steps divided by the beautiful central carved block of the "Spirit Stairway" leading to the "Hall of Great Perfection," and take their places on the platform under the direction of a master of ceremonies. The musicians stand beside the quaint orchestral instruments. Here is the sonorous jade, the *t'êh ch'ing*, a hanging stone gong which has been used in China since the highest antiquity, in fact ever since music was first decreed for religious services as "the Harmony between Heaven and Earth." The Emperor Yao (2300 B.C.) left a hymn in praise of it:

"When I smite my musical stone,
Be it gently or strong,
Then do the fiercest hearts leap for joy,
Then do the chiefs agree among themselves.
When ye make to respond the stone melodious,
When ye touch the lyre which is called *ch'in*,
Then do the ghosts of the Ancestors, having pleasure
  in sweet sounds,
Approach to hear."

The *ch'in, or* "Moon Lute," to which His Majesty refers, has silken strings fixed to turnable screws. It is this primitive dulcimer with a limited range and small sweet sound that we have developed, through the harpsichord, into our grand piano.

Another ritual instrument, the little *shêng*, with its sixteen short bamboo tubes fixed in a case of wood, is the ancestor of our mighty organ. Inspired by one of these "Clustered Bamboo Pipes," carried back to Europe by an early Russian embassy, a musician of St. Petersburg invented the accordion and the harmonium.

More typically Oriental are the stone chimes, or *pien ch'ing*, poetically called the "Engaging Jades," a set of sixteen jade slabs each sounding a different note; the *po chung*, or pointed oval bell; the *pien chung*, or set of sixteen barrel-shaped bells hung within a wooden frame; the *ch'ih*, or bamboo flute transversely blown, the *ti*, or horizontal bamboo pipe, with six finger-holes, a modern invention dating only from the Han dynasty; the *shun*, or "mud gourd," a clay ocarina with several blow-holes; the embryo clarinet, blown through a small reed in the manner of the chanter of the bag-pipes; the *chü*, a wooden tub struck with a mallet; the *yü*, or "wooden tiger" crouching upon a pedestal, its back a row of iron teeth; the *shou pan*, or "black wooden clappers;" and the various drums, the *lung hu*, the "Dragon Drum," the *chin ku*, or "Great Drum," six feet in diameter, the *t'ung ku*, or metal drum, the barrel-shaped *ying ku*, the *p'ai ku*, the small skin drum, the *têh shêng ku*, or "Victory Drum", the *yün loh*, or "Cloud Cymbals" suspended ten in a row, and, finally, the "Little Stars," a group of tiny metal castanets. Many of these beautiful old instruments are mounted upon carved and gilded stands which add much colour and elegance to the orchestra.

The whole setting for the archaic ceremonial accentuates its simple solemnity. Overhead a sky so blue that it puts even Italian skies to shame. Around us a hush of reverent silence. We seem to be miles away from the city of Peking, this vast city of the old East, with its warring politics, its Oriental fatalism and intrigue. We seem to stand upon a spiritual magic carpet that carries us back into the past Confucius himself so loved, "that golden age of China with its semi-divine rulers who threw a pall over his imagination." The centuries look down upon us in the cypresses a thousand years old, in the yellow tiles of the great hall that seem to burn against them, in the tablets that record the victories of mighty Emperors like Ch'ien Lung and K'ang Hsi and the triumphs of great scholars, and in the stone drums inscribed with the oldest records of Chinese writing that have come down to us.

When the President's representative ascends the Dragon Stairs, the chant of the choir boys rises from the marble terrace:

"Confucius, Confucius! How great is Confucius!"

The master of ceremonies invites the officiant to come and worship; the First Hymn is played by seven different pairs of instruments. The accompanying chant, sung slowly and in unison like the early Christian plain chant, has a twisted melody running through it like a thread binding a bundle of gold, and ends abruptly on a note which brings no feeling of *finale* to our ears. The spirit of Confucius, in life a great lover of

THE SECOND MOON, OR "BUDDING MOON" 209

music, is supposed to arrive and take part in the ceremony as this hymn begins, hence its name: "Receiving the Spirit."

When the music ceases, the officiating magistrate enters the great hall where the spirit tablets of the Teacher and his disciples stand, and bows his head in reverence. At the actual ceremony, he approaches the altar-table spread with pigs and sheep and spotless calves, with grain and fruit, with oil and wine. The droning notes of the prayer specially composed for each sacrifice are heard, and the fragrance of incense floats out through the open door. Again the sound of music breaks the solemn silence as the viands are presented before the tablets, once, twice, and then carried out into the courtyard to be burned in an iron brazier. Finally, symbolic offerings of jade and silk are presented with another series of deep bows and the prayer is burned. The magistrate leaves the hall, descends the dragon stairs. Again the boys' voices cry:

"Confucius, Confucius! How great is Confucius!"

and "the rites are over for this, the two thousand and twentieth year since the great ruler of the Han house, Emperor Wu, inaugurated the rites of the meat offering."

This same service, but with a less elaborate ritual, is held by local magistrates in every city in China. Moreover, even in modern schools, pupils in foreign-style uniforms have a service twice a year in honour of the Supreme Teacher. Each class, led by its preceptors, bows before his spirit-tablet and burns incense to his spirit, proving that Confucianism is still the code of the learned, notwithstanding the transition period through which scholarship in China, like so much else, is passing. Even in Christian colleges it is becoming the custom to have a special celebration on the birthday of Confucius, when men outside the Church, but prominent in Government and official life, are invited to address the students. Christian leaders admit Confucius as one of the world's great philosophers, and Christian converts are not debarred from membership in those Confucian societies formed to commemorate the great Teacher of Ethics.

* * *

Ch'ü Fu   If further proof be needed of the vitality of the Master's principles and the respect still accorded to him, we may have it abundantly by visiting the Sage's tomb, one of the most impressive pilgrimages in China.

From the nearest point on the Tientsin-Pukow Railway, it is a six mile journey by Chinese cart, or the still more primitive wheelbarrow, to the quaint old town of Ch'ü Fu, where the seventy fourth lineal descendant of Confucius lives. The present holder of the title of Duke Kung—the only hereditary title in China, conferred on the Sage and his descendants to perpetuity, and confirmed by each succeeding dynasty,—is a little boy still under the tutelage of his uncle. There is something very impressive about the idea of this child, living in a magnificent palace built upon the very site of the Master's modest house, as his descendants have lived for twenty-four hundred years— still receiving the revenues of vast grants of land

CONFUCIAN TEMPLE, CH'Ü FU.

THE "DEW WELL" USED BY CONFUCIUS, CH'Ü FU.

held tax-free like his liegemen, the farmers who till his fields, because he still fulfils the duty of guarding the tomb of the Sage and sacrificing in honour of his memory.

One third of the little old-fashioned country town, grown static through the ages, is occupied by this palace and by the temple dedicated to Confucius. The original shrine built here in 47 B.C. to honour "the Princely Man, uncrowned and yet a King," was a small affair of but three rooms. The beautiful halls of to-day, though planned in their present proportions as early as A.D. 739, have been constantly repaired and embellished by a long line of Emperors till, as they now stand, they are one of the noblest examples of architecture in the whole of China. Nowhere are these sweeping roofs, hung with bells on which the wind plays a ceaseless dirge, surpassed,—nowhere are the nineteen high stone pillars wreathed with dragons equalled,—nowhere do the Chinese find as many holy remembrances gathered within a single enclosure. The visitor on his sentimental pilgrimage is conducted reverently by an attendant through courtyards filled with splendid cypresses that date from the Han dynasty. He is shown the "Apricot Pavilion" where the Master taught, the "Dew Well" from which he drank, the group of purple "dust-repelling" stones, the pathetic stump of the tree he planted with his own hands, the ink-box that he used and other treasures of remembrance, the scores of monuments, big and small, recording Imperial and princely tributes to the Sage, the hall containing his portrait, the famous set of slates depicting his travels, finally, within the main building, the imposing Spirit Tablet of the "Master Example of all Ages."

Thence one goes on to the graveyard about a mile beyond the north gate of the city. Sitting within the bumpy blue cloth-covered cart we follow a road lined by hoary old cypress trees, dusty white, cross a stone bridge over an imaginary stream to the gate of the outer enclosure where distant descendants of Confucius lie. A shrine on the right is dedicated to Yen Hui, known as the "favourite scholar (*see* "The Ninth Moon"), who died prematurely and was mourned sincerely by the Master. Admirable indeed was the virtue of Hui," said the Master. "With a single bamboo bowl of rice, a single gourd of fresh water, and a hovel in a mean, narrow lane, while others could not have endured the distress, he did not allow his joy to be affected by it."

Beyond, there is a second wall enclosing the tombs of the Kung Dukes, direct descendants of Confucius, and an avenue of stone animals leading to the inner cemetery. In the eastern corner lies buried the Master's only son and, on the south, his grandson Tzŭ Ssŭ, author of the "Doctrine of the Mean" and known as the "Transmitting Scholar."

The Sage himself occupies the central place of honour, his tomb marked by a simple tumulus of earth a few yards high. Before it is a stone incense-altar and a simple stone tablet bearing the inscription: "Grave of the all-accomplished saintly Prince Wen Hsüan." No grandiloquent epitaph, no display of splendour, only an unpretentious mound shaded by old trees whose trunks reflect a sheen of steel and platinum and, over all, solemn repose and dignity.

Kings and princes in many lands have built themselves magnificent sepulchres, but what other grave of a giant among men can compare in impressiveness with this quiet unadorned resting place, the centre of a great nation's thoughts. We visit the little stone hut marking the spot where Tzŭ Kung mourned the death of his beloved Master for six long years, long after all the other disciples had dispersed. A few blue butterflies, like flakes of sky, settle upon the stone slab,—the monument of faithfulness.

Towards sunset we return again by special favour to the tomb, drawn back by the solemnity of worshipful reverence. Effacing colours and obliterating distances, the dusk turns the pines to ghostliness. Their gnarled trunks and twisted branches throw shadows like rams' horns. The new moon, a crooked finger with a long silver nail, reaches out of the sleeve of night and touches the tablet. And we realize that this single grave symbolises the highest Chinese ideals. Many things have passed away, but the tradition of the Master and his influence are still alive, radiating from this holy province of Shantung. And as we turn away to journey back to Ch'ü Fu across the bridge of silence through a land of mists and legends, we feel that weird, sad, delicious thrill which accompanies the "sudden backward flowing of the tides of life and time," the sensations of an epoch summed up in the emotional feeling of a moment.

CHAPTER VII.
THE THIRD MOON, OR "SLEEPY MOON."

HIRD Moon, "sleepy moon," so runs the popular saying in China. It means that the drowsiness of spring is in the air, the true spring, not the *Li Ch'un*, or calendar spring which comes while the weather is still cold. On the great northern plains that all winter long have looked hopelessly dead and barren, a miracle takes place. Tree-branches, dry and bare one day, burst into bud overnight. Fragile catkins appear on the willows—pale, faceless things that can scarcely be called flowers. Brown orchards suddenly blush pink, and shivering poplars put on new liveries of green and silver, coaxed by the tender *ku yü*, or "corn rains." Wild geese fly in fan-shaped formations across the sky. Swallows return from the south to build their nests under the eaves of city gate-towers, and remind us that to them the Chinese capital owes its name, *Yen Ching*—the "Swallow Capital."

Now that the frost is gone, the farmers "open the ground." At dawn, they leave their villages for the long day's work in the fields, fathers, sons, and grandsons, carrying the primitive farm implements that have not changed since the days of the Perfect Emperors who reigned many centuries before Christ. Even their ploughs are light enough to be borne shoulder-high. They must be so in a land where cattle yoked to them are small, and, failing cattle, little grey donkeys, or even men, are harnessed to dig the shallow furrows.

Yet, despite their primitive tools, these Chinese farmers are unequalled in the whole wide world. Long before us, they learned the secret of crop-rotation—a secret which enables them to work the same land for five thousand years without exhausting it, and to support themselves on an incredibly small acreage per capita. But the price of life is incessant drudgery so long as daylight lasts, with no conveniences to lighten labour. Each member of each family is a slave to the buds, shoots, and roots that pierce the earth. The men plough, or guide the donkey-drawn harrow like a big comb with wooden teeth, or pull the little round stone roller, the size and shape of a lady's muff; the women do the lighter work, such as weeding and cotton-picking, while even tiny children cut and carry dry grass or stalks for fuel.

All of them farm tightly and carefully, well knowing "that the image of a cake is not nourishing"; that sufficiency, let alone comfort, means unremitting toil and self-denial, means turning every copper earned this way and that. But the fields they cultivate are their own, and no absentee landlord grinds them down. This is one reason why the canker of social discontent is unknown, and unfelt, among Chinese peasants.

Besides, even for them, life has its lighter moments, brief snatched leisures packed with simple delights, impossible of purchase by those whose daily habit is sated idleness. There are local fairs and feasts to vary the dull routine. Then, there are also the great holidays of the year (*see* "The Twelfth Moon"), which none are too poor or too heavily burdened to observe and enjoy.

* * *

The Han Shih   Of these, the *Ch'ing Ming*, or Spring Festival (literally—"Pure and Bright"), falls early in the third moon though, occasionally, late in the second —a movable feast according to our calendar but, by Chinese reckoning, fixed one hundred and six days after the Winter Solstice (*see* "The Eleventh Moon").

On the eve of *Ch'ing Ming*, everyone used to observe the *Han Shih* ("Cold Food") Feast, when nothing hot was eaten, and no fires lighted for twenty-four hours. The custom is dying out in Peking but, we believe, is still followed in other parts of China. Its origin, long since forgotten by the common people, was connected with the solemn rite of kindling new fire once a year— an old, old rite, probably a relic of tribal times, when the *Han Shih* marked the interval between the extinction of the old fire and the lighting of the new.

Historical records prove that, as late as the T'ang dynasty, new fire was obtained by rubbing two willow-sticks together. The children of courtiers performed the ceremony in the open space before the Imperial palace, he who first set his sticks alight receiving a golden cup and three pieces of silk. The picturesque custom, which died out under the invading barbarian dynasty of the Yüans (A.D. 1260-1368), was, undoubtedly,

THE THIRD MOON, OR "SLEEPY MOON"

a survival of sun-worship in dim, distant ages, and may claim a vague kinship with the solar rite of the new fire adopted, and still observed, by the Catholic Church as a Paschal ceremony.[1]

In course of time, when the deeper significance of these mysteries became lost in China, a myth was invented, as so often happens, to account for them. Thus, the *Han Shih* is explained as a memorial feast in honour of a patriot of the Chou dynasty (350 B.C.), to whom the first ancestral tablet was erected (*see* "The Twelfth Moon"). Once upon a time, this hero accompanied his Sovereign Lord of Tsin on a journey. Misfortune befell the travellers. Food supplies failed—whereupon the devoted retainer cut off a piece of his own flesh to feed his starving lord. The latter, desirous of rewarding his faithful servant who had fled to the mountains, commanded the underbrush fired to chase his henchman out of hiding. "Thus," said the lord, "modesty shall have no excuse to escape just gratitude." But rather than stain his disinterestedness, the loyal liegeman preferred to burn alive. "Greater love hath no subject for his ruler!" cried His Majesty, deeply grieving and rending his robes. "Let none forget his noble example, and let the people honour his memory each year at this season by lighting no fires in their homes for three days, and eating cold food as a sign of remembrance."

* * *

---

[1] The popular habit, which persisted in England not so many years ago, of allowing the hearth-fire to grow cold on Easter Sunday, and re-kindling it on Easter Monday, is one more memory of older cults existing in regions as far apart as China and Mexico, Egypt and Peru.

The Ch'ing Ming— First Feast of the Dead

The feast of *Ch'ing Ming*, which follows the semi-fast of *Han Shih*, was originally an orgiastic festival of life-renewal, such as appears in every religion to celebrate the spring mating-season. Before the dawn of history, it was observed in the Far East by youths and maidens dancing together garlanded with flowers on river banks. As increasing standards of culture forbade the licence that accompanied these Saturnalias when the evils of a dead year were ritually got rid of by various means, they gradually altered completely in character and intention till, nowadays, the *Ch'ing Ming* has become an All Souls' Festival, on which it is the universal custom to visit the ancestral tombs.

Further proof that the *Ch'ing Ming* is a direct descendant of the antique Spring Festival is the alternative name, still in use, of *Chih Shu Chieh*, or "Tree Planting Festival," which exactly corresponds to our Arbor Day. The connection is too obvious to need explanation. In bygone years the Emperor, or a prince deputed by him, planted trees in the palace grounds on this particular day. Though many people imagine the custom was brought back to China by returned students from American universities, they actually revived and adapted a historic Chinese custom. Nowadays, the President of the Republic, or his deputy, does the planting in some public place.

How the dead came to be connected with what was, originally, a festival of life, is interesting to trace. As we have already seen, the basic idea underlying all Chinese religious feeling is the ancestral cult. Now in some mysterious way which we cannot fathom, and which the Chinese themselves cannot clearly explain, one of man's several souls remains near his grave, hovering about this world, and maintaining contact with his descendants through a subtle connection with his mortal remains. Such a *kuei*, or spirit, has definite requirements of worship and sacrifice at stated intervals—the annual festivals of the dead. It has, also, definite powers for good and harm. Indeed, it is not too much to say that the Chinese believe the dead still dwell in this world and rule it, influencing not only the destinies of men, but the conditions of nature.

Hence the natural desire of an agricultural people to chose the period of the year's re-birth for tomb sacrifices. Craving the blessing of the ancestors on the first fruits of the harvest sown in the true spring of the year (much in the same spirit as they invoked the blessing of the official gods at the calendar spring), men propitiate the ghost-gods—for the souls of the departed do take on, to some degree, the powers of gods—at the *Ch'ing Ming*, with offerings of agreeable food, of music, of burnt sacrifices, in short—of whatever is likely to put them in a good humour.

Such ideas are not confined to the Chinese. The Jews, according to St. Matthew's Gospel, were also in the habit of embellishing their tombs in the neighbourhood of Jerusalem at the return of spring, a little before the celebration of Passover. We find similar practices elsewhere connected with the once universal worship of the dead. Even in details, the Occidental and Oriental rites resembled each other. For example, "the

Chief Magistrate of Platea, clad in a purple robe, washed with his own hands the tombstones of warriors who had fallen in battle with the Persians, and anointed them with oil. He also slaughtered a black bull over a burning pyre, and called upon the spirits to come and partake of the banquet of blood. Funeral games and combats to soothe the ghosts of the departed were common occurrences among our own ancestors . . . also horse races, foot races and shooting matches." Exactly as in China, "in primitive Ireland great fairs, such as Emain and Carmain, originated in honour of the dead, and the belief was held as firmly as in the East that the welfare of the living depended upon the welfare of the departed, that so long as homage in the shape of funeral games and food (probably the first fruits of the harvest) was given, so long the spirits would bless their people by causing the earth to yield plentiful crops, the cows to give milk, and the waters to swarm with fish, whereas if they deemed themselves neglected or slighted, they would avenge their wrongs by cutting off their food supply, and afflicting the people with death and other calamities." (See *The Golden Bough*, by Frazer).

This conception led to a desire for the careful preservation of the bodies of the dead.[2] The Chinese did not, it is true, attempt to mummify like the Egyptians. But they strove for a similar

---

[2] In an attempt to arrest the decay of the body, jade—an emblem of vitality—was often buried with a corpse. Mortuary sets for placing in the orifices of the body, such as the nose, ears, and mouth, may still be found occasionally in curio-shops. The symbolism of jade in China, and its connection with the rites for the dead, is a subject of unusual interest. (See De Groot, *The Religious System of China*, and Laufer, *Jade*).

ideal in their preparations for burial, extending over many months; in their elaborate funeral ceremonies, involving vast expenditure and heavy personal sacrifices (men have been known to sell themselves into slavery in order to bury a parent with suitable dignity); and in the careful laying out and upkeep of burying grounds.

Owing to the close connection supposed to exist between soul and body, and their vague belief that one of a person's several souls may attain re-birth in human form, the Chinese have a great horror of descending to their graves with any physical mutilation. Decapitation, for this reason, has always been much more dreaded as a punishment in China than strangulation. The amputation of a limb, even to save life, is seldom undertaken by foreign surgeons without the written consent of the patient's family, since many prefer to see their dear ones die rather than risk their re-incarnation as cripples. Some who agree to the operation attempt to avoid, or lessen, the risk, by burying the limb which has been cut off in the grave where the sick person will ultimately lie. Before condemning such superstitions, let us remember that we ourselves have had analogies to them in Europe. They were widespread in Ireland not so many years ago, and an authenticated case is cited fairly recently "of a woman in Wiltshire who had her leg amputated, got a little coffin made for it, and caused it to be buried in the (Christian) churchyard."

Except for eunuchs, priests, and courtesans who, according to the polite euphemism, have "left the home," and those who have accepted *Chinese Graves* foreign religions, there are few public cemeteries in China. In some of the southern provinces, however, where the population is dense and the land terribly overcrowded (near Canton, for example), we do find common burial-grounds in which each family has a private plot. Too poor to afford individual monuments for their dead, these humble folk (crowded in death as in life) erect one large stone tablet, where those who are unable to distinguish the graves of their own forefathers may worship them vicariously.

As a rule, however, each Chinese family has a private burial-ground outside the city walls, often in the midst of cultivated fields, sometimes with arable lands attached to it. The crops grown on such lands provide a living for the caretakers, and funds for the repair of the graves.

The site of a Chinese graveyard is so important that an astrologer is consulted to insure it be "lucky." A southern aspect with mountains behind or, failing these, a horse-shoe wall, contribute to "good influences." A stream or moat is also desirable, and trees, especially cypresses, are supposed to attract the vital essences, the *Yang* and the *Yin,* and thus preserve the bodies against dissolution and the attacks of evil spirits. Pines, too, with their branches bending like sheltering arms, are also favoured. The Chinese call them the "dukes" of the hierarchy of the trees, because the character for pine is a combination of "wood" and "duke." "Wood," originally a rude picture of a tree, now consists of a cross and two strokes, thus making a similar ideograph to the one used for the number eighteen ("ten" and "eight").

Hence, the classical comparison of "a high wind blowing through funereal pines" to the noise of "eighteen dukes leading their retainers to victory."

The graves themselves, with slight variations in different provinces, are usually in the shape of mounds or hills, a traditional remembrance of days when the original Chinese settlers buried on mountain-slopes as a protection against floods, robbers, etc. Typical examples are the enormous tumuli covering the remains of Chinese sovereigns, with underground grave-chambers arranged in a manner strikingly reminiscent of the Egyptian pyramids. Princely tombs, though smaller, have the same form; likewise, even humble peasant graves which are simply cones of heaped-up earth. Farmers bury in their own fields, because they can not afford to do otherwise. Nevertheless, Chinese geomancers still—theoretically at least—consider flat lands unlucky for sepulchre.

Each grave-plot is managed by the elders of the family to whom it belongs. Their duty is to see that, even in death, the clan is not divided. Furthermore, each member must lie in his appointed place, according to the status he had in life, and "certain parts of the graveyard are always more honourable in the hierarchic sense than others . . . The back, centre, and front portions of the ground are reserved for married couples who have left children (man and wife being frequently buried under the same mound) and, therefore, take an honoured place in the family pedigree, whereas those who have died unmarried, or in babyhood, are either not accommodated in the family graveyard at all or, if admitted, are buried close to the right or left boundary." "A villager was once brought before me," says Johnston in *Lion and Dragon in Northern China*, "on the charge of having buried his dead infant, a child of two years, in a part of the graveyard that was reserved for its dignified elders. As it is advisable in such matters to uphold local custom, I felt reluctantly obliged to order the man to remove his child's body to that part of the graveyard which is regarded as appropriate for those who have died in infancy."

Practical, as well as moral, considerations require that a special effort be made to visit tombs at least once a year in China, as such care constitutes a renewed right to the property. An unswept grave indicates that a family has died out. After a considerable lapse of time if, at the Feast of Tombs, none came to offer sacrifice according to established custom, the abandoned cemetery would gradually melt into the surrounding fields, and the plough of the farmer level earth-mounds to earth, and dust to dust. Finally, when all traces of the neglected tombs were gone, the plot might be sold to others as a burial-place. So long, however, as the right of ownership is asserted, heavy penalties are enforced for damaging graves, for encroaching on graveyards in the hope of getting a few more feet of arable land, and for cutting trees, or pasturing cattle thereon. Village elders promptly punish such offences, compelling the offender to make an expiatory offering to the dead whom they have insulted. In the case of an Imperial burial-ground, the penalty was death.

Those who find themselves far from their native province at the *Ch'ing Ming* and, therefore, unable to make a personal visit to their family burial-ground, will send a deputy, and themselves perform certain rites in the house where they happen to be. For this purpose, they buy paper bags about a foot square, decorated with two human figures surrounded by symbolical printed flowers. In the space left between these decorations, the name of the ancestor for whom the bag is intended is inscribed. It is then filled with paper money and placed on the *k'ang*, or brick bed-platform, with food-offerings. After these preliminaries, the father, as high priest of the family, makes his bows with hands uplifted, just as he would at the actual grave. Later, the offering-bags are carried outside the street door, and burned. It is important that this ceremony be performed before sundown, as the spirits of the ancestors, who are believed to come and spend the day with their descendants, must get back to their graves before the city gates close at nightfall.

But, if possible, people try to be in their own homes at the spring festival of the dead which is an occasion of family re-union. A few days beforehand, rich people send out old retainers (if they have no caretakers on the spot) to sweep and repair their graves, while the poor man, shovel on shoulder, goes himself to plaster fresh earth on mounds that winter winds and snows have crumbled.

Early in the morning of *Ch'ing Ming* day, the countryside is alive with people wending their way to serve the dead. The Chinese expression is significant: *pai shan*, to "worship at the hill." Though in some provinces men only visit the graves, in Chihli, at least, the entire family troop out,—children astride donkeys whose pack-worn hides show their everyday activities, women hobbling painfully on bound feet, men carrying incense sticks and paper money tied up in cloth bundles. One sees no signs of grief or mourning, except among those recently bereaved,—here and there a woman, in white "keens,"—but the majority of the peasants in clean blue cotton clothes have rather a holiday air, as befits those who are on their way to a pleasant ceremony of remembrance.

There is hardly a family whose land is not dotted with the little brown tents of sleeping ancestors, bound up, in death as in life, with the inherited ground. "Man belongs to the soil, not the soil to man, it will never let its children go . . . and when they die, these peasants, they return in child-like confidence to what to them is the real womb of their mother." Death in China, according to Chinese conceptions, is too natural to be horrible. Our gloomy doctrines of everlasting punishment for sinners are unknown. Nor are their saints shut into a hard golden heaven where the living can not follow them, but remain quite near and caring. This sense of the nearness of the dead strongly affects the attitude of Chinese mourners, though, here as elsewhere, there is genuine grief when loved ones die. It permits

CARRYING PAPER MONEY TO BURN AT THE TOMBS.

PILGRIMS AT A TEMPLE.

Photo by the Asiatic Photo Publishing Co.

relatives round a sick man's couch to speak with perfect propriety of his approaching end, and to discuss with him the elaborate funeral preparing in his honour. It even allows his wife to sit beside him, sewing on mourning dresses for the family, and the carpenters to put finishing touches on his coffin within earshot.

There is no false shyness, either, about graveside offerings, and we may watch, if we wish, the simple yet touching service. Among farmer folk it consists of a few bows, as dishes of coarse grain are held between uplifted hands with the whispered invocation: "Your children have come to-day with gifts of food. Disdain them not, we beg you, for we are poor and cannot offer you a rich repast. Therefore, forgive and come and eat." Then two squares of paper, stamped to represent cash, are put on each grave and held in place by a stone, or a lump of clay. For days afterwards, these papers flutter in the breeze, giving the effect of white doves settled on the plain.

Chinese gentlefolk set out more elaborate offerings on the stone altar-tables of their clan-cemeteries in the twilight of the trees. Many will serve a complete feast. But worshippers are always careful that their sacrifices to the spirits be offered in even numbers, just as sacrifices to the gods are offered in odd numbers, the reason being that gods correspond to the *Yang*, or uneven principle, and ghosts to the *Yin*, or even principle.

At each grave the same rites are performed, but lest any distant ancestor be neglected (which may well happen when the inscriptions on the oldest monuments become illegible), a general sacrifice is made to all clan-ancestors in addition to the personal sacrifices to the last few generations.

The simple reverence of the service dignifies the humblest offering. It is touching evidence of a mutual relationship of love and gratitude. Love indeed makes the dead divine to men who feel that spirit-eyes are watching them,[3] that spirit-ears are listening to the prayer never said aloud: "Forefathers of our family, of our kindred and of our race, deign to accord us your protection. All that we have is your gift, all that we know is your knowledge bequeathed—about laws of life and death, about the things to be done and the things to be avoided, about ways of making life less painful than Nature willed it, about right and wrong, sorrow and happiness, about the error of selfishness, the wisdom of kindness, about the need of sacrifice. Unto you, the founders of our homes, we humbly utter our thanks."

Meanwhile, there may be, indeed there often used to be, musicians hidden among the trees, playing old airs on shrill reed-pipes for the pleasure of the souls. Even nowadays one occasionally hears the notes of a flute among the graves, softly piping a mystical serenade full of melancholy, like the shepherd's melody in Wagner's *"Tristan and Isolde."*

---

[3] On old bronze sacrificial vessels formerly used at the grave-rites you will notice a curious design which rudely resembles a pair of eyes. It is intended to represent the eyes of the ancestors watching their descendants at the sacrifices. This primitive picture proves how literal is, and has been for centuries, the Chinese belief that ghostly presences gather near their tombs at the festivals of the dead.

After worship, the family sometimes dines near the tombs *al fresco,* or returns home for a feast of re-union, when the offerings, of which the dead have already absorbed the spiritual essence, are eaten by the living.[4]

In one way, and one way only, the Chinese grave-offerings used to differ from those of almost every people, including even their neighbours, the Japanese. No flowers were placed on their tombs, or planted near them by private individuals, since in ancient days this was an exclusive privilege reserved to members of the reigning family who, as a rule, used artificial flowers specially manufactured in Amoy for this purpose.

* * *

Willow Myths

The only plant permitted for general grave-decoration was the willow. One still sees willow-branches stuck in grave-mounds, for the willow has a mystical connection with the spring festival, when sprigs of it are also hung under the eaves of the houses. Popular legend connects this latter custom with a certain rebel of the T'ang dynasty who took the willow as his badge. Sympathisers were ordered to hang out a branch of the chosen tree. On *Ch'ing Ming* day, when the signal for revolt was given, those who displayed the mark

---

[4] The offering of food to the dead seems to have been made by all nations from very early times. In some Christian countries funeral meats are still placed on graves at "All Souls'," and in Russia a special meal is taken by families near their tombs on the day known as "Parents' Saturday."

of loyalty to the new regime were spared in the general massacre that ensued. Their descendants gratefully continue the habit.

As a matter of fact, the true significance of the willow as a spring-emblem lies much deeper. First of all trees to respond to the sun's kiss by putting out tender leaves, it has become identified with the principle of light. Moreover, it is extremely hardy and will grow almost anywhere. Popular imagination therefore sees in it a symbol of vitality and an enemy of darkness.

A logical step further in superstition makes the willow a general protective talisman against evil spirits,—"they that walk in darkness." The third emperor of the T'ang dynasty bade his suite wear willow-wreaths on their heads to guard them against the venom of scorpions. Other well-known men used amulets of willow-wood to ward off sickness, owing to the connection between this "fortunate tree" and the Sun, the Great Physician. In Kiangsu, women still stick a sprig of willow in their hair at the *Ch'ing Ming*, lest evil spirits harm them, and in other provinces young people wear willow-sprouts, called "willow dogs," all day, because of the old saying: "Those who wear no willow at the *Ch'ing Ming* will be re-born as yellow dogs in a future life"—not a pleasant fate in China.

Not only does this ever-useful plant rout the forces of evil. It also has power to attract good influences. It can draw the spirits of ancestors back to their homes, while at the same time repelling stranger and, possibly, mischievous ghosts. It is helpful to those possessed by devils and, finally, as we shall see later (*see* "The Sixth Moon"), it is a powerful rain-charm.[5]

*   *   *

Apart from the solemn duties connected with them, the *Ch'ing Ming* visits to the graves are looked upon as a pleasant holiday picnic—the first real excursion of the year. More will follow. With the spring weather, temple festivals begin. These serve not only as excellent pretexts for an outing, but also bring contributions to impoverished coffers. *Hsi Wang Mu*

In Peking, the first important religious anniversary of the third moon is the birthday of the Hsi Wang Mu, mother of the Western Heaven, celebrated by a three-day fair (from the first to the third) at the P'an T'ao Kung, a little temple situated on the banks of the canal near the Tung Pien Mên.

The identity of the Goddess honoured is rather confused, as often happens in China. Now she appears under one name, now under another. But her simplest title, "Mother of Heaven," best expresses her place in popular affection.

The Hsi Wang Mu and her husband, the Mu Kung (or Tung Wang Kung), are supposed to represent the original pair of beings, born respectively of the Quintessence of the Eastern and the Western airs, progenitors of the Masculine and Feminine forces, the *Yang* and the *Yin*, which play such a large part in Chinese sym-

---

[5] The Chinese are not alone in their belief in the power of the willow. This tree is revered by the oldest inhabitants of Asia, such as the Ainus, the Manchus, etc.

bolism. Mu Kung, also a personality wrapped in doubtful legends, is God of Male Immortals and the Ruler of the Eastern Heaven where, clothed in violet mists, he dwells in a palace made of clouds, whereas his wife, in modern fashion, lives apart from him in the Western Heaven supposed to be situated in the K'un Lun mountains of Chinese Turkestan.[6] Her abode is beautiful beyond belief. It has ramparts of gold three hundred miles around, with twelve jade towers and battlements of precious stones. Her gardens contain fairy fountains spouting jewels, and in her magical orchard the peaches of Immortality grow.[7]

This enchanted abode is the home of the Immortals, those wonder-men and women, divided into grades according to the colour of their robes, —blue, black, yellow, violet, and fawn,—who keep the Hsi Wang Mu company. Rarely, very rarely, a human visitor is admitted to this paradise. Prince Mu Wang came once in the Xth century

---

[6] Their mystical reverence for the distant K'un Lun Mountains, situated in the direction of the setting sun, is possibly due to a hazy, yet persistent, tradition of western origin for the Chinese race. It is significant that many peoples of divergent culture imagine their Paradise—or their Inferno—to be situated in the West. Here the Greeks located their fabled Gardens of the Hesperides. Here Buddhist literature places the Heaven of Amida (*See* "The Eleventh Moon").

Polynesian ghosts went towards "the turning place of the sun," and in our modern expression "going west" is a strange survival of a widespread myth originating in the Sun cult.

[7] Herbert Giles, in the *Adversaria Sinica*, compares the Hsi Wang Mu with Juno. One lives in the K'un Lun mountains above the clouds, the other on cloud-capped Olympus. One has a garden where the Peaches of Immortality grow, the other a garden of equally magic apples. Juno's attribute is the peacock, and the Hsi Wang Mu is accompanied by the phœnix, who may well have been inspired by the Indian peacock.

Parallels like these suggest that more than one member of the Taoist pantheon can be traced to myths of the distant Occident.

THE THIRD MOON, OR "SLEEPY MOON"

B.C., with his eight favourite horses—a popular theme of Chinese artists. (Some believe this visit symbolises early Chinese relations with the West). The mighty Han Emperor, Wu Ti, also received from her own hands, in his palace at Lo Yang, seven of her magical peaches that ripen every three thousand summers.

But regularly once a year, on her birthday, the third of the third moon, the gods gather at her palace to offer congratulations. The God of Happiness, Fu Shên, comes in his blue official robes, the God of Riches, Tsai Shên, with his arms full of treasures, the Dragon King, Lung Wang, Ruler of the Seas and the Jasper Lake, rides thither on a thundercloud. These, and many other popular Taoist divinities, hasten bearing gifts, and remain to a feast worthy of the illustrious company. The rarest meats are served, bears' paws, monkeys' livers, the marrow of phœnixes, and, for desert, the peaches that renew the immortality of immortals, while music from invisible instruments and songs sweeter than any heard by mortal ears delight the banqueters.

It is in memory of this, most wonderful of birthday parties, that the festival of the P'an T'ao Kung is held. Alas! the magical atmosphere of the Western Heaven is sadly lacking. The "Palace of the Trained Peach-Trees" in Peking is a tiny sanctuary, cramped and tawdry. It contains an insignificant clay statue of the "Mother," dusted for the occasion, and a trellis of faded paper peach-blossoms whose branches are bent like a crouching dragon. (Hence the name of the temple, from *t'ao*, "a peach," and *p'an*, "to bend"). Nevertheless, the shrine is crowded with worshippers, especially women and children. In matters of religion, the Chinese imagination is capable of performing miracles, just as in matters theatrical. A riding-whip is as good as a horse when it becomes necessary to indicate a stage cavalry-general. A small temple is as good as a big one, if it enshrines a popular goddess. Dusty altars are no discredit, lack of ventilation is no disadvantage. Indeed, the more people jostling one another in a confined space, the more incense-fumes pouring out in a suffocating cloud, the more noise, the more children and chickens under-foot, the more tea served, and spilled, in temple courtyards, the greater the success of the festival. Let us add that most of these conditions are fulfilled at the P'an T'ao Kung and, consequently, "country potatoes" (peasants) and town-dwellers alike make happy holiday here.

* * *

The Tung Yo Miao   Another equally important pilgrimage of the third moon, which appeals more especially to the men folk, is that to the Tung Yo Miao (a Taoist temple near Peking, on the round to T'ung Chou) dedicated to the Spirit of Mount T'ai Shan.

There things are better organized. Three guilds contribute to the success of the festival. Dusters dust the images, the Lamp-makers donate a lantern whose flame must not be allowed to go out, and all the artificial flower-makers put an arch of paper blooms before the entrance. The Tung Yo Miao is a rich temple, and large, having many halls dedicated to many divinities. With gods to suit all tastes, it is in a position to cater to several varieties of "religious custom." Lovers come here to pray to the "Old Man of the Moon" (*see* "The Eighth Moon"), debtors to settle disputed accounts with the aid of the magical abacus hanging on the wall just inside the gate, invalids to touch the famous bronze mule supposed to cure all maladies, and scholars to worship the God of Literature.

Though China has been torn and wracked by invasions throughout her long, long history, her people, nevertheless, still believe that the "pen is mightier than the sword." Kingdoms may disappear, and empires fall, but the glories of wisdom remain and conquer in the end. Therefore, the deepest allegiance of the Chinese has ever been not to a government or to the land (feelings which constitute patriotism in our sense of the word), but to their ancient culture. It is a wonderful conception, though diametrically opposed to our Western militaristic ideals based on Roman expansion and Germanic conquests by force of arms. We poke fun at the bookworm and praise the warrior—the "officer and the gentleman" point of view.

But the Chinese despise the military caste. They consider scholarship the test of ability to govern, because scholarship implies morality, and morality, in their eyes, is the primary force of the world,—far above physical force.

The logical outcome of this point of view was the old-fashioned system of examinations held by the State and based on the classical books, for the most part moral treatises intended to promote good behaviour and, therefore, good government, —that is to say as little government as possible, because well regulated people need few laws and fewer police. "If a father is a good father, a husband a good husband, a neighbour a good neighbour, a ruler a good ruler, and all classes are assiduously practising the five heavenly virtues, justice, magnanimity, politeness, understanding and the faithful execution of duty, what need for the State machinery to interfere?"

In theory, the result was a perfectly regulated world where all men were equal, since hereditary titles (with one or two exceptions) did not exist —a true democracy, as we dream of democracy, though under the headship of a sovereign, where not birth but scholarship was the passport to honour and to high office, and the doors of learning stood wide open to high and low alike.

The underlying principle of the old Chinese State examinations was useful and good. It built up a moral distinction, a conception of human dignity unfettered by class prejudice, and a mighty and unbroken chain of culture of which any people might well be proud. But, in practice, it strangled progress and was, therefore, doomed to die of senile decay. Modern Chinese scholars declare that the cut-and-dried literary pursuits imposed by the classical system were bent upon curbing and crushing the politically dangerous development of free thought, much as our own classical education, with its study of dead languages, has been accused of doing in Europe. Actually, both methods were intended to train young minds to logical thinking, to protect them

THE THIRD MOON, OR "SLEEPY MOON"

against the vagaries of unscholared intellects, to serve as a mental discipline and provide the historic background against which the present appears in proper focus.

To-day, the mantle of Chinese culture, albeit often torn and ragged, has fallen upon the shoulders of students whose ideals are so different and so antagonistic to those of their forbears. But even these schoolboys retain a measure of popular respect and are permitted a semblance of authority, just because they study books. What they study, and how deeply, is a secondary consideration. As descendants of the *literati*, some sympathy is considered due to them, even if their learning has faded and their manners toughened—due more especially at this crucial period when a new machinery of learning must be devised to replace the old ideals of culture without losing touch with race traditions.

Gods of Literature

With scholarship a passport to rank and honour, it is no wonder that Gods of Literature have always been worshipped in China by those who were, at the same time, seekers of knowledge and seekers of office. Chief among these divine patrons is Wen Ch'ang, who lives in the constellation of six stars, also called Wen Ch'ang, connected with *Ursa Major*. His personality is vague, his history confused. Now he appears as a hero of the T'ang dynasty, a brilliant scholar in his lifetime, and a resident of Szechuan province, where he is still specially revered. Again, we find him incarnated as a serpent, according to a legend where Taoist fancy has free rein. In other myths, he is a saint with seventeen re-incarnations, lasting over a period of three thousand years. Soldier, scholar, prophet, take him as you will, he ends his chequered career by official deification, and gets himself identified in the mysterious Chinese way, which allows spirits of mortals (whose merits have qualified them) to rule a portion of the universe from a star.

Wen Ch'ang is usually portrayed in a long blue gown, a sceptre (*ju yi*) in his hand. He rides a white horse and is accompanied by two servants known as "Deaf as Heaven" and "Dumb as Earth,"—convenient followers who do not give away their master's secrets when he distributes intellectual gifts.

Associated with him also, by a process too long and too complicated to explain here, is Kuei Hsing, likewise a star in the Great Bear. There is no real connection between the two, but the "convenient 'elasticities' of dualities" among the gods in the Taoist scheme of things allowed them to "borrow glory from one another" and, finally, combined their cults by a stroke of priestly cunning which obtained Imperial sanction. Nowadays, at the Tung Yo Miao and elsewhere, the stately Wen Ch'ang shares his shrine with Kuei Hsing, a tiny person with the face of a devil. He is so represented because of a pun on the character for *kuei*, spirit, or devil. One leg kicked-up roguishly helps the resemblance to the ideograph. Little Kuei Hsin is rather a pathetic figure. In life, he was a deformed scholar, as repulsive physically as he was brilliant mentally. First academician at the metropolitan examinations, he appeared before the Emperor to receive the golden rose due to the prize winner. But the sovereign turned from his ugliness and refused what was the dwarf's right. Thereupon, the miserable rejected one threw himself into the ocean. Fortunately, a sea-monster raised him on its back and brought the despairing scholar to the surface again. The legend gave rise to an expression formerly used by Chinese to describe one who did well in an examination: "To stand alone on the sea-monster's head."

As for Kuei Hsing, he got his revenge. Ascending to heaven, he took up his abode in the star "kuei hsing," his namesake, got identified with Wen Ch'ang (owing to a confusion in the character *kuei*, which has several meanings), took a share of his worship, and in heaven obtained the official recognition he was denied on earth.

In the same literary group, for these various figures often appear together in Taoist temples, we find Chu Yi, "Mr. Redcoat," the inseparable companion of Kuei Hsing, and Chin Chia, "Mr. Golden Cuirass." Chu Yi was once the beloved of an immortal princess who conducted him to heaven. He returned to earth, because he preferred to study the classics, and attain the dignity of a Minister. The story goes that once he appeared in a long red robe—hence his nickname, "Mr. Redcoat"—before an examiner, and commanded him by a gesture to pass an essay. Hence the expression: "Perhaps Mr. Redcoat will nod his head," with which a poor scholar's friends would endeavour to hearten him on the eve of an examination.

Chu Yi came to be regarded as the purveyor of official posts, while his friend Chin Chia, "Mr. Golden Cuirass," was the terror of wicked scholars. At the same time, the latter could and did reward the righteous. He holds a flag which he waves before the homes of the elect, thus warning families that one of their members or descendants will win high literary honours.

The cult of these combined literary gods is still very much alive, as witness the offerings of pen-brushes, ink-slabs, and writing-paper made at the Tung Yo Miao during the festival.[8]

* * *

The pilgrimages described so far are all city pilgrimages but, as the warmer weather comes, the pious begin to go further afield. A favourite excursion is to T'an Chê Ssŭ, one of the oldest, finest, and certainly the best preserved monasteries near Peking. A bluish confusion of roofs, beautiful as the curved flight of arrows, interspersed here and there with yellow-tiled pavilions like pools of sunlight, show sharp against the green of wooded hills as we approach the temple. Beyond the main gate with its fine archway, painted in butterfly colours, open a series of splendid courtyards with great halls of prayer. Here, in perpetual twilight, the Buddhas are enshrined amid soft silken hangings, on gilded pedestals, behind altars enriched by bronze incense-burners, by multi-coloured enamel flower-vessels, and lacquered lotuses with a purplish sheen.

---

[8] There are other patrons of letters, including one of the Eight Taoist Immortals, Liu Tung Pin.

THE THIRD MOON, OR "SLEEPY MOON" 241

Further up the hill-side are other edifices belonging to this princely establishment of the Buddhist Church, secondary shrines to lesser gods, each with its treasures of painting, weaving, or metal work,—a lovely little round treasure-house like a jewel-casket, a hall of assembly, a *chieh t'ai,* or platform for consecrating monks, and the "Bamboo Courtyard," one of the many suites of guest-rooms reserved for important Chinese pilgrims who make large donations to the temple, as so many sovereigns and statesmen have done since its foundation when Peking was still a provincial town.

Of peculiar interest is a magnificent ginkgo tree with an honorific *p'ai lou* in front of it, and a special altar where prayers are said. This tree was canonised by Imperial edict, and came to be considered a patron saint, so to speak, of the Manchu dynasty, since at the accession of each Emperor it put forth a new shoot from the parent stem. *Tree Worship*

This is the only case we know of where a tree has actually been included in the official cult, though tree-worship among the people is not uncommon in China. In fact, it is "one of the recognised by-paths of Chinese religion," inherited from primitive settlers, and connected only accidentally, as it were, with Confucianism, Taoism, or Buddhism. The cult is particularly interesting, because it leads to a region of folklore and myth common to China, India and Europe. Thus, in Rome, we have the sacred Fig-tree of Romulus, an object of popular devotion; in India we find the Bo tree and the Jambu tree,

16

revered for their connection with the Buddha, and traces of tree-worship actually exist in popular tradition and local custom right across Europe. There are legends of holy trees and talking trees who pleaded for their lives with wood-cutters in England, of tree-spirits in Switzerland and in the Tyrol, where Frazer, in *The Golden Bough*, speaks of a tree that, as late as 1859, was thought to bleed when cut. "With a varnish of Christianity," he says, "and the substitution of a saint's name, tree-worship, like water-worship, holds its own to this day."

It is not astonishing that in China, where mountains, lakes, rivers, even stones, are believed to be the abode of spirits, trees should also have their "ghosts." Some are good, some bad. Therefore, to cut down a tree is not without risk, lest the spirit be offended. One such angry *kuei* has been known to rush out in the form of a blue bull. We ourselves know of an instance where disasters that befell a certain village were traced to a magnificent elm possessed of a demon in the form of a centipede living in its trunk. The country folk feared to take action against their ghostly enemy. But the God of Thunder, in answer to their prayers, shot his bolts and freed them.

In southern Fukien men dare not fell large trees. Their spirits are considered too dangerous, and may visit a whole neighbourhood with disease and calamity if irritated. In Amoy, the farmers even dislike planting trees, because they believe that when the trunks become as thick as human necks, they will be strangled.

Sometimes, peasants will apologize to a tree before cutting it down, asking pardon for the pain they are obliged to cause: "Our children are cold and we have no wood to heat our cooking pots." For trees, like men, feel pain. Even in standard Chinese histories we read of trees that utter cries of agony and bleed when cut. The "Lone Pine" on Spider Hill, near Hei Lung T'an in the Western Hills, near Peking, is an example, and one of the six junipers in the Temple of the God of Mount T'ai Shan, in T'ai An Fu, is another. From the latter, at the first blow of the woodcutter's axe, it is said "blood-red sap gushed out in streams, and the workmen fled before the wrath of the offended tree-spirit."

The idea of propitiation doubtless underlies a custom common in certain parts of North China, notably in north Shensi, north and west Shansi, and on the borders of the Ordos desert, where the Chinese peasants "always hang the heads and feet of their goats and sheep that die, or are killed for food, on trees as propitiatory offerings to the gods." The same custom prevailed among the Teutons, also tree-worshippers. Each of their clans had one or more specially sacred trees, a pine, or spruce, or some similarly-shaped conifer, upon which its members hung offerings to the Sun during the Yule Tide. These offerings consisted of the heads, skins, and various other parts of the animals (either domestic or wild) they killed and ate, and from this custom we have derived our Christmas tree.

Fortunately, all Chinese tree-spirits are not revengeful. Many are kindly disposed towards men and grant petitions, especially for recovery from sickness. To this category belong the sacred trees whose branches are hung with votive paper-scrolls, inscribed by grateful worshippers, or bright-coloured rags—an immemorial custom inherited from the oldest inhabitants of the Far East—and those that have incense burned to them, and offerings of food set out as before the image of a god. We know of one curious instance, in the courtyard of a secondary temple at T'an Chê Ssŭ, where a miniature shrine is built against the trunk of an old chestnut, the home of two serpents, supposed to be incarnations of the spirits of the Universe like those kept in the main hall of the temple. Naturally, the destruction of such sacred trees is considered a calamity by the entire neighbourhood.

A few of the Chinese tree-legends are well worth re-telling. There is, for example, the story of the pine set upon a hill. One day, a peasant met a traveller with a white dog near by. "Where do you live, elder brother?" the countryman inquired. "In yonder tree," the stranger answered, and proceeded on his way up the mountain. The peasant, curious yet unbelieving, followed him when, lo, and behold! he saw both man and dog disappear into the tree. Then he knew that the man was indeed the spirit of the pine, and the dog the ghost of the white fungus growing at its root.[9]

---

[9] This fungus is of great value and importance in China. It is called the "fungus of immortality," and appears in Chinese ornaments as a symbol of longevity. Laufer says that it "is a species of agaric and considered a felicitous plant because it absorbs the vapours of the earth . . . A delicacy at Chinese feasts, it also has valuable medicinal

Generally, pines in legend stand for morality, as witness those that "changed colour"—blushed in sympathy—with the pious ruler of legendary times who used to weep at his own unworthiness, taking upon himself before Heaven the responsibility for natural calamities which were actually beyond the power of mortal man to prevent. Another conscientious tree, near the birthplace of Confucius in Shantung, showed an interest in good government by refusing to put forth new branches "until moved to do so by the advent of a virtuous sovereign. The beginning of the Manchu dynasty and the birth of the Emperor Yung Cheng . . . . were marked by this wonder."

Maples also were associated with old Chinese rulers. Under the Han dynasty, the Imperial palace was known as the "Maple Halls" because some of these trees shaded its courtyards. A certain "demon maple," famous for granting the requests of its worshippers, originated the popularity of maple wood for making "prayer-answering" images of the gods.

The willow has a gentler spirit, as we have already seen. Light and graceful legends, concerning scholars and ghostly maidens, tales of meetings in sunny gardens, of visions and vanishings, are connected with it. Of such is the story

---

properties as a tonic, and western Hupei, Szechuan and Kuangsi, where climatic conditions are most favourable to the development of this fungus, export nearly five hundred thousand taels' worth each year." Apart from its power to restore health, and even bring the dead to life, this magical plant suggested the form of the *ju yi*, or sceptre, presented to and by officials, because it supposedly ensured the "accomplishment of wishes."

about the youth who, while studying the classics under an old willow-tree, heard the sounds of a dulcimer softly played. Inquiring the name of the unknown musician, he was answered by a soft voice: "I am the spirit of the willow," while an unseen hand sprinkled him with a soft rain of willow-sap. Then the voice continued: "You will gain the highest degree of the Empire without fail. And, afterwards, you must make me offerings of date cakes." The young man promised he would do so and, sure enough, he received the foretold degree.

Another willow-legend is charmingly depicted on the Chinese stage. This tale also concerns a scholar, young and handsome too, who possessed the picture of a lovely maiden. In the midst of his studies, his eyes often wandered to the painting and, finally, one day the beautiful face smiled at him. Fascinated, he closed his books, knelt down before the scroll, and begged the image to speak. Of course, at the sound of a voice sweeter than the perfume of the lotus, he straight-away fell in love with the picture-maiden. "Who may you be?" he inquired, trembling with emotion, and she, blushing, replied: "I am the spirit of the willow-tree which grows yonder in your garden." Humbly then he begged her to become his wife, and she agreed if he would prepare suitable garments for her. When all was ready, she stepped out of her brocade frame to receive the blessing of his parents, was dressed in the marriage robe, and became a mortal wife.

Every woman, according to an old Chinese superstition, is represented by a tree in the spirit-world. If this ghostly counterpart of her mortal self blossoms, it means she herself will have children. Red flowers show that girls will be sent her, white flowers, boys, while bare branches mean an empty cradle. In the latter case, a child may be adopted with the certainty that it will acquire all the family characteristics. As in nature one tree is grafted on another to get fruit, so this child is considered a human graft on a barren family-tree and will resemble that tree.

Perhaps the prettiest Chinese tree-legend we know dates from the eighth century of our era. In those days, a certain scholar lived alone with his books for companions, and flowers, which he dearly loved and tended, for friends. One beautiful spring evening, as he was strolling in his garden, a maiden in trailing garments flitted towards him. She greeted him softly, saying: "My companions and I are on our way to visit an aunt, and we beg your gracious permission to rest a little while here." Of course, the scholar gave his consent, so she called her friends who were hiding in the bamboo-grove where the fire-flies gleamed like tiny lamps, and introduced them. "This is Miss Plum, yonder—in crimson—is Pomegranate, and I am called Peach-blossom." The youth was charmed with these graceful girls who came towards him shyly, carrying flowers and willow-branches. Their slender, graceful figures swayed like wisteria-tassels in the breeze. When they moved, a delicious perfume filled the air, and when they spoke, there was a tinkling as of tiny bells.

As host and guests chatted together, the aunt, Lady Zephyr, arrived with a fluttering of rainbow silks. Thereupon, wine was served in her honour, and a feast of dainties spread in that enchanted garden. The party soon grew merry, enlivened by songs and dances in which the maidens posed as gracefully as a flight of butterflies touched by the blue fingers of the moon. Unfortunately, in the midst of the gaiety, Lady Zephyr carelessly spilled some wine over the robe of her niece, Miss Peach. Losing her temper, this young lady reproached her aunt, and in her turn, was rebuked by Lady Zephyr. A quarrel ensued. In the midst of it, suddenly the whole group vanished into thin air.

Now, the very next evening, the offended maiden appeared to the scholar again and said: "My companions and I, who visited you last night, all live in your garden. Alas! we can not dwell there in peace, because every year we have a visit from the cruel East Wind. Usually, we ask our aunt to protect us. I should have done so yesterday evening, but she spoiled my dress, I grew angry and forgot to beg assistance. So, that is why I have come to ask you to help us."

"How may I do this?" the scholar inquired.

"We desire you to prepare a crimson flag, embroidered with the moon and the planets in gold upon it, and to set it in your garden. Pray, do this early in the morning, as soon as you see the East Wind coming, for the flag will guard us against his rough embraces."

The young man readily agreed to do as they wished, and the girl, thanking him, said: "We are grateful for your help, and will certainly repay it."

Next morning a furious wind blew. It bent bending willow-branches, made harps of the pines, and tore the tenderer shrubs to pieces. But against the scholar's garden, protected by the magic flag, the gale was powerless and not a blossom stirred.

Then the scholar realised that his guests, the flower-maidens, were tree-spirits, and their aunt, who felt herself insulted and sought revenge, was in reality the wicked East Wind.

That same evening, the girls appeared to him once more with blossom-garlands as thank-offerings. "You have saved us," they said, "and we have nothing else to give you. But if you eat these flowers, you will never grow old. Pray, do so! And at the same time we beg of you to put up the flag in your garden every year, on the anniversary of this day, that we also may renew our youth."

This quaint myth is the origin of two popular superstitions. In memory of the flower maidens, the country folk gather peach-blossoms at this season, dry and powder them, mix them with well-water, and drink the decoction as a cure for heart disease, or, in other words, to renew life. At the same time, they remember the revenge of the wicked Miss Wind in the saying: "If the wind whirls up the dust on the tombs on *Ch'ing Ming* day, it will blow incessantly for forty-five days thereafter," a prophecy which, in our experience, is inevitably fulfilled, at least in Peking.

CHAPTER VIII.

THE FOURTH MOON, OR "PEONY MOON."

UMMER in China begins in the fourth moon. In Peking, street-hawkers sell little red cherries like children's rosy cheeks, and mauve roses, dewy wet, from which Chinese housekeepers make a delicious jam.

Then every garden is at its loveliest, wide awake at last. Nature rewards the Chinese love of flowers with a gift for growing them. Bushes, shrubs, even trees, forget to follow their original intent and bend to the will of the Oriental gardener, yielding him an almost wifely submission. The geranium in an old tomato tin blooms generously for the poor boatwoman who lovingly tends it. The New Year narcissus unfolds its star-petals from the cracked cup of the humblest coolie. A clump of asters grows in the stony ground beside his sentry-box for the policeman who waters them in full uniform with his tea-pot.

The Chinese calendar is marked with flower-anniversaries, and every moon is hostessed by a flower-fairy presiding over a long chain of flower-fêtes. The Rose Fairy presides over the first moon; Apricot Blossom over the second; Peach Bloom over the third; Mistress Climbing Rose over the fourth; the Pomegranate Maiden over the fifth; and the Saintly Lotus Lady over the

THE FOURTH MOON, OR "PEONY MOON" 251

sixth. In the seventh comes the perfumed Balsam Fairy, and in the eighth the Fairy of the Cassia Flower, so small but so sweet. The ninth sees the reign of the Chrysanthemum Queen, the tenth the Golden Lady of the Marigolds, the eleventh the cold and virginal Camellia Fairy, and the twelfth is in charge of the pale Winter Blossom.

The "birthday of flowers" in general is on the twelfth (in some places on the fifteenth) of the second moon. If no rain falls on that day, they will be plentiful throughout the year. The 19th of the fourth moon is the day known as the "Washing of Flowers" and also the birthday of Wei Shên, Protectress of Blossoms. People gather in their gardens to celebrate this double anniversary.

Throughout China the fourth moon begins the fashionable season for garden parties, and the great event everywhere is the opening of the peonies, for the peony in China is the King of Flowers, and used to be called "the ornament of Empire." When these favourites reach perfection, the President of the Republic himself sends to privileged guests invitations for a "peony-viewing."[1] They enter the Sea Palaces, where His Excellency is lodged, by that "Home-Looking Gate" from whose balcony Ch'ien Lung's "Stranger Concubine" gazed with lonely eyes towards her birthplace in distant Kashgaria. They are conveyed to the audience hall, across a lake of lapis-lazuli, in one of those barges, with rowers

---

[1] We are speaking here of Chihli province and, especially, Peking. In Shensi the birthday of peonies is on the fifth of the third moon, and the date of their blossoming varies in other provinces. See *La Pivoine*, by Henri Imbert.

standing to their oars, that formerly transported Sovereigns and their suites. Imagination calls up pictures of majesty seated in golden robes on golden cushions, surrounded by dainty ladies with flower-crowned head-dresses, and appropriate garments of ceremony, of purple-robed attendants bowing low. Alas! when this setting belonged by right to kings and courtiers, they were never able to hold it. But now that they have lost it, we feel that it is theirs' forever. "The living are passing travellers, the dead are men come home."

Peony-viewing

Though ghostly presences linger, much poetry and picturesqueness have vanished since thrones are empty, and Imperial pageants things of the past. The Presidential party of to-day, though dignified, is commonplace. Guests gather in a modern hall, greeted by an Executive in uniform, regaled on sandwiches provided by a local caterer. Only a few of the older officials appear as survivors of a vanished regime. The majority wear frock-coats, and among these "moderns" is a sprinkling of young Chinese "returned students" from Western universities,—men of different provinces speaking different dialects, strangers in their own land who exchange banalities in English.

When formal greetings are over, guests are free to wander through the Palace gardens where the flowers appear as beautiful pictures most beautifully framed. We cross zigzag bridges over ponds with shimmering gold-fish, like streaks of fire. We pause in open pavilions whose roofs are draped with wisteria. Here and there, long purple-blue sprays hang as a curtain, staining the sun-

THE FOURTH MOON, OR "PEONY MOON"

light as it passes through. A mauve carpet of fallen petals covers the marble floor. Above our heads, the bees in the blossoms make a sound like the drone of the sea in a shell.

Slowly, stopping often to admire each new vista, we proceed to the throne of the King of Flowers, the grey rock-gardens where fantasy runs riot in stone, the terraces faced with yellow-glazed tiles splintering into sunbeams. Here the plants are set out in stately rows showing how well those Chinese gardeners of long ago understood the value of contrast: pink against green, grey against rose-colour, the grouping of feathery bamboos as a background, the dark note of twisted pines, all arranged to enhance the perfect blooms, sun-drenched through the trees,—themselves sun-filled cups, "Holy Grails at a garden party."

"Wonderful, are they not?" says a Chinese gentleman at our elbow. "Remember, the peony has long been a cult in my country, and these plants you see here have been carefully tended for hundreds of years to bring their blossoms to perfection. With us they are a rich man's hobby, symbol of old ideals,—riches, power, and stately elegance. Whereas in your honourable country I understand but one or two varieties exist, we have peonies of many kinds and many colours,—plants expanding in wanton luxuriance with blossoms large and round as summer fans, and dwarf specimens whose flowers are scarcely larger than a tea-cup. Five varieties are classed as remarkable. But the two most famous are the yellow peonies of the "Yao family," and the purple peonies of the "Wei family." These grow only in Lo Yang. It is considered a great honour for the town, an honour gained in a rather curious way. Once upon a time, according to legend, the Empress Wu Tsê T'ien, concubine of T'ai Tsung and herself one of China's strongest rulers despite her passionate weaknesses, was walking in the Imperial park at Hsi An Fu. Suddenly, a whim seized her. 'I desire,' said she addressing the flowers, 'that when I return to-morrow every blossom shall open to greet me. Let none wait for the soft spring rains. Let none linger for the summer sunshine. Hear, my beauties, and obey.'"

"At this autocratic command, the tiger-lilies hastened to unfurl their banners, the honeysuckles opened their tiny trumpets, the jasmines unfolded their waxen cups. Only the royal peonies considered it beneath their dignity to bloom. The Empress was furious at their disobedience, and indeed, in those days of authority, such insubordination was intolerable. 'Let them be exiled from my garden for *lèse-majesté*,' she commanded. 'Send them out of my sight, the rebels.' So to Lo Yang they were banished, and in Lo Yang they have remained, proud holders of the title 'the unbending and the loveliest in the land.'"[2]

Not everyone can afford the luxury of possessing such kingly plants at home. In Peking, the general public goes to "taste the flowers" in the

---

[2] Lo Yang in Honan province, in recent years the headquarters of the soldier-poet Wu Pei-fu, was for centuries the capital of China. Under the T'ang dynasty, to which the Empress Wu belonged, it was superseded by Ch'ang An, the modern Hsi An, capital of the province of Shensi.

temple gardens of Tsung Hsiao Ssŭ, the reputed resting place of the Immortal Lü Tung-pin, or in dreamy old Fa Yüan Ssŭ, both in the Chinese City and both founded by the great T'ai Tsung himself. The priests, especially in the former temple, tend their gardens more reverently than their altars, thereby attracting visitors who are expected, and urged, to make donations. The first general impression is a rich pattern of flowers on the grey carpet of the stone courtyards. The finest plants, many of them over two hundred years old, are enclosed in low walls of open brick-work or tiles, and literally covered with magnificent blooms of different varieties,—wine-red, pink, plum shading to black, golden yellow, pure white, and the rare green only a shade lighter than the leaves. True peony-lovers come at sunrise, hour of perfume. Stirred to poetry, they greet their favourites in verse, calling them by name: "Water that sleeps in the Moonlight," "Maiden's Dream," "White Robe crumpled by the Son of Heaven," "Black Robe stained with wine," or "Pale Cheek touched with a Rainbow Flush." But even ordinary folk appreciate their beauty, as witness the solemn admiration of a group of school-girls who will stand for half an hour looking at a perfect bloom.

* * *

When the last flowers have faded, and the temple courtyards are once more shut off from the disturbing currents of the world, the visiting monks also take their departure, about the time of the *Li Hsia*, or Summer Solstice. Alms-bowl in hand, they beg their way back to their mother-temples where they remain during the hot months, nominally under strict monastic rule, following the example of the first Indian disciples. The Master himself began the custom of a yearly retreat at this season, and preached some of his most famous sermons in one or other of the garden-monasteries to which it was his habit to retire. "Let the Brotherhood," he advised, "do likewise; let them gather together when the rains begin for a period of quiet meditation and religious training,"—a period answering in some respects to our Lent.[3]

Chinese Buddhist monks, and also Buddhist nuns, as a class, have sunk far below the example of the Master and his followers in practising the faith. Though a few men have real conviction and zeal, living holy lives and devoting themselves to serious study, the communities, as a whole, are composed of indolent priests attracted to the cloister by a peaceful, assured livelihood. They mechanically fulfil the low expectations of the populace who hold them in slight esteem, asking of them only reasonably orderly conduct instead of the pure, unworldly lives enjoined by the Blessed One.

But while they take their vows lightly enough, Chinese Buddhist monks do, on the great festivals, celebrate masses with some fervour, the convinced

---

[3] In China, as in India, indeed throughout the East, the rainy season is very different from the wet weather in Europe or America. The rains "break" about the same time every year and continue, with few interruptions, for a definite period. Travel at this season is practically impossible. Therefore, the founder of Buddhism, like all great religious leaders, made allowances in his quest of the ideal for the requirements of practical life.

through faith, the indifferent for the sake of "face" before the public, or in the hope of donations.

One of the greatest of their church festivals is the birthday of Gautama, or Sakyamuni (8th of the fourth moon), the Teacher of the World, historical founder of Buddhism. *Birthday of Buddha*

Few stories are more appealing than that of this preacher of a universal faith still professed by hundreds of millions—the first faith in which principles common to the whole of humanity superseded racial gods.

As Edkins says, "the best key to the understanding of Buddhism is to be found in the study of the life of its founder . . . for, in Sakyamuni himself, humanity is first seen, then divinity." The Buddha legend is so rich in incident and grandeur, and so well known through the recitals of scholars who have given their lives to the study of it, that we shall attempt no more than a brief outline of this beautiful story.

Prince Siddharta, "Sage of the Royal House of Sakya," was born in the VIth century B.C. at Kapilavastu, the capital of a little state in modern Nepaul, in the shade of the mighty snow-crowned peaks of the Himalayas. His mother, Queen Maya, a young bride, dreamed of his coming from Heaven. Miraculously conceived, he was miraculously born from her side after she saw the vision of a many-tusked white elephant.

As the holy child entered the world, the god Indra appeared in the form of an old woman to assist his mother. Brahma and other older deities gathered to greet him, and the "Four Heavenly

Kings," standing at the four corners of the earth, respectfully saluted the new-born Buddha. All nature rejoiced. Cool breezes gently stirred the trees, springs of water burst from the dry ground, and a great light illumined the whole world. The Devas[4] made offerings of flowers, while four mighty dragons (according to the Chinese version) poured streams of water from heaven over him in baptism. Thereupon, the Wonder-Child immediately stood upon his feet and took seven steps to the east and seven steps to the west.

Son of a king, the little prince was brought up as became his rank, taught every accomplishment and surrounded by every luxury. A seer predicted for him a choice between two kinds of greatness. "If," said this wise man, "he stays in his royal home, he will become a mighty ruler, such as only appears once in ten thousand years. His conquests shall extend to the far corners of the earth, and all nations shall bow down before him. But if he chooses to renounce the world, and wander away into homelessness, he will then become a great saint and teacher of mankind."

As the King longed to keep his son near him and see him crowned with material glories, he filled the boy's life with every possible joy and interest, lest the Prince desire to leave his home. By royal command, all sad and ugly sights were hidden from the beautiful and beloved youth. But, inevitably, incidents of earthly misery did come before his eyes. While driving to some pleasure-gardens in his gilded chariot, drawn by

---

[4] Saints—mostly old Brahmanist divinities included in the Buddhist Pantheon.

Photo by Yamamoto, Peking.

BUDDHA SAKYAMUNI, GANDHARA TYPE, CHENG TING FU.

A BEGGING PRIEST.

milk-white horses covered with trappings of gold, the young Prince had four meetings—with poverty, misery, pain, and death, four omens that decided his fate. Once he saw an old man, white-haired and bent, begging for alms. Again, a man lying by the roadside, grievously ill. A third time he met a corpse followed by weeping women and, finally, he encountered a poor monk begging scraps of food. The Prince, deeply impressed by human grief, was struck by the peaceful and happy expression of the mendicant priest. As they conversed together by the roadside, suddenly all his doubts and difficulties cleared away... "*I* will do as this man has done," he said to himself. "I will give up all I possess and follow the way. Abandoning the world and its pleasures, I shall find peace and learn the wisdom which will teach mankind to overcome the miseries of mortal life."

That very night, when all in the palace were asleep after a great festival, Siddharta arose determined on the Great Renunciation, for riches, power, and high position, could hold him no longer. Softly he crossed the perfumed apartments of his drowsy harem, and the women looked in their sleep so like the dead that he was filled with loathing. Silently he bade farewell to his wife and infant son dreaming on a couch of jessamine flowers. He felt a great longing to hold his son in his arms but, mastering human emotion, he stepped out into the palace courtyard "where the moon shone with so bright a light that it seemed as if the snow of the mountains covered the land."

His faithful servant was waiting with his favourite horse, Khantaka. As the Prince rode through the silent streets of the sleeping city, Devas scattered flowers so that no sound of hoof-beats should be heard, while the heavy gates of the town swung open miraculously to permit his passage.

The royal pilgrim travelled far, and only when the moon had set and the eastern sky gleamed golden with the light of day, did he dismount on the bank of the river Anoma. Here he parted with his horse which reverently kissed his feet. Here he took off the jewelled crown and heavy ear-rings of a royal prince which he entrusted to his servant, bidding him carry them back to his father's house. Siddharta then drew his sword, cut off his long hair and beard, exchanged his dress of fine muslin for the robes of a beggar, and set forth in the perfumed silence of an Indian night to seek the way of Deliverance.

Long and hard was the road trodden by the homeless wanderer. For six weary years the Prince, accustomed all his life to being served, to soft couches, dainty food, and fine raiment, was without a roof to cover his head, with only the melting snows to quench his thirst, and no pittance save what charitable passers-by put into his begging-bowl made of leaves. Yet, this life of asceticism, of fasts and penances, brought him no nearer to the truth.

At last came the Day of Enlightenment. Siddharta, or Gautama as he is more usually called after the beginning of his wandering life, retired to the jungles of Uruvela and, seating himself under a peepul tree, determined that he would not stir from this spot until he had grasped the highest wisdom. There Mara, the Spirit of Evil, tempted him with the forces of Hell, and assailed him with all the terrors that affright the human mind. But the Holy One sat unmoved while the fiends attempted to do him bodily harm, their deadly weapons falling as harmless lotus-petals at his feet. Mara sent his own daughters, Desire, Pleasure, and Delight, to stir the earthly passions of Gautama, but the beautiful maidens changed into ugly old hags. "When the night was far-spent, and all the Evil One's devices had failed before the steadfast mind and innate holiness of the Buddha, Mara owned himself defeated . . . and the angels and archangels and all the hosts of heaven rejoiced at his victory, while Gautama himself received the peace which passeth understanding."

The Holy One was thirty-five years of age before he attained Buddhahood. As soon as his message was clear to him, "like a man who comes forth from a dark prison into the glorious light," he returned into the world to preach the truth, because of his great love for humanity. His first sermon on the "Turning of the Wheel of the Law" was preached in the Deer Park near Benares to all who came to hear him—not, like the Brahmins, only to privileged people, but equally to rich and poor, old and young, men and women. Thereafter, he and his disciples went their separate ways to spread the doctrine. Even when age and infirmities were heavy upon him, the Buddha still travelled from village to village, teaching the people until, finally, when he reached almost four-score years (he died at seventy-nine), he knew his ministry was ended. Calling his disciples around him in the garden monastery where he lay sick with a mortal sickness, he said to them: "Your Master's body is now bent and infirm, and, just as an old worn-out cart which is bound-up with cords can with difficulty be kept going, so it is only with care and trouble that the flesh continues to exist. I am old. My journey is nearly ended, but sorrow not and let truth be your refuge." So saying, he passed away into the bosom of the Nirvana, and his mighty spirit merged again into the eternal and the impersonal, as a drop of water descends from the clouds in dew upon a lotus leaf, glides from its green cup into a pool, from a pool escapes into a river, and thence is carried to the Ocean.

In the Buddha's lifetime, his teachings spread far and wide through Hindustan. Other peoples soon sought relief in his tranquil doctrines from the uncertainties of change and the burdens of earthly life. Over land and sea the Faith was carried by the devoted disciples of the Master, even as far as Syria and Malaya, Corea and Japan. To China also, the Buddhist *Chen T'an*, the Law travelled with preacher and merchant, over the glaciers of the "Roof of the World," across the sands of Central Asia. Neither difficulties nor dangers could hold back the Great Thought carried with the

"Great grey caravans on great grey wings......
 Plunging over trackless wastes where trails can never meet,

THE FOURTH MOON, OR "PEONY MOON" 263

> Spraying noiseless gravel as they crowd on out of sight,
> Swift silent caravans on swift silent feet.
>
> Soft slow caravans swaying through the night,
> Tinkling bells and padded feet, and spices that the traders brought,
> Easing through the moonlight over sands dull white,
> Soft slow caravans of soft slow thought."

In that long passage of time and distance, the simple faith of the Blessed One evolved into an elaborate theological system. Schisms occurred in the Church, and each new sect made its own efforts to recover the Divine from the maze of an exaggerated polytheism that clouded the pure Doctrine. "As the Christianity of Rome differs from that of a sect like the Quakers, so does the Buddhism of Thibet, with its complicated rites and ceremonies, differ from the simpler form of faith practised in Ceylon and Burma."

The personality of the Perfect Example, the patient and loving Teacher, has likewise undergone an evolution. When the story of the Buddha's life was told again and again "in countries far removed from the scenes where the events had taken place, tales of wonders and miracles were added to the first simple narrative," until the figure of the Master is now shrouded in many veils and presented under many incarnations, his sacred precepts overlaid by many commentaries.

Distant is the gaze of Gautama seated on his golden throne in this far land, and strange his first message must have appeared to the Chinese. But the waters of universal compassion swept away all obstacles to understanding, and the influence of Buddhism, as we have already seen in the chapter on "The Hundred Gods," has deeply affected Chinese thought and culture. Even to-day, it remains a living faith in the land, a faith revered not only by the simple folk but by many highly educated men and women.

* * *

Throughout China innumerable temples have been erected to the Buddha, and in every one, on his "day of remembrance," the *Sutras* are chanted musically with an accompaniment of booming drums, the tinkle of small bells and large bells, the thud of the wooden fish-head[5] struck with a small stick, the clang of brass cymbals, the whirr of the *t'ang tsi*, the tiny gong held by a half-cross to which it is tied by strings, and the clang of the *yin ch'ing*, a flat metallic plate cut out in flower-shapes and struck with a thin iron rod to mark time for the chanters.

As for the shabbinesses and absurdities that too often accompany the mass—the soiled robes of the priests, the inattention of the acolytes—an Oriental believer ignores them in his appreciation of the grandiose intention of the general artistic effect of the service. If the Chinese priests themselves are slack in the performance of symbolic ceremonial, it is because, to them, the invisible thing symbolised is so much more important than the symbol, though they would not consciously argue so.

An abbot we know invites us to see Buddha's Birthday Service at one of the big temples in Peking. "It is a pity you could not have come

---

[5] Symbol of watchfulness, as the fish never closes its eyes.

last night," he says as he greets us, "you would have been present at the ceremony of washing the images of the gods, which takes place once a year. All the statues, whether of wood or stone, are carried out into the courtyard and sprinkled with water which must be exceptionally pure. That is our custom." In private shrines the Buddhas are not removed from their places, but strips of paper are carefully pasted over their eyes, so that one may not see when another is bathed. They are, likewise, blindfolded whenever the sanctuary is dusted. Some place an image of Gautama in a big jar of water, and the faithful, as they pass through the courtyard where it stands, sprinkle the head with a spoonful of water and then further baptise it with a handful of copper coins, afterwards used for buying incense.

Like all night-services, the "Washing of the Buddhas" is very picturesque because each monastery, glowing with lanterns, appears like a torch against the darkness of the night. The accompanying prayers are chanted by grey-robed priests kneeling on the grey pavement, each celebrant a ghostly statue. The images themselves, not ranked by hierarchies as in a temple but mingled without order, fill the shadowy space; Kuan Yin of many forms; Buddha under many names; armoured effigies of Wei T'o; P'u Hsien Pusa riding his elephant, and Amida on his throne of gold,—shadowy shapes growing more and more shadowy to one another as the lights burn low, while incense from a wrought-bronze brazier perfumes the water that trickles musically over crowned heads and aureoles of dull gold.

Excusing himself, for he has a service to attend, the abbot bids us wander round the temple. It is a good opportunity to study at our ease, and under the competent guidance of an intelligent young monk detailed to show us everything, the general arrangement which, even as regards architecture, holds good for any ordinary Chinese Buddhist temple. Moreover, we see here and shall meet everywhere the same gods, in the same order, so that, once accustomed to their positions and attributes, we can always recognise the more famous divinities.

Outside the entrance-gate stand two stone lions, guardians of Buddha,—probably converted demons. Passing this gate, we notice that all the important buildings face south. Great stress is always laid in China on the points of the compass, and proper orientation has its bearings even on social relations, as one discovers in that ancient guide to ceremony and propriety, the Book of Rites. Buddhist halls in China are also built in accordance with this principle.

The Four Diamond Kings The first hall contains the gigantic statues of the "Four Heavenly Maharajahs," or "Four Diamond Kings," *Chin Kang*. (already mentioned in connection with Buddha's birth), two on either side guarding the entrance.

Although of Hindoo origin, they are also to be found under different names, and with different legends attached to them, in Taoist sanctuaries. "Authorities, both English and Chinese, disagree in minor details in connecting the Buddhist *Chin Kang* and the Taoist Heavenly Kings . . . but, in general, they may be taken to represent the

TWO OF THE HEAVENLY MAHARAJAHS.

Photo by Yamamoto, Peking.

IMAGE OF PU TAI, HANGCHOW.

four seasons and the four directions . . . They are but seldom worshipped, an occasional incense-stick being placed before them by the devotee of other idols." (See *Chinese Religion Seen Through the Proverb*, by C. H. Plopper).

Vaiçravana,[6] or *Pei-Sha-Mên,* Lord of the North ("He Who Has Heard Much"), holds in his hand a rat (mongoose) or reptile that possesses magic power. Dhritarashtra, or *Chih Kuo,* Lord of the East ("Protector of Kingdoms"), has for his attribute a giant guitar that controls the gales. Virudhaka (*Tsêng Chang*), Lord of the South ("Increased Grandeur"), is distinguished by his magic sword, "Blue Cloud," that gives birth to the wind, and Virupaksha (*Kuang Mu*), Lord of the West ("Large Eyes"), by his umbrella which, when opened, brings universal darkness.

All four are impersonations of hurricane power, described as actively interfering in the affairs of the world. When, for instance, kings and nations neglect the Law of Buddha, they withdraw their protection. At the same time, despite their ferocious faces, painted blue, black, red, and white —"they bestow all kinds of calm happiness on

---

[6] The language from which the Buddhist sacred writings were originally translated was Sanscrit and Chinese Buddhists use transcriptions, or transliterations, of Sanscrit names to designate their gods. For the same reason "manuscripts, inscriptions, charms cut on copper mirrors, and lucky sentences under eaves and over doors in monasteries, are often in Sanscrit and, in polyglot books printed in Peking, Sanscrit is one of the languages employed."

As Baron Staël-Holstein points out, Vaiçravana, also known as Kuvera, is in some countries identified with one of the Seven Gods of Riches, his treasures being protected by a race of giant ants. His wife is Hariti, originally a female demon and the Goddess of Smallpox. (*See* "The Second Moon,"—Kuan Yin).

those that honour the *San Pao,* or 'Three Treasures,': Buddha, the Law, and the Priesthood."

The Marshals Hêng & Ha

Between them and the south wall are sometimes placed the figures of two other guardians in military uniform and with fierce countenances, called *Hêng-Ha Erh-Chiang,* the two Marshals Hêng and Ha. These worthies are supposed to have been war-lords during the bitter feuds which resulted in the Chou dynasty wresting the throne from the Shang dynasty, in 1122 B.C. Both Tuchüns received supernatural powers from the great Taoist magician of their day, at least so the legend runs, ignoring the fact that Taoism was not born till many centuries later. Hêng, Chief Superintendent of Military Supplies of the Chou Wang (last of the Shang line), was able to eject from his nostrils two rays of light with a booming sound like a bell struck heavily,—while Ha could blow out of his mouth a great gust of horrid yellow gas which annihilated all that it touched. Hence their popular nicknames of "Snorter and Blower." [7]

Ha finally perished in an encounter with Hêng who, in turn, met his death at the hands of another paladin. Both Marshals were canonised by the dynasty against which they fought so heroically. Thus the Chinese rewarded loyalty even in their

---

[7] Can the Chinese claim the invention of poison-gas, as they do of the aeroplane and the taxi-cab? The roots of many modern inventions are found in Eastern mythology, and even in our own fairy tales. "Did Icarus anticipate Wilbur Wright? We do not believe so. But many old-fashioned Chinese scholars do—or at least did—accept such flattering legends at their face value. While tradition had the force of fact, this viewpoint helped them build up their sense of race-superiority or, rather, cultural superiority, and confirmed them in the belief that all other nations are barbarians."

THE FOURTH MOON, OR "PEONY MOON" 269

enemies, and propitiated mighty foes in the Spirit World. That is why the "Blower" with dilated nostrils and the "Snorter" with open mouth, giving vent to all the old high imprecations, stand in temple vestibules.

In the same building with these guardians, but opposite the front door, there is always an image of the priest Pu Tai, a historical figure, popularly supposed, in China, to be an incarnation of Maitreya, the Merciful One, Buddha of the Future (anniversary 3rd of the sixth moon). The traditional Hindoo conception of the Buddha-To-Come is a kingly figure, sixty ells in height—the size the virtuous will appear after the Resurrection, but such representations of him are rare in Chinese Buddhism, though not in the Thibetan Church. A fine example of the Lamaist "Maidari" is the huge wooden figure at the Yung Ho Kung in Peking. The usual incarnation of Maitreya (*Mi Lo Fo*) as Pu Tai is represented as a very stout monk with robes slipping off his breast and naked paunch exposed to view. Plump as a fruit ready to burst, often surrounded by the children beloved by him in life, this "little fat gentleman, who sits in *négligé* costume with a broad smile on his face, has earned among foreigners the sobriquet of the 'Laughing Buddha.'" We judge him as the embodiment of carnal virtues. Yet in reality he is revered as the Saviour of the World three thousand years hence, though his future mission contrasts oddly with his gross appearance.

Pu Tai

Mi Lo sometimes shares his shrine with an image of Kuan Ti (*see* "The Fifth Moon"), but

Wei T'o

invariably stands back-to-back with Wei T'o, a warrior and the Protector of the Law, facing the inner courts on the other side of a wooden screen. Originally a *Chin Kang*, advanced to the Pusa state on account of his zeal and goodness, Wei T'o, in full armour, holds by right the "diamond club," "ready to avenge instantly any slight on the Buddhist faith." It seems incongruous to find mailed figures protecting a religion of tranquility and love, whose command "thou shalt not kill" extends to the lowest forms of animal life. But Wei T'o only wars on demons and evil spirits who are, until successive rebirths purify them, outside the pale. Moreover, Wei T'o has other less warlike characteristics. He is an attendant upon Kuan Yin, and often found in her company. He appears also to be the business agent of monasteries (see *Peking*, by Juliet Bredon) since begging monks, when collecting offerings for temples, carry a portable shrine of Wei T'o on their shoulders, and religious communities appeal to him for their material needs. "Wei T'o," says Johnston in *Lion and Dragon in Northern China*, "is often depicted on the last page of Buddhist books. This prevents their destruction by fire and insects and (it is confidently asserted) compels their borrower to return them to their owner. A private Wei T'o," he adds, "would be a most welcome addition to the furniture of many an Englishman's library."

Wei T'o looks towards the main halls of the temple. To enter one of these principal sanctuaries is to realise how much Buddhism contributed in China to the arts of carving, painting, sculpture, and mural decoration. We sometimes find the walls embellished with splendid frescoes, reminiscent of the Italian Primitives, or scroll-paintings suspended side by side, showing the incidents of a soul's journey to the realm of judgment, and all the horrors of the various hells. Often these pictured terrors are off-set by pictured consolations—by the beautiful figure of Kuan Yin, White Goddess of Mercy, by the compassionate smile of Ti Tsang Pusa, or by a vision of Paradise in which the devout artist opens to simple fancy the delights of heaven, with its jewelled trees and that lake where the souls of the righteous are re-born as lotus blossoms.

On the altars, carved reliquaries of lace-like beauty recall the work of our Renaissance, while the sculptured entablatures of pedestals, with their resemblance to Greek and Roman monuments, remind us that Alexander's conquest of Persia and invasion of India was a "signal for a host of new thoughts to originate in the countries conquered," and of the influences which came with Buddhism to China.

Three Great Buddhas

In the first great hall incense-burners, candlesticks, and the "Eight Precious Jewels"—the conch-shell, the brazier, the Wheel of the Law, the umbrella, the *stupa*, the pair of fishes, the net, and the banner, stand before images of gods smiling in gold, a triad of immense figures known as the Buddhas of the Present, the Past, and the Future. The central figure is Sakyamuni, the historical Buddha and the Buddha of the present. Seated on his lotus pedestal, his figure glows dully like a dying fire in the midst of little sparks, the votive lamps lighted for the festival. Art suggestions of Indian origin are still recognisable in his stately draperies, his classical features, and his inscrutable smile with a touch of sadness, a smile reflecting the pathos of a beautiful exile. When bitter winds blow, as they often do in Northern China, that touch of melancholy on the face of the Lord Gautama seems to hint at the sacrifices he endured, the material sacrifices of his kingly home in a land of warmth and luxury, the spiritual sacrifices, too, for worshippers who often give him but a hasty lip-service.

The other two figures of the trinity are similar to the figure of Gautama, except for the position of the hands, since all belong to the highest rank of wisdom and power. The triple presentation is simply an attempt to give a larger conception of the Buddha, all-knowing, all-embracing, all-powerful. "Countless the Buddhas appear," say the *Sutras*, "yet there is truly but one Buddha. The many are forms only."

The Trinity of the Three Precious Ones (Past, Present, and Future), always together and easily recognised, among other attributes, by their short curly hair formed of snail-shells,[8] are by no means the only three-fold group of gods to be found in Buddhist temples.

Kuan Yin, P'u Hsien, Amida, the *Wu Liang Shou* Buddha, "Buddha of Long Life," and others figure in such triads, and Sakyamuni himself may appear in a second great hall with his favourite

---

[8] A pretty, but disputed, legend gives the faithful snail the honour of protecting Gautama's head from the fierce rays of the Indian sun, and thus explains the skull-cap of shells.

THE FOURTH MOON, OR "PEONY MOON"

disciples on either side,—Kasyapa, the aged, first patriarch of Buddhism, and Ananda, the youth, the Beloved on whom the Buddha set his heart, "as the sun, on rising, sheds his light straight upon the western wall."

A place in this same sanctuary is sometimes given to *Jan T'êng*, teacher of the Buddha in a former life, or to *Yo Shih Fo*, the Divine Healer. Devas like "Brahma" (*Fan T'ien*), and "Shakra" (*Ti Shih*), also appear there. The central halls of a complete Buddhist temple in China contain "representatives of all the four ranks above the range of the metempsychosis . . . . the four highest classes of beings recognised by Buddhism . . . . " Disciples of the lower ranks who are, however, delivered from the world of life and death, and are called *Shêng Wen*—listeners, are represented in Ananda and Kasyapa, the Devas, or the Arhats. Though they have not yet attained to the rank of Bodhisattva[9] owing to some overpowering attachment to the Wheel which binds them to the passions and pursuits of the world, they are still permitted to penetrate the mysteries of earth and heaven, owing to the power and knowledge they have gained through listening to the teachings of the Buddha. The rank above them in holiness, that of Bodhisattva, is represented by Wên Shu and P'u Hsien, crowned with their golden crowns.

Sakyamuni's disciples, the Arhats, or *Lo Hans*, stand on the east and west sides of the sanctuary, usually nine in a row, and each with his own

Lo Hans

---

[9] For the meaning of the term *Bodhisattva*, or *Pusa*, see "The Second Moon," Kuan Yin, footnote.

emblem. These saints sometimes include one thousand two hundred figures, sometimes five hundred, but in most places only the eighteen most prominent followers appear. According to the holy books, when Buddha was about to die he entrusted his religion to sixteen of the great Arhats who were delegated to watch over the "welfare of the lay believers and the spiritual interests of Buddhism. They are to remain in existence all the long time until Maitreya appears as Buddha, and brings in a new system. Then these saints will collect all the relics of the Buddha and build over them a magnificent *stupa*. When this is finished, they will pay their last worship to the relics, rise in the air and vanish into the remainderless Nirvana." The Chinese have added two Lo Hans to the original number, thus making eighteen in all.

Their present spiritual powers are indicated by the wild animals crouching submissively beside them and by the popular names: "Subduer of the Dragon," attached to the figure with the dragon coiled around it and with the head in its hand, and "Repressor of the Tiger," used for the image with the tiger on its back, or at his feet.

Like the saints and martyrs of the West, to whom they vaguely correspond, the Lo Hans act as intercessors for suffering humanity before the throne of the Almightly. Twenty Devas (whose status in some ways corresponds to that of the Lo Hans) include incarnations of the older Hindoo gods and even of Nature deities like the moon, but all more or less disguised with Chinese names and costumes. They stand ten on a side

THE FOURTH MOON, OR "PEONY MOON"

at the south end of the two rows of Arhats that line the eastern and western walls.[10]

Central positions are given to the two important figures of Wên Shu and P'u Hsien, two of the four great Pusas of Buddhism who, from their mountain seats, guard the Law in China, as the Heavenly Maharajahs guard every Buddhist sanctuary. While many of the fabulous beings mentioned in the literature of Northern Buddhism have no image or shrine in the temples of the present day, the above-mentioned Bodhisattvas are seldom absent.

* * *

Wên Shu Pusa

Wên Shu, the Indian Manjusri, is one of the most famous of the lesser divinities. Incarnated in the Living Buddha of Urga, according to the Yellow Lamaist sect, he is generally considered by the faithful as the personification of wisdom, the "Enlightener of the World." By the sword in his right hand, and the book resting on a flower in his left, we shall know him. The sword stands for pure intellect, sharp enough to pierce the deepest recesses of Buddhist thought. The book is symbolical of Buddha's sacred teaching.

According to the Canon, Manjusri was bidden by the Great Teacher to seek the instruction and salvation of the Chinese by making his home at Wu T'ai and there cause the Wheel of the Law to revolve incessantly on the five pagoda-crowned mountains.

---

[10] Their appearance, comparatively rare except in very large Chinese temples, accords with the description given in the *Sutras* of the congregation of disciples which gathers round Buddha to hear his preaching. Fine examples of the Deva figures exist at Ta Chüeh Ssŭ and at Hsi Yü Ssŭ, near Peking.

Legend has ever since identified him with this place, one of the Four Sacred Buddhist mountains, on the eastern border of Shansi, a wonderful winged peak where, once upon a time, there was a collection of monasteries constituting the richest religious establishment in the Empire. Persecutions in the IXth century A.D. somewhat reduced the magnificence of its shrines. Nevertheless, it remains to this day an important place of pilgrimage. Manjusri himself appears at the mountain retreat from time to time, sanctifying again and again the misty peak so often veiled in clouds. But on those days when the air is bright and sunny its grey and silver outline can be seen from afar, and when the sun sets behind it, the holy hill glows like a purple pyramid. Then it speaks of peace and perfection, of remoteness from the worries and futilities of earth, of another life, better and nobler than ours. This life the pilgrims, toiling slowly upwards, hope to attain through patience and perseverance.

The temples on the mountain are, for the most part, served by Lamas. Mongols, Buriats—and even Kalmucks from distant Russia—are in the majority among the pilgrims, so that nowhere in China do we find a more colourful and picturesque group of worshippers making the round of pagodas that rise like a prayer into the blue sky, or bowing awkwardly with a clinking of turquoise ornaments before the figure of Wên Shu, seated upon a lion, with a blue lotus in his hand and his golden-green eyes looking down on them, lustrous with wisdom.

Photo by Owen Lattimore.

BRONZE PAGODA, WU T'AI SHAN.

LAMA PRIEST IN FULL DRESS, PEKING.

P'u Hsien Pusa (*Samantabhadra,* "The All Gracious"), whose chief seat of worship is imposing Mount Omei, in Szechuan, stands scarcely less high in popular esteem than Wên Shu, and thousands of pilgrims, including Thibetans and Nepaulese, yearly climb this magnificent peak to do honour to their patron saint. The whole of this noble mountain, a colossal vision of beauty which is a religion in itself, was once church property with temples every few *li* all along its slope. The golden summit was crowned by a shrine of pure bronze, a Wan Li temple destroyed by lightning in 1819, and so expensive to rebuild that no abbot has yet succeeded in collecting sufficient funds.

P'u Hsien Pusa

Legend says that as Wên Shu visited his mountain riding upon a lion, so P'u Hsien reached his upon the back of an elephant. In proof of this wonder-tale there exists a bronze elephant at Nan Nien Ssŭ, one of the seventy monasteries served by two thousand priests that cover the mountain. Baber (the first foreigner to ascend the mountain in 1877) affirms, in a record whose accuracy has not been questioned, that this temple, with its square base and ingenious arrangement of triangles, segments of circles, and projections arranged to support a dome, is actually, after the Great Wall, the oldest Chinese monument in existence. As for P'u Hsien's elephant, now surrounded by a wooden cage reminiscent of the Zoo, he declares it to be the oldest cast-bronze figure of any great size in the world. Powerfully modelled, it stands over twelve feet high. Upon its back, seated in a huge lotus, is the Bodhisattva with a crown of glory.

Yet even more remarkable is a natural phenomenon observable on the very summit of Omei Shan. When the sky is clear, and clouds of fleecy mist float in the chasm below the peak, and the great holy stillness is not disturbed by so much as the cry of a lonely bird, there appears a spectre akin to the Spectre of the Brocken, which takes the form of "a golden ball, surrounded by a rainbow floating on the surface of the mists." Not to everyone is the vision given, but the pure in heart know it as the *Fo Kuang*, or "Glory of Buddha," and believe it a reflection of His halo, "and an outward and visible sign of the holiness of the mountain." So impressive is this miraculous apparition that pilgrims, beholding it at dawn, have been known to throw themselves over the cliff in an ecstasy of religious fervour. Therefore, nowadays, the mighty precipice of black rocks spattered with ruddy patches as though drops of blood had oozed in places from its stony heart is guarded by a barricade of chains.

* * *

Returning to the typical arrangement of a Buddhist temple, we find other chapels at the side of the main halls. One is surely dedicated to Kuan Yin the Merciful (*see* "The Second Moon"), and another to Ti Tsang (*see* "The Seventh Moon"), who stand, together with Wên Shu and P'u Hsien, as the four favourite Pusas of Chinese Buddhism, and like four brilliant torches illuminate the Doctrine. Omito Buddha (*see* "The Eleventh Moon"), he of the Western Paradise, is either associated with Kuan Yin or has his own hall, unless he is represented in one of the major trinities.

The same may be said of divinities like the Yo Tsang Pusa and the Yo Wang Pusa, but the arrangement of these gods is subject to so many exceptions that there is no hard and fast rule.

Even non-Buddhist gods appear in Buddhist sanctuaries, personalities like Tsai Shên, "He Who Presides over Riches," Kuan Ti, also the Lung Wang, and some of the lesser Medicine Gods. The little shrines of the latter, seldom visited, are draped with spider webs that ghost against our faces.

* * *

Pilgrimages

Advanced or philosophical Buddhists will tell you that the function of all these images is simply to remind the devout of something higher and better, and direct their thoughts towards the Eternal Verities. But the common people, unable to grasp abstract ideas, see an actual god in every image.

To this category belong the group of "incense guests" gathered for a festival—mostly old ladies —kneeling stiffly before the Buddha, calling his holy name, and making an offering of a white bean with each invocation. The beans will be cooked later with young carrots and willow-buds. Great cauldrons full of them are steaming already in the temple kitchen. "We call them happiness-beans," our guide explains. "First, they are offered in plaited baskets to Buddha and, afterwards, some are distributed by the grand-dames to passers-by. Others are reserved for the children. Thus we attempt to establish a religious link with the little folk of the neighbourhood—a link we believe and hope will lead them later to our altars and the path of virtue."

A favourite method by which Chinese priests keep in touch with the people, and draw contributions for the maintenance of monasteries, is by organising and encouraging pilgrimages. Just as in Chaucer's England May was the month when "longen folk to go on pilgrimages," so in China the fourth, or corresponding, moon is the favourite season for the pious to visit the holy places of the hills, not that "the lamp of piety burns more brightly then than at other times, but because nature herself lures men out of doors."

In every province in China, indeed in almost every district, there is a sacred mountain to attract the faithful, whether Buddhist or Taoist or both, and whole volumes would be required to describe all the great pilgrim-centres of China. The T'ai Shan (*see* "The Tenth Moon") and P'u T'o (*see* "The Second Moon") are national goals drawing the devout from every part of this vast land. But the "Tea-Burden Procession" to Miao Fêng Shan, a peak in the Yang Shan chain of mountains to the west of Peking (abrupt and fantastic hills that appear so unreal in Chinese paintings, yet are so characteristic of the Chinese landscape) is sufficiently typical of the little local pilgrimages to warrant detailed description.

This cluster of small sanctuaries, most beautifully situated on a mountain-top overlooking the Hun River valley, is dedicated to three goddesses: the T'ien Hsien Niang Niang whom the Taoists have identified with the grand-daughter of the Spirit of Mount T'ai Shan, sometimes called the "Jade Lady," but more often the *Pi Hsia Yüan Chün* (*see* "The Tenth Moon"), or "Princess of the Coloured Clouds"; the Yen Kuang Niang Niang, who cures the eye-diseases so common in China, and is a patroness of the blind; and the Tzû Sun Niang Niang (*see* "The Eighth Moon"), or Giver of Sons, sometimes represented with a mask over her face and a sack over her shoulders, in which she carries the babies she intends to bestow upon her worshippers. Thank-offerings of shoes and red eggs are made to her by grateful mothers, and those who have no children will steal these, thinking thus to obtain the good fortune of their sisters. The statue of this kindly goddess is often taken from her altar and carried into the room of a woman during childbirth.

The Niangs Niangs belong to a group of nine protecting divinities: 1. The *T'ien Hsien Niang Niang*, the All Highest, 2. The *Tsai Shêng Niang Niang*, the Heavenly Midwife, 3. The *Pan Chen Niang Niang*, the Goddess of Scarlet Fever, 4. The *Nai Mu Niang Niang*, Patroness of Wet Nurses, 5. The *P'ei Yang Niang Niang* who watches over the birth, feeding, and general upbringing of children, 6. The *Yen Kuang Niang Niang*, "Our Lady of Good Sight," 7. The *Tzû Sun Niang Niang*, She Who Grants Posterity (*see* "Marriage Gods,"—"Eighth Moon"), 8. The *Yin Mêng Niang Niang*, Protectress of little Children, —and 9. The *T'ou Chen Niang Niang*, Goddess of Smallpox, who has two attendants. People beg the last to keep away from their homes, but once she visits them a tablet to her is set up, and she is daily beseeched for the recovery of the patient, with great care and courtesy, as she is easily offended. If the patient recovers from the scourge of a "Thousand Flowers," a paper chair or boat is made, her tablet placed within it, and the whole mounted upon the back of a phœnix. The effigy is then burned on a pile of straw. Thus she is escorted away from the place. But if the sick person dies, she is cursed off the premises.

In South China more Niang Niangs are added to this group. Sometimes we find sixteen or twenty, all with similar functions, all occupied with the care of women and children.

Of the three chief guardian goddesses, the Jade Lady is the best beloved and the most powerful. Like the mythical Jade Emperor (*see* "The First Moon"), she came late to the Taoist Pantheon, a figure invented under the Sungs to answer the need for a goddess rivalling the Buddhist Kuan Yin who had already appeared and was making a strong appeal to the public mind. Abstract thinkers are content with the doctrine of the Tao in its early purity, but simple folk need personal divinities to hear their prayers, and women desire a feminine intercessor, one who understands their ardent maternal instincts, and is worshipped as wife and mother, equally with the masculine ideal of hero and father.

One of the most lovable divinities of Taoism, the appeal of the Jade Lady was instantaneous, persistent, and widespread. She flourished under the Mings, and still retains her popularity, though under different names in different provinces. In

Fukien, for example, they simply call her "Mother" identifying her with the antique patron-divinity of that province (*see* "The Second Moon"). In other places she is disassociated from all rival goddesses and given her own temples, her own assistants, and her own attributes connected with children and childbirth.

Wherever her shrines are found there is no scarcity of worshippers, and all four roads leading to Miao Fêng Shan are crowded with devotees going to beg her favours from the first to the eighteenth of the fourth moon. What could be more natural "in a world where duty so seldom coincides with pleasure," and keen physical and mental enjoyment of good air and beautiful nature can so rarely be combined by busy folk with "an exhilarating sense of religious fulfilment."

The pilgrims that frequent the sanctuary of the Three Niang Niangs are of all kinds and all classes,—retired officials and their ladies, sleek monks, peasant men and women. Few pious folk in China tread the pilgrim-road alone, unless they are fulfilling some special vow, like men who do the journey in chains borrowed from a temple where such instruments are kept, or childless women whose pilgrimage is accomplished wholly in prostrations. Standing, such penitents sink to their knees and fall straight forward, flat on their faces in the dust, then rise again to stand once more on the extreme point their finger-tips have touched, repeating the painful and laborious process till the bourne is reached. Or, again, those would-be-mothers who try to earn the right to bear a son by dressing in red robes, like condemned criminals, believing thus to wipe away past sins.

But, as a rule, pilgrims band themselves together for the journey, and in nearly every province societies are organised for people who desire to visit certain shrines.[11] Where humble toilers who, after all, form the major portion of temple visitors, cannot afford the cost of several days' journey individually, even though their wants are few, the pilgrim-fund permits a certain number of subscribers, selected by lot, to represent even the poorest hamlet. Winning tickets entitle their holders to draw all expenses for inns and cart-hire from the common chest, though tips to servants and alms to beggars—matters in which Orientals are very generous—each person defrays out of his own pocket.

Members of the richer associations travel a little more comfortably than the poor farmers who pack their own beds upon their backs. Where the latter walk, the former use chairs or carts. One class carries rubber cushions and thermos bottles, the other sits contentedly on wooden benches.

---

[11] Arthur Smith, in *Village Life in China* speaks of two kinds of Travelling Societies, the "True Travelling" (*Hsing Shan Hui*) and the "Stationary Travelling" (*Tso Shan Hui*). "The former lays plans for a visit to a sacred mountain . . . whereas the latter, or 'Sitting Society,' is a device by means of which the actual results of a pilgrimage are accomplished without the trouble and expense necessary to the visit of a distant and inaccessible peak. A 'Sitting Society' pilgrimage includes a theatrical performance, jugglers, and a feast. Between these amusements, the pilgrims find time to worship the mountain goddess at a paper mountain which, by a simple fiction, is held to be for all intents and purposes the real Great Mountain. That the people feel they have less merit in this vicarious pilgrimage is proved by the contempt of those who actually climb for the stay-at-homes—a contempt expressed in the nickname 'Squatting and fattening societies.'"

The poor buy provisions in the villages, the rich carry their own, or have them carried in boxes gay with yellow pennants and bells. The tinkle of these bells, as the bearers trot across the plain, is a characteristic sound announcing that the "mountain is open."

Yet, on the whole, a perfect democracy exists among the travellers. Along the road, people share alike, and snobbery is non-existent. The gentlemen in silks are not offended at the good-natured jostling of the cotton-coated artisans and both share their holiday pleasures.

In olden days, the Acrobats' Club used to give performances by the roadside. Professionals hoped, while they enlivened the weary climbers, to attract Imperial patronage and the reward of a yellow flag proudly displayed at their town performances. It was a good advertisement, like the label "by special appointment to His Majesty." Young amateur sports, who enjoyed dare-devil stunts like dancing on bridge railings, riding two ponies at once Roman-fashion, balancing heavy weights, or trick-conjuring with earthenware jars, also performed for the amusement of the populace.

When a stop is made at a village to burn incense at the local temple, where soft-toned bells are continuously ringing, and slender pencils of smoke rise through the perforated lids of old incense-burners, the rich merchant's wife mingles quite freely with her poorer sisters. At the rest-houses, chair-coolies sit and sip tea beside their employers with perfect equality, yet without impertinence, because the Chinese thoroughly understand the harmony of human relations, a hard lesson learned by generations of living close together. Their manners are, rather let us say were, so perfectly regulated that absolute freedom guarantee no encroachment of one man upon another. Imposed barriers are not necessary where society long ago set its own limits where it needed them.

In addition to the ordinary code, pilgrims have a special manual for their guidance, just as the pilgrims in Europe did before there were Baedekers. Such manuals give information concerning roads to the shrines where "they are to bend the knee and open the purse." Chinese pilgrim handbooks likewise have secular hints on travelling, interspersed with precepts for proper religious behaviour and careful injunctions as to how to conduct themselves when meeting superiors, equals, and inferiors, what thoughts to think on approaching the heights, what gods to worship, and what tips to give.

Materially speaking, all necessities are provided for the pilgrims to Miao Fêng Shan. Matsheds are erected for their entertainment by a committee of benevolent persons, and here hot tea is served. Every village offers fresh eggs for a consideration, and bowls of water are set out for thirsty animals, the carters dropping a copper in the receptacle after their mules have drunk their fill. The local people offer for sale fans and hair-ornaments of plaited straw, and "happy flags" made of red chenille in the shape of bats, butterflies, or swasticas. Both men and women buy these because it is the custom, and means "carrying happiness home." Indeed, one can never fail to recognise pilgrims by these ornaments stuck in the bright cloths tied about their hair as a protection against dust.

On the mountain, a series of little temples mark the regular halts and, in addition, open-air restaurants spring up at every point where the view over the plain inclines weary climbers to sip a cup of tea and eat a bowl of macaroni. Peddlers install their booths near by, selling pilgrim-staves with carved dragon-heads, baskets, etc.—all the simple necessities of simple folk when they take the road.

Once over the high passes, most people linger in the pretty valley where the roses that perfume the tea are pouring out their incense, before attacking the last steep bit of road partly shaded by gnarled vines whose bent branches throw a network of shadows across the path, like the wrinkles on the face of an old, old peasant woman.

Crowning the peak, perched on the rocky height, are a few small shrines of scant historic interest; legend says they were built in 1622 by a Taoist monk, a Buddhist monk, and a pious layman, who together collected subscriptions for the purpose. Nor are the good goddesses, lately regilded, works of art. Of greater interest is a secondary shrine with the statue of an old woman in a blue peasant's coat, reputed as a healer and canonised unofficially. She is a good example of how persons quite undistinguished in life can sometimes, in favourable circumstances, attain the dignity of local deities or saints. Though she belongs to no pantheon and has no official titles, women bow before her, whispering phrases at once occult and familiar like an amulet worn on the heart.

Gathered together in the name of the Princess of the Coloured Clouds, the pious neglect none of the other shrines on the peak. Every act of merit is a *fu yüan*, "cause of happiness," hence the endless prayers, the offerings of copper coins that fill large pits several feet deep, the waving torches of incense, the endless kneelings, bowings, and head-knockings even on the stone flags of the courtyard, where the brown out-cropping of the peak itself—the holy boulder which is the actual *Miao Fêng* ("The Marvellous Peak")—pierces through.

After performing all their meritorious acts, the pilgrims crowd the little terrace overhanging the valley to enjoy the view of the Hun River. The sunset gives haloes to everyone, aureoles for every head. The old women have their nimbus, and the young willow-trees below, and even the weary farmers toiling up are crowned with light. So beautiful is the setting of the shrines poised on the rocks like birds, and surrounded by the blue silhouettes of pointed peaks, that the pilgrims linger till dark, and then descend the mountain in picturesque procession by the light of torches flaring among the rocks.

* * *

The Eight Immortals

If back in time from Miao Fêng Shan, some of the faithful will hasten to burn a candle at the shrine of Lü Tung-pin (whose birthday is on the 14th) and Han Chung-li (fêted on the 15th), two of the most popular Immortals. Did the Chinese pay reverence to them all, their religious duties would never be done, for, in addition to the Hundred Gods of their Pantheon, the Immortals are a vast group of figures of both sexes, some of whom are patron-saints of various crafts and, therefore, not to be neglected.

These Immortals, or Saints of the Taoist Church, are a peculiar invention of that peculiar creed—a conception combining the idea of everlasting life in a better world with the retention of an earthly bodily form. Its appeal is obvious. Who would not prefer to be one of these "eternal children, eternally healthy and happy, rather than a hazy *kuei*, or creature of spiritual essence, dependent for a joyless existence in the sunless nether world on offerings made by relations in this?"

The principle of cheerful immortality once established, the Taoist priests thereby much strengthened their grasp on popular imagination. They—and they only—hold the key to such a state of beatitude. They make the rules for attaining it. People have to come to them if they want to reach this blissful condition.

Theoretically, anyone is eligible. Practically it is not so easy to become an Immortal. Candidates must follow a strict regime of body and soul, while at the same time actively searching for the "True Essence." Attempts to do this led to the most varied experiments to obtain the "Elixir of Life" and the fabulous "Pill of Immortality." Finally, in the perfecting of the superman, there are three recognised stages. First the ascetic who comes out of the shell of his old body, like a cicada and, while keeping his earthly semblance, has perfect health and the power to travel throughout the universe, eat and drink copiously, and is completely happy without fear of disease or death. Next the "hero," or "perfect man," whose entire body has grown spiritual, who can fly through the air seated upon the clouds, free of all natural laws. Last the Saint, or Immortal of the First Class, who can make transformations in full daylight in contradistinction to those on the lower rungs of the ladder of perfection, who can be metamorphosed only in darkness. Such candidates as succeed in changing themselves at will night or day "become the rulers of the world."

Many of the Immortals, whose names are legion, live in the Palace of the Hsi Wang Mu (*see* "The Third Moon") and are misty figures that make little appeal to the man in the street. But the group known as the "Eight Immortals" remain immensely popular to this day. We find these fantastic personages, some of whom were historical figures and others pure creations of fancy, represented on every kind of art work, from valuable screens and porcelains to common teapots and posters sold for a cash. Their adventures are still told by story-tellers at country fairs, embodied in modern novels, and represented on the stage. "Born with a gift of laughter, and a sense that the world is mad," they are still the heroes of dramas, romances, and myths.

Like Taoism itself, which originated in a revulsion of thought and feeling against the prevalent order of things,—an attitude to which this faith has often reverted throughout the ages,

—the Eight Immortals were born during the Sung dynasty from the reaction of the vibrant Chinese sense of humour against the stilted official cult with its cut-and-dried classical examples who always did the right deed in the approved and conventional way. Shocking in appearance and habits, this boisterous, roystering band is a group of Chinese Don Quixotes who attained the highest bliss not by following the usual righteous road but by the devious paths of weird and fantastic adventure, thanks to the formula of the magic Elixir of Life. Not for them are the solemnity of things long reverenced. The universe was their box of toys and they themselves enchanted goblins at play. That is why they live when so many others, more virtuous than they, have died—because of their everlasting appeal to the little boy buried deep in every man's heart, joined to the Oriental love of make-believe. Many are those who do not take them seriously, but everybody loves them for their storied antics and their droll conceits. They folded up donkeys and tucked them in a girdle, they travelled on clouds, set fire to the Ocean, rode the waves on a crutch, met in drinking dens, and danced in and out of the gates of heaven.

No wonder the Chinese get a thrill from the absurd juxtapositions, the happy incongruities and the naive humour of the happy Eight. Dull folk see in these figures an ideal of perfect if imaginary happiness and, necessarily abstemious themselves, envy their heroes' orgies of food and wine. Tradition-bound, they enjoy the delightful spectacle of this band of national cut-ups kicking their heels like colts in a pasture. Moreover, the Eight Immortals have types appealing to all classes, old and young, rich and poor, male and female, ignorant and cultured, sound and sick. The complete story of their adventures would fill a whole book. Only the chief characteristics distinguishing each of these quaint figures can, therefore, be given here.

Lü Tung-pin, the most famous of the band, is an actual historic personage born in Shansi under a T'ang Emperor to whom he was vaguely related. From the hour of his birth, when a great light shone and a stork flew through the room, he was plainly marked for no common destiny and seems to have grown up to expectations, an impressive figure eight feet tall with a sparse beard. A gentleman and a scholar, he rose to an exalted position holding high office until the usurping Empress Wu seized the throne. Then he and his wife fled to the mountains. In exile he changed his family name of Li to Lü Tung-pin, "Guest of the Rocks," studied the mysteries of alchemy and, living on air alone, slew dragons and "rid the earth of diverse kinds of evils for upwards of four hundred years." His usual symbols are his "devil-slaying sabre," and a fly-whisk, or "cloud sweeper," the latter indicating that "he is able to fly at will through the air, or walk on the clouds of Heaven." More rarely, he bears in his arms a little boy, thus appearing to promise his devotees male children who shall grow up to be *literati* or famous officials. For this reason scholars pay him honour, and his small temple in the Chinese city of Peking is filled with votive tablets presented by some of the most important men of modern China. In the midst of these serious tributes, the itinerant barber and the ignorant soothsayer come to burn incense. How did Lü Tung-pin gather this doubtful *clientèle*? Owing to a slovenly mispronunciation of his name, he is confused in the popular mind with Lü Tsu Ta Hsien, a disciple of Lao Tzŭ who behaved so badly that he had to renounce the ideal of perfection and return to the world. Here, necessity being the mother of invention, he became a barber in order to earn his living. It is this ne'er-do-well who is actually the patron saint of barbers. But fate, by a ridiculous error, allows the Razor Guild to meet yearly in Lü Tung-pin's sanctuary to burn incense, enjoy theatricals, and discuss the price of a Chinese shave and hair-cut. Both the occasion and the company seem ill-suited to the dignity of the Imperial Patron of Letters.

Lü Tung-pin's teacher, friend, and fellow wine-bibber, was Han Chung-li, the typical soldier of the Group. Yet, there are so many versions of his life and adventures that he sometimes figures as a warrior, sometimes as a Taoist priest, and sometimes as a beggar receiving the Pill of Immortality. His military virtues are eclipsed by his alchemistic talents. All his pictures show him in mufti, sometimes holding his famous feather-fan, or a peach of long life.

If Han Chung-li and Lü Tung-pin are the most famous of the Eight Immortals, Li T'ieh-kuai, "Li with the Iron Crutch," has the honour of being the first among them to gain immortality, and the Hsi Wang Mu herself taught him the doctrine, after curing him of an ulcer of the leg. Poor Li is a pathetic figure. Losing his parents in early youth, he was ill-treated and half-starved by a cruel sister-in-law, whereupon he fled to the hills. There, while wandering in a wood, he had a most distressing adventure. His soul left his body to visit the Hua Shan, one of the Sacred Mountains. Some say that it ascended into the heavens to see his Master, Lao Tzŭ. In any case, it was on a perfectly respectable errand, but on returning Li's soul found that his body had vanished and it had to enter the first body it could find from which the vital essence had not yet completely departed. This happened to be the untenanted corpse of a beggar—a poor wretch lately dead of hunger with a lame leg, matted hair, and bulging eyes. When he discovered this, Li wanted, of course, to exchange the vile shell against a better one, but it was too late. Lao Tzŭ advised him from the skies not to attempt the impossible, but gave "him a gold band to keep his hair in order, and an iron crutch to help his lame leg." Thus his picturesque, contorted figure is unmistakable. Chinese druggists sometimes use his portrait as a shop-sign because of the medicine-gourd he carries, and exorcisers look upon him as their patron.

Another member of the "happy band" is old Chang Kuo of Shansi who, with commendable worldly wisdom, refused all invitations to the brilliant court of the T'ang dynasty and preferred to remain obscure. Though he appears to have been old from birth, and liked to pretend to be a re-incarnation of the "White Bat that appeared after the First Chaos," he had a taste for sport. He rode thousands of miles a day on a white mule which, at the end of the journey, he folded up like a sheet of paper, resuscitating his magical mount, when wanted again, by sprinkling a few drops of water on him. A picture of Chang Kuo sitting on this mule, but mounted backwards (a peculiarity of this long-distance rider) and offering a son to a newly-married couple is often placed in a bridal chamber. It seems rather inappropriate that the aged ascetic should be concerned with marital happiness and the birth of children. But even in life he was famous for conjuring tricks, and the phœnix-feather in his hand marks him as a worker of miracles.

If Chang Kuo stands for old age among the Sung Immortals, youth is personified by Han Hsiang-tzŭ, grand-nephew of Han Yü (A.D. 768-824) the great statesman, philosopher and poet of the T'ang dynasty. The child, educated by his cultured uncle, grew up to be a lover of flowers, producing miraculous plants with poems written in letters of gold on their leaves. This refined, gentle youth, supposed by some to have been the pupil of Lü Tung-pin, either carries a basket of the blossoms he so loved, or a jade flute, and is the patron of Chinese musicians.

Officialdom is represented in the "Merry Eight" by the person of the stately Ts'ao Kuo-ch'iu who was connected with the Imperial family of the Sungs, and seems to have been let into the company of Immortals after leading a most dissolute life on earth for no better reason than that, the eighth grotto of the Upper Spheres, where they lived, happened to be vacant, and they nominated him because he "had the disposition of a genie." Excess of temperament involved him in the murder of another man's wife. He was imprisoned, but released, profiting by a universal amnesty granted by the Emperor, and then gave himself up to the practice of perfection. Nowadays, we find him pictured as a most respectable figure holding in both hands the kind of tablet formerly used at Imperial audiences—a tablet emitting miraculous light. This is a passport to audiences with the superior divinities, and becomes Ts'ao particularly well since, in life, he was allied to the reigning family and, though looked upon as the black sheep of the clan, had access to the Palace.

One woman only is a member of this strange company,—Ho Hsien-ku, supposed to have been born in Canton about A.D. 700. "A maiden holding in her hand a lotus blossom, the flower of open-heartedness, or the peach given her by Lü Tung-pin as a symbol of identity, playing at times on the *shêng*, a reed organ, or drinking wine"—such is the usual picture of this young lady, who gained immortality by eating powdered mother-of-pearl presented to her by a friendly ghost, and henceforth spent most of her time wandering on mountain-tops or floating on coloured clouds.

Finally, the maddest of all the mad hatters is Lan Ts'ai-ho, whose sex even is not clearly determined. This doubtful individual is the semi-crazed strolling musician or mountebank of the Immortals. No one knows whence Lan Ts'ai-ho came, nor whether he was as lunatic as he pretended to be. "A man, yet not a man," he called himself as he earned his meagre livelihood singing in the streets. His verses denouncing "this fleeting life and its delusive pleasures" would seem to prove his sanity. Yet, when money was given him, he strung the cash on a string and dragged it after him, never looking back to see whether it was still there or not,—a sure proof of lunacy to the frugal Chinese mind. Nor was his costume less peculiar than his behaviour; a tattered blue gown held together by a wide wooden belt, one foot shod, the other bare. In summer he wore wadded garments, in winter he slept in the snow, his "breath rising in a brilliant cloud like the steam from a cauldron." His final disappearance from this world was in harmony with his wild life. He rose lightly to Heaven on the fumes of wine he loved, throwing down to earth his shoe, his belt, his robe and castanets, with the grandiloquent gesture of one who goes to pluck the stars. Why this queer individual has come to be represented with a basket of flowers, and is styled patron of gardeners, is one of the mysteries of this mysterious band of picturesque gentry.

Finally, as a group, the Immortals who are, essentially, above and beyond the law, may serve to remind the respectability-loving Chinese of the old adage: "no man can be sure to avoid prison or the beggar's bag,"—the classical *"per aspera ad astra."* They embody, and subtly suggest also, the Oriental reverence for madmen and their admiration for the type that renounces the vanities of earthly honours and possessions, which are so dear to the majority of men. Indeed the wisdom, the humour, and the moral lessons, to be drawn from the lives of the Eight Immortals still keep them alive even in the XXth century—sharp figures of broken dreams with which to cheat the conventions.

CHAPTER IX.

THE FIFTH MOON, OR "DRAGON MOON."

THE fifth moon. Summer is here; bright brassy skies, dust where the oxen pass, and the ceaseless clamour of cicadæ shrilling together in the trees.

Peking streets echo with the song of the ragged peddler:

"Soothing syrups cooled with ice,
"Try them once, you'll take them twice,
"A copper a glass to forget the heat,
"And the taste is sweet, the taste is sw-e-e-t."

Customers gather round the stall sipping these water-ices flavoured with "driving-away-heat ingredients,"—dried orange-peel, magnolia-rind, cardamon and hibiscus,—from cracked glasses, and stirring them with spoons no thicker than tin foil. In bygone days, when the Emperor watched over his people with fatherly solicitude, jars of cool water for free distribution were placed near police stations with the inscription: "The Imperial mercy is all-embracing." Rich and charitable folk continued the custom for a while, but their generosity was sometimes ill repaid, if we may believe the cynical saying: "Some give the ice water, others steal the drinking spoons." In South China it is still a meritorious act to provide hot tea in big kettles for passers-by, as many southerners consider unboiled water dangerous to drink.

The Peking populace, evidently, has no such fear of germs, if we may judge by the crowd of coolies clustering round the fruit-stalls where pomegranates with wide-open crimson mouths, and slices of watermelons covered with flies, find ready purchasers.

Trade is brisk now, customers plentiful. Even the poorest may indulge in a little extravagance, because a holiday approaches—"the notorious fifth of the fifth," celebrated all over China, even in distant Yünnan where it has been adopted by aboriginal tribes, like the Miaos and the Lolos.

* * *

The Dragon-Boat Festival

This universal festival is called simply the *Wu Yüeh Chieh*, or "Feast of the Fifth Month," but is known to foreigners as the "Dragon Boat Festival."

Economically speaking, it is one of the three settlement days of the year; astronomically, it marks a turning point in the seasons, for till this day nature has been gradually ripening, and, from now on, she gradually declines. Here then we have a milestone in the calendar of growth, celebrated in different ways all over the world, since the dawn of civilisation, as the "Festival of the Summer Solstice." Look back far enough, and we catch glimpses of magical feasts with propitiatory rites in connection with hunting, fishing, and tribal wars—the three important events in the life of primitive peoples.

In China, these rites appear to have survived chiefly in the cult of the River Gods upon whom, in the old days, a great part of the population depended for food in the form of fish. The first Water Divinities included the ghosts of the drowned, deprived by the manner of their death of sanctified sepulchre and forced into the category of wandering, therefore unappeased and mischief-making, ghosts. De Groot hazards the theory that, as in China every man-eating tiger is animated by the soul of his victims, so the man-eating alligators of former ages, who devoured the drowned, became possessed of their spirits and their power. If it was from them or their saurian ancestors that the dragon was evolved, we have the link connecting the Dragon with the Midsummer Festival. As the Controller of the Waters and Dispenser of Rain (*see* "The Sixth Moon") is invoked in the season of drought just prior to the breaking of the rains in the sixth moon, so in midsummer—season of the harvest—and in the fifth month—the month of plenty—people placate the Spirits of the Drowned embodied in the monsters they animate, lest neglect lead them to spoil the ripening crops and send famine and her gruesome sister, pestilence.

Of course, such complicated and distant origins for the Dragon Festival were long ago forgotten by the Chinese people. They needed a semi-historical peg on which to hang the older myths. Therefore, according to popular tradition, the holiday commemorates a high-minded statesman and poet, called Ch'ü Yüan, a minister of the State of Ch'u, who lived in the feudal period in the IVth century B.C.

An honest and upright figure in a troublous and dishonest age, he vainly urged reforms on a Prince who turned a deaf ear to his good counsel. Those were the days when loyal patriots believed in the duty of suicide as a moral protest—a suitable "remonstrance against shameless conduct on the part of one's lord," imperative when all other means of persuasion had been tried in vain. Thus, when he found himself powerless to check the abuses of his age, Ch'ü Yüan calmly composed the famous poem *Li Sao* detailing his anxieties, and jumped into the T'ung Ting Lake (in modern Hunan Province) clasping a great rock in his arms. Some fishermen, who witnessed the act, hastily rowed out to save him. They could not even recover his body. Then all the people wept for admiration of his sacrifice and threw rice upon the waters to feed his ghost. Nevertheless, the spirit remained unsatisfied, until one day it appeared to a group of worshippers on the bank and said: "I have, hitherto, been unable to avail myself of the offerings which you and others have so graciously presented to me because of a huge reptile (evidently a dragon is meant) which immediately seizes and devours all things that are cast into the waters. I request you, therefore, to wrap such offerings in small pieces of silk, and to carefully bind the same by means of five threads, each being of a different colour. Offerings which are in this manner enclosed the reptile will not dare to touch." This request is the origin of the triangular rice-cakes that, nowadays folded in leaves, are still offered to Ch'ü Yüan.[1] In a way, it is laughable that a disembodied spirit should trouble to return and make practical suggestions about food but, on the other hand, there is real pathos in the idea that virtuous souls are still concerned about their dinners.

As it is obviously impossible that the death of a rather obscure minister of State should be the real motive for one of the great festivals of the moon year, the virtuous Ch'ü Yüan may be taken to represent in his own person all the drowned, and share his offerings with them—the little rice-cakes, the lanterns set afloat in order that the hungry ghosts may see their way to them, and the Dragon-boat races which, though held in his name, honour all those lost in the waters from time immemorial. According to de Groot, these boat races "represent fighting dragons, in order to stimulate a real dragon-fight (in the Heavens) which is always accompanied by heavy rains."

Regattas are rare in North China,[2] but in the south, where rivers and lakes are numerous, they still remain a popular amusement. At

---

[1] In the inland districts peasants eat these cakes, appropriate to the day (made of rice or millet, with a bit of sugar or candied fruit inside), often without knowing their meaning.

[2] In Peking various land-sports once marked the festival day—polo in the Palace and shooting at willow-wands in the open spaces of the capital. Picnics to various places outside the town amused the populace, while officials, bearing banners, went on a curious frog-catching expedition to the Nan Hai Tzŭ, or Southern Hunting Park. The frogs, when netted, were pricked over the eyes, for the sake of a liquid they exuded, known as a valuable heart-tonic and still used by Chinese doctors. The "white tear" has been discovered by Western science to be more potent than digitalis, one more proof that Chinese empiric methods long ago reached results only lately obtained in our medicine.

Amoy, Foochow and Canton the water festival is particularly brilliant, and the races sometimes last several days. Thousands of people crowd the shore where matsheds are erected for their convenience, or hire *sampans* parked along the short course. Little maids, like brilliant moths and butterflies in their bright holiday costumes, play on quaint instruments. The pointed notes of the serpent-bellied *san hsien* float over the water in a plaintive love-song that dates from the Chou dynasty. Families picnic on the decks of brightly-painted red and purple junks, friends toast each other, or gossip squatting on their heels. Everywhere there is the bustle of thickly-thronged life, a kaleidoscope of colour and sound, of lights and shadows, of moving boats and people, an ever-changing grouping on land and water in the tawny sunshine with its fierce, prowling splendour.

Suddenly, the attention of the crowd is fixed on the starting point. The rowers are about to take their places in the boats. Amidst wild applause and shouts of "*hao! hao!*" from the spectators, they paddle slowly down the course, giving the crowd plenty of time to admire. Enthusiasm is justified. They are a fine sight, these huge boats resembling dragons, each over ninety feet long and so gracefully slim that two men are crowded as they sit side by side. High sterns with long steering-paddles rise many feet above the gunwales, high prows are shaped like a dragon's head with open mouth and cruel fangs, and the long body between is gaily painted to represent scales, and touched up with brilliant gilding.

One man stands in each bow, as if looking for the corpse of Ch'ü Yüan, and throws his arms about as though casting rice upon the waters. Others, interspersed among the rowers, wave brilliant flags or beat gongs and cymbals, so that the deafening clamour may frighten away the monster that Ch'ü Yüan feared.

Now the boats gather at the starting point. There is one tense moment of breathless silence. Then the signal is given and they are off. The rhythmic splash of two hundred paddles, kept to time by a coxswain with a bright waving banner, sends the slim dragons with arrow-like speed through the water. White and silver jets shoot up from the bows, foam pours like cream round the blades of the oars. The play of muscles glistening in a mighty effort, the howling of cymbals, the clashing of gongs to excite the rowers, the life-like movements of the dragons' heads controlled by rope lines, the bitter rivalry between the boats of different guilds, the loud partisanship of the spectators and then, at last, the deafening applause that greets the winners, all form part of an unforgettable spectacle.

In Canton, in 1926, according to the *North China Herald,* the procession held in honour of regatta-day was distinguished by modern characteristics and had as its *pièce de résistance* a man dressed as a mandarin of the old regime with queue and formidable moustache. He was walking between two Europeans (Chinese dressed in paper clothes that looked exactly like the latest thing from Bond Street) who were pestering the official to sign some paper. Another member of the procession was carrying a placard on a bamboo pole, and as this apparently gave the meaning of the scene, a bystander was asked what it was all about. "The foreigners," he said, "are asking the poor official to sign a treaty." The mandarin and the "Europeans" played their parts well. They looked very glum and serious, and the crowds thoroughly enjoyed the spectacle.

Next came some Chinese generals, also in paper uniforms with epaulettes, swords, and paper top-boots. They passed without much comment, but afterwards there was wild cheering and *"hai yahs"* as the "peasants" appeared,—coolies armed with brooms, poor boat-people with oars over their shoulders, and farmers clad in tatters carrying spades and other implements. Their placard, borne aloft on a broom, stated that these were the poor workers who always got the worst of the bargain in the world. "They work most and get least" was the slogan. Then, coming down the French Cathedral Road, was a big, formidable looking black gun. Cleverly made with bamboo framework, covered with black paper, when seen from a distance it looked exactly like the real thing. It was on a platform with wheels, and ten men had a hard struggle, or pretended to have a hard struggle, to drag it along. On the gun was a poster stating: "A useless weapon." Following the gun were many strikers, and upon their placard was written: "This is the weapon we use." Chinese explained all these things, but they did it without any sign of ill-feeling. The whole procession, indeed, was regarded simply as great fun.

THE FIFTH MOON, OR "DRAGON MOON"

Alas! the joy of the Dragon Day sport is often marred by fatal accidents as, owing to their shallowness and their peculiar construction, the dragon-boats may capsize, causing loss of life.

In Foochow and in Canton there were so many accidents due to sudden floods, or boats fouling, or faction fights, that the police forbade the races for some years, though they are now allowed again. Many fatalities might be avoided but for "the curious and cruel Chinese superstition which sometimes prevents a Chinese from helping a drowning comrade, even when he could save the man without danger to himself. The apparent callousness is due to the fear that he will incur the revenge of the water spirit cheated of his prey, or even be required as a substitute for the person saved. This superstition was once widespread and existed in places as remote from one another as Ireland and the Solomon Isles."

After nightfall the dragon-boats are often taken down the course again in slow procession and outlined with lanterns. The effect is fairy-like in the warm waters alive with phosphorescence. Oars gleam with light at each stroke, and at the prow ripples trimmed with fire flee away to right and left into the darkness, brightening as they run till, suddenly, they go out like a blown lamp-flame. The crests and wavelets burst into showers of sparks, and every patch of spray catches fire and smoulders for a moment. The lanterns launched on the waters add picturesqueness to the scene. To the superstitious they are not only an offering but also an insurance against plague.

* * *

The "Ministry of Medicine."

Ever since 500 B.C., the fifth moon has been called the Pestilential, Evil, or Wicked, Moon and regarded as the most poisonous month of the year. Not without reason. Outdoors is like a furnace. Dry winds and droughts come first—sirocco winds that try the nerves. Then steamy heat hangs over the land like a hot, clammy blanket. Pestilential vapours exude from the earth. Disease lurks in the odours that arise in undrained Chinese cities, and the Five Venomous Animals appear: the Snake, the Scorpion, the Lizard, the Toad, and the Centipede.

In order to repel the noxious influences rampant at this season, people pray to the Gods of Medicine. Orthodox Buddhists tell their beads in honour of the Yo Shih Fo, the "Healing Teacher" who relieves suffering and prolongs life. His figure appears in most of the temples, and his heaven, a rich invention of the Oriental imagination, rivals the Paradise of Amitabha (*see* "The Eleventh Moon") with its jades of all colours, "blue-green as the feathers of the kingfisher, yellow like bees' wax, red as the cock's comb, and black as ink."

The Taoists have a regular Ministry of Medicine, presided over by the mythical Emperors Fu Hsi, Shên Nung and Huang Ti, who represent a "sort of ancestral triad of medicine gods," with Fu Hsi, their leader and the discoverer of the Eight Trigrams. These combinations of a straight and of a broken line symbolise in China the deepest wisdom and the foundation of all knowledge. Without attempting to quote any of the innumerable commentaries on these figures, we may say that the great philosopher Leibnitz thought them worthy of consideration, and the early Catholic fathers—men of science—discussed them at great length. Chinese physicians of the old school still believe that their mystical power controls the maladies and fates of their patients.

Fu Hsi, moreover, is credited with having established marriage (as the human race before his time bred like animals), with having planned the basis of civil administration, inaugurated writing and the computation of time. His wife, or sister, (the vagueness of the relationship is characteristic) Nü Kua is, nevertheless, still thought of as a semi-woman, semi-serpent or fish, like certain early Near Eastern idols.

Fu Hsi's successor, Shên Nung, added to the invention of agriculture that of astronomy, arms, and medicine properly speaking, as distinguished from divination. Under him, the patriarchal organisation of the Chinese began to separate into clan-groupings which later became the Kingdoms of the "middle ages."

This sovereign, who was a contemporary of the Great Pyramids and of the Sumero-Akkadian empire, was succeeded on the throne by "The Yellow Emperor"—Huang Ti—who started sericulture, the working of metals, the coining of money, etc.

Whether or not any historical truth lies behind this triad of inventors, they symbolise with remarkable historical accuracy the slow progress of their race from barbarism to civilisation. Thus, the first two are popularly represented clad in leaves, Fu Hsi in the middle, holding the symbol of the *Yang* and the *Yin* surrounded by his Eight Trigrams, Shên Nung on his right, while Huang Ti, on the left, is the first to be clad in State robes. This group is frequently found in village shrines.

The people, however, pay little attention to these misty Emperors. They ask relief for their aches and pains from the Yao Wang, King of Remedies. He is the Celestial Doctor whose shrine is in every hamlet, where, often enough, he seems to have power to save others long after he is able to save himself and his temple from crumbling to pieces. Often we have been touched by the spectacle of a peasant-mother praying for the sick child in her arms at his tiny half-ruined sanctuary which the community can not afford to repair. Her faith is undimmed by the woe-begone structure with the plaster gnawed away by winds and rain, with its walls the same colour as the dust from which they rose. The paper ceiling hangs in tatters, the altar-table has one leg awry, and the figure of the god, swathed in a long weather-stained cloak, has a flattened face, faded and sodden like blotting-paper left out all night in a drizzle.

This beloved Patron of the Healing Art, specially worshipped in the "dangerous Fifth" (though his birthday is in the third moon), is identified with various persons, human and divine. He absorbed old medical celebrities into his own person in the mysterious way Taoist heroes can, including a very famous leech from Kashgar, or India. Accompanied by a black dog—is there a hint of witchcraft here?—this worthy travelled to China, and "reduced his lanterns"[3] to a minimum or, in other words, cured more people than he killed, and became a specialist in fevers.

Assisting the Yao Wang to watch over the people's health are numerous other members of the Heavenly Ministry of Medicine. The Nine Lady Healers, the Niang Niangs, of whom we have spoken in "The Fourth Moon," have an important place in the group of specialists, and the Gods of the Five Epidemics (*see* "The Tenth Moon") have charge of the seasonal plagues. In addition, there is a separate "Ministry of Smallpox," significant of the prevalence of this disease in China. Most Chinese children "put out flowers," and though, nowadays, a primitive form of vaccination is common, "the Ministry of Smallpox has not been abolished and, possibly, its members, like those of some mundane Ministries, continue to draw large salaries for doing little or no work."

Besides many spiritual patrons of the healing-art, numerous historical and famous doctors have been deified and swell the retinue of the Yao Wang. One of the most striking is a physician called Hua T'o, who attended the famous usurper Tsao Tsao, conqueror of the Han dynasty. Tsao Tsao suffered from dreadful headaches, and an operation akin to modern trepanning was advised. Hua T'o was allowed to begin, but not to finish a preliminary treatment, because Tsao

---

[3] An old rule obliged physicians to hang a lantern over their door for every patient that died.

A hundred years ago, in England, doctors were supposed to follow in the funeral processions of their late patients; in Japan, both doctors and nurses still do so.

Tsao grew suspicious of the physician, believing him a tool of his enemies. The poor doctor, suspected of designs to kill his patient, was himself condemned to death. Before his execution he bequeathed to his jailor, as a reward for various kindnesses while in prison, his wonderful library. But when the jailor went to fetch his legacy, he found Hua T'o's wife had used most of the priceless volumes to light the kitchen fire. "What was the use of them?" said she, "they indirectly caused my husband's death. However, if you want the one book left—and even that has a few pages missing—take it, and welcome!" This single treatise, all that remained of Hua T'o's works, is an important surgical manual, still in use. The last physician to be included in the Ministry of Medicine, to our knowledge, was a successful practitioner deified, according to Father Doré, as late as 1913.

* * *

The "Ministry of Exorcisms"

In the "dangerous Fifth," the Chinese supplement their prayers to the medical divinities by appeals to the Ministry of Exorcisms, a Taoist conception,[4] including seven chief Ministers whose duty it is "to expel evil spirits from dwellings, and generally to counteract the annoyances of evil demons."

Two of these worthies are of particular interest —P'an Kuan and Chung Kuei. P'an Kuan, Guardian of the Living and the Dead in the

---

[4] Couling, in his *Encyclopædia Sinica*, points out that exorcism and the arts of healing and of divination have long gone hand in hand with Taoist doctrines. "The elixir of life and the genii originated as medical ideas."

Underworld, has been supplanted in popular favour by Chung Kuei, "Prince Protector Against Evil Spirits." Once, in a fever-dream, the Emperor Ming Huang (A.D. 712-756) was tormented by a small devil dressed in red who broke into the palace and disported himself mischievously. When His Majesty angrily inquired who he was, the Red Imp replied flippantly: "I am called Hsü Hao, or 'Emptiness and Devastation.'" Further irritated by his disrespectful manner, Ming Huang was about to call the guard and have him arrested, when a tall figure dressed in a blue robe appeared, went up to the imp, tore out one of his eyes and ate it. When asked who *he* was, the hero replied: "A physician of Shensi province, unjustly defrauded of his rightful honours in the public examinations." Shame and indignation led him to commit suicide, as a protest, on the steps of the Imperial Palace. The Emperor of the day, the first of the line of T'ang (A.D. 618-627), ordered him buried in a green robe, a special honour reserved for members of the Imperial clan, whereupon, in gratitude, the scholar swore to protect the Sovereign, whoever he might be, from the annoyances of "Emptiness and Desolation." On waking, Ming Huang found that the fever had left him. He then called for a celebrated portrait-painter to paint the person he had seen in his dream. The likeness proved excellent, the artist was rewarded with a hundred taels of gold, the picture was hung in the palace, and Chung Kuei canonised with the title of "Great Spiritual Chaser of Demons for the Whole Empire." Furthermore, copies of his portrait have ever since been pasted on house-doors in the "dangerous Fifth."

* * *

Talismans

Because of all the evil influences rampant during this month, people are careful of everything they do. For example, no one will start building a new house or begin any undertaking lest it prove unlucky. Nor will housewives put out their bedding in the sun.

In towns and villages one notices a bunch of mugwort, *artemisia vulgaris,* and leaves of the sweet-flag, sprigs of garlic, and other aromatic plants, hung over doorways. Each has a special significance as a preserver against evil influences. The leaves of the flag are pointed; they represent a demon-killing sword. The mugwort has historical associations. There is an old saying that he who on the Tuan Wu festival (the fifth of the fifth Moon) "hangs out no mugwort, will eat no new wheat," a saying explained by the legend of a famous rebel, named Huang Ch'ao, who gave orders to spare any family that exhibited this plant at its door. The custom would appear, however, to be much older than the legend. In fact, from very ancient times the fragrant artemisia was considered efficacious against troublesome ghosts. Magistrates recognised this when they stepped over burning mugwort after an inquest, in order to keep the ghost of a corpse from following them.

It is significant that nearly all the talisman plants used are strong smelling—capable of keeping off summer odours and insects. Chinese have long believed that the stronger the scent—and the bigger the pill—the better the remedy, whether moral or physical. Buddha himself was cured of a sickness by the perfume of a lotus. It is still more curious that the custom of hanging up fragrant herbs in midsummer, and even the choice of plants, is the same in the West as in the East. The Herb of St. John put up on the Feast of St. John in Russia, Norway, Sweden, Belgium, England and France—indeed all over Europe—on houses and stables, as a protection against sorcerers and their spells, is none other than the *artemisia vulgaris* that hangs over the Chinese farm-house door. The coincidence is as peculiar as it is exact.

Moreover, a connection with the universal rites of the Midsummer Festival is suggested by the old French belief that the herb must be gathered on Midsummer Eve—the day before the Feast, when it has power to preserve man and beast from ulcers, fever, plague, and also from lightning and fire. Pliny remarks that magic plants must be picked on this day and no other.

In China there is a well-known custom "to rise early and walk exactly one hundred paces into a field without turning the head, then to pluck one hundred blades of grass which must be carefully taken home. The grass is put into a pot of water and thoroughly boiled. The water, into which all the virtues of the grass are now supposed to have passed, is strained, boiled a second time, and bottled as a remedy for headaches, wounds and nervous diseases. It is called the *pai tsao kao*, 'hundred grass lotion.' Wise men who hand down this valuable recipe from generation to generation are careful to explain that the medicine will be of no use whatever if any of the prescribed conditions have been neglected. It is absolutely necessary to walk neither more nor less than one hundred paces, to pluck neither more nor less than one hundred blades of grass, and to boil and strain the water in the manner laid down. Above all, everything must be done on the fifth day of the fifth moon, as it is only on that day that ordinary grass possesses *ling,* spiritual, or health-giving, properties."

Anxious mothers of only children rise with the sun to collect dew, and use it for moistening ink. With this ink little dots are made on the foreheads of precious sons to preserve them against illness. Or else, someone is begged to write the character "Wang" (王), meaning "prince," on the baby face, as this ideograph is supposed to resemble the wrinkles on the forehead of a tiger, the terror of evil spirits. Children who have reached seven years are allowed to plait their hair in a single braid, instead of the many little pigtails seen on small babies. Older people drink sulphur-brandy to safeguard themselves against magicians. This superstition is founded on the legend of a youth who transformed himself into the wife of a good man. One day, having drunk sulphur-wine, he took back his original masculine form, to the great consternation of his mother-in-law, whereupon he perished on the site of an ancient temple still shown at Hangchow.

The list of charms to combat wicked influences is endless—too long to quote here, especially as every province has its own set of superstitious talismans. A few may be mentioned, however, to show the kind of fetish that appeals to the Chinese mind.

Red threads are tied to the wrists of little boys to insure them long life and strengthen the memory, or five-coloured threads, symbolic of the "Five Poisonous Ones." This is an old custom, dating from the VIth century A.D., and an infallible antidote against illness and wounds. Friendly families make gifts of these threads to one another's children. Each mother puts them on her offspring with perfect faith, but is careful to take them off again after noon. This is called "throwing away evil." Another custom, prevalent in South China, is to make or buy paper dolls, one for each member of the family. These are known as "substitutes for the body." A prudent housewife places all her puppets together in a basket, as a sign that the family shall remain united. Later, she takes them out one by one, and begs each please to take upon itself any ill-fortune which may be coming to the individual it represents. Finally, in the presence of the assembled relatives, the dolls are burned. To conclude the little ceremony, everyone drinks a concoction of sulphur and cinnabar dissolved in wine, as a further guarantee against malign influences. What is left over may be used to smear the noses and ears of the children to preserve them from the pimples and eruptions so common in hot weather, or sprinkled through the house as a protection against vermin and the Five Poisonous Animals. In the north, old-fashioned folk, in addition to hanging out pieces of green called *nai tzû*, place a cloth boy-doll on their gateposts, as a guarantee that sickness will not visit the home during the next twelve months.

To protect little girls, paper flowers are worn in the hair, with pictures of the "Poisonous Five" on their petals, or sachets filled with aromatic plants are worn round the neck, hung in their clothing, or hidden in their hair. Such sachets are prettily embroidered, and good needlewomen pride themselves on making them very, very small, as the tinier they are, the greater credit to their skill. They may be of any colour, but preferably red, because red is supposedly the colour of the peach-blossom, in itself a powerful protector against demons.[5]

In villages, we sometimes find curious amulets that are survivals of very antique beliefs. Of such are the miniature swords made of willow. Obviously a sword is a good defence against enemies, mortal or supernatural. But why the willow? Because the willow is sacred to the sun, the Great Healer and Physician of Nature and the light that by its essence combats darkness (*see* "The Third Moon"). Likewise, the gourd is used because it too suggests the idea of healing, being the Chinese horn of plenty used by apothecaries as a receptacle for their drugs and simples.

---

[5] For the power of the peach-tree against bad influences, see the story of the Door-Guardians in "The Twelfth Moon." Note also that exorcising priests use seals made of peach wood. It is with a club of peach-wood that the attendants of the Kings of Hell kill unrepentant souls, and fever patients are struck with peach-branches to drive out the bad spirit that is making them ill.

One might think that there were enough anxieties to be guarded against in the dangerous "double Fifth" without the "Five Poisonous Animals." But these creatures never fail to appear as scheduled, and must be carefully outwitted, like the devils and the diseases. For this purpose, charms of yellow paper with pictures of them are sold to be hung over doors and windows. A popular talisman is a bronze medal with the picture of the "Chang T'ien Shih" (or "Chang, the Master of Heaven,") richly robed and riding a tiger, vanquishing these insect pests with his magic sword. Thereby, as usual, hangs a tale—an old legend still alive in popular imagination. A sovereign of the Han dynasty, the Emperor Hsiao Ho (A.D. 89-105), once requested a Taoist sage, Chang Tao-ling, who lived in the "Dragon-Tiger Mountain" in Kiangsi, to capture the Five Animals and put them in his magic box in which he used to distil the Elixir of Life. The sage did so, and within this casket, on the fifth of the Fifth Moon, the expected miracle happened. Each creature gave up its essence, contributing to form the true Elixir, called the "Dragon-Tiger Elixir."

* * *

The Chang T'ien Shih

The invention of written charms, so popular in China,—charms once inscribed on bamboo slips but nowadays printed on paper.—to keep off devils and all malevolent creatures, visible and invisible, is attributed to Chang Tao-ling, the first Taoist "pope."

Connected, as Taoism has often been through passing centuries, with movements of opposition and rebellion against the existing authorities, especially in times of trouble, Chang Tao-ling is credited with a direct contact with the "Yellow Turbans," the powerful revolutionaries of the end of the Han dynasty. As regards Taoism itself, he appeared at the psychological moment when the abstract mysticism, or "nihilism," of early days reverted to the quest of immortality in this life by means of the "philosophical stone," so much sought after in the Middle Ages of Europe.

In this movement Chang Tao-ling stood foremost, and even if he was a charlatan, he must certainly have been a very remarkable man—perhaps a great scientist for his age.

Legend states that he received a mystic treatise from Lao Tzû himself, by means of which he was able ultimately to compound the "elixir of life." Then, a spirit directed him to the hills where he was to find a stone hut within which were concealed writings of the Three Perfect Emperors and an ancient canonical book. He went as directed; he searched; he found them, and he learned how to "discipline himself for a thousand days." After that, he could leave his body and walk among the stars. He could fly,[6] and distinguish sounds inaudible to mortal ears. He could divide the mountains and the seas, and command the thunder and the rain to obey him. He could fight the King of Demons single-handed, and the Hosts of Evil fled before him, "leaving not a trace of their retreating footsteps." In a

---

[6] Taoist magicians are often called "feathered scholars," because they are able to fly.

THE FIFTH MOON, OR "DRAGON MOON"

word, he became a great sorcerer, the greatest that ever lived in China.

At the ripe age of 123, Chang Tao-ling swallowed the Pill of Immortality in his mountain retreat, and ascended to Heaven in broad daylight in A.D. 157, accompanied by his two disciples with whom he generously shared the Gift of Everlasting Life. Even now the name of the Supreme Wizard, who must have been shrewd enough, but to our way of thinking suggests little of the Lord's anointed, remains a household word in China. He is the founder of the second stage of Taoism, a practical interpreter to the simple folk of Lao Tzû's vague teachings,—the earthly Viceroy of the Jade Emperor who rules the skies (*see* "The First Moon").

His sixty-second descendant in an unbroken line, Chang, "The Master of Heaven" (a title granted in A.D. 424, rescinded, as irreverent, under the Mings but still in popular use) continues to live on the "Dragon-Tiger Mountain" to which his illustrious ancestor retired nearly two thousand years ago, and even now rules a great spiritual empire maintaining, in these republican days, a state which mimics the old Imperial regime. Till lately, he conferred "buttons" like the Emperor, and he receives visiting priests like a sovereign, granting them titles and seals of office.

His palace stands in the middle of a little town. It is new and of no special interest, having been re-built after the T'ai P'ing Rebellion. Nevertheless, when he grants an audience, even in these semi-modern surroundings, it is impressive.

A band of pilgrims approaches. They have journeyed through picturesque natural scenery. They have paused at the foot of the mountain to worship at the temple on the very spot where the first "Pope" ascended to Heaven. They are rich invalids bearing costly gifts, because the services of the "Master of Heaven" as a demon-exorciser are expensive. They enter his reception-hall through high doors thrown open by attendants, and find the Magician seated on his throne in canonical robes, with gold and silver embroidery. They claim his power of exorcism. One old man has a daughter possessed of a devil, and buys a devil-chasing charm prepared on thin yellow paper to be pasted on the sick girl's forehead. A rich widow complains that her house is haunted. The Chang T'ien Shih promises, for a suitable consideration of course, to chase away the demon, with his "sword" which all demons fear. When he wields it, the Great Magician can catch them and put them into jars which he seals with a charm. A visitor to the Dragon-Tiger Mountain remarks that somewhere on the mountain there are many rows of such jars, although he personally did not see any.

When a Taoist pontiff departs this life, his soul passes into another male member of the family of the ancient House of Chang. In order to determine the chosen vessel, all the male members of the clan assemble at the palace, their names are engraved on tablets of lead, the tablets are then thrown into a vase full of water, and the one which bears the name of the new "incarnation" floats on the surface.

TAOIST MOUNTAIN SHRINE.

Photo by the Asiatic Photo Publishing Co.

GATHERING FOR A TEMPLE FAIR.
Photo by the Asiatic Photo Publishing Co.

THE FIFTH MOON, OR "DRAGON MOON" 323

The present incumbent is a very normal man, with two wives and several sons, for Taoism, in the persons of its wizards, retains marriage. Indeed, it is difficult to imagine the tall, handsome, middle-aged gentleman who receives visitors so courteously and cordially, who seems so sane and so sensible, in his rôle of Magician, and to picture him seriously bottling-up evil spirits.

Powerful as he is, legend tells us that on one occasion (in A.D. 1014) the hereditary head of the Taoist Fraternity had to call for help in exorcising a noxious demon. This demon, who turned out to be no other than a famous rebel of the XXVIIth century B.C. who plagued Huang Ti—the "Yellow Emperor"—lived in a pond near a temple recently erected to that Sovereign. Apparently, the holy shrine annoyed the demon. Yet it could not be destroyed, because this would have meant a "loss of face" to the Imperial spirit. The "Heavenly Master," called to Court to settle the dilemma, either could not, or would not, take on the task of ousting the devil, and suggested that a loyal soldier of past ages be summoned to vanquish the trouble-maker. Accordingly, the spirit of Kuan Ti was called up and arrived in a black cloud that hung over the haunted pond. A great aerial battle raged, with the clashing of arms and the charging of invisible horses. When the sun appeared once more, the demon was defeated and the throne ordered a thanksgiving service in honour of the hero.

* * *

Kuan Ti

The 13th of the fifth moon is the day still fixed for the official sacrifices to Kuan Ti, sacrifices which persist where so many others, to gods seemingly more essential to the welfare of the State, have been abandoned. The "golden warrant," or special patent commanding this official recognition, dates from A.D. 1614. That the Republic should continue to recognise an old Imperial mandate is a sign of the times. Stronger than democracy is militarism. The leaders of modern China are soldiers, and Kuan Ti is the patron of those "who take the sword in hand to perish by the sword." [7]

Among a peace-loving people like the Chinese, who have long despised the warrior caste, it is strange that Kuan Ti, the canonised hero generally described as the Chinese Mars, or God of War, should be beloved by the masses as he is, and that his shrine should be found in most villages, as often as that of the Lung Wang (*see* "The Sixth Moon"), the T'u Ti, or local god (*see* "The Ninth Moon"), and the Yao Wang, or God of Medicine. As a matter of fact, Kuan Ti, in his true character, is not a fire-eating divinity, delighting in battle and bloodshed, but a deified general whose own martial successes taught him that "war is hell," and whose spiritual powers are used rather to prevent war and protect people from its horrors. In proof of this, note among his many honorific titles those of: "The Prince of Peace by War," or "Prince of War-Won Peace."

---

[7] The 13th of the fifth Moon is one of the disputed dates of Kuan Ti's birthday. It is also the day on which, according to popular tradition, he sharpens his sword and, invariably, during the twenty-four hours, rain falls. We have here a curious illustration of his dual powers, as a war-god and as a rain-god.

Unlike so many gods of the Chinese pantheon, Kuan Ti was not invented out of whole cloth, but was once very much alive. A native of Shansi province, he was born in A.D. 162, in an age of perpetual intrigue, of fights for power, of fear, of wearying petty internecine wars, of general discomfort, uncertainty and unrest. The stage was set, much as it is at this very day, for a popular hero. The country needed his knightly figure, tall as a tree, with unusual strength, both eyes glaring like a menace in righteous anger,—a romantic personality in one of the most romantic periods of Chinese history, a hero creating a sound and noble impression, always doing the manly thing, and, what's more, the knightly, chivalrous thing,—a warrior "bold enough to stroke the tiger's beard." This paladin character was marked for adventure from the first. He began his career by succouring a damsel in distress—an honourable escapade which involved the murder of a tyrannical magistrate, the adoption of the name of Kuan to conceal his identity, and flight from his native place.

Thenceforward he began a stirring career, "not too uniformly successful to be monotonous." Chance or Fate, call it what you will, associated him with two other celebrated characters of the period, Liu Pei and Chang Fei, and these three became lifelong and devoted friends, swearing blood-brotherhood by the famous "Oath of The Peach Orchard" near Cho Chou (on the Peking-Hankow Railway)—an oath sealed under the blossoming trees by the opening of their veins, by the sacrifice of a black bull and a white horse, and the solemn words: "One to another we shall be brothers, now, henceforth, and until death." [8]

The romantic adventures of these friends are told at great length and with many fantastic additions in the "Story of the Three Kingdoms," the Chinese Iliad, a book that has done marvellous press-work for Kuan Ti and helped "raise him to a pitch of popularity which is really independent of his position on the official roll of divinities, or saints, and would hardly be affected by the total withdrawal of official recognition."

Nevertheless, we regret no cinema existed in those days to give us exact portraits of Liu Pei, a descendant of the founder of the Han Dynasty, who eventually carved his way to the throne of one of the Three Kingdoms, largely through the devotion and generalship of Kuan Ti; of Liu Pei, the nine-foot hero, with eyebrows like silkworms and a beard two feet long; and of Chang Fei, eight feet tall with a panther's head, a swallow's chin, eyes round and sparkling like jewels, and a voice like thunder—a butcher and wine-merchant by trade, and owner of the historic orchard.

Between such giants, that brotherhood oath was no light thing. It meant these voluntary relations pooled their resources and their talents. It drove them to battle together in defence of their Sovereign with conspicuous success to their combined arms. Unfortunately, their most stunning victory, when, at the head of three thousand men, they defeated three hundred thousand rebels known as "Yellow Turbans" under Ling Ti (A.D.

---

[8] Such oaths between comrades are not uncommon in Chinese history and are not unfrequent to this day.

THE FIFTH MOON, OR "DRAGON MOON"

168-190), was not appreciated by that effeminate monarch who continued his life of idle pleasure till, finally, saddened by a new revolt, he died of grief. Again, a second time, under the Emperor Hsien Ti (A.D. 190-220), Liu Pei took command of the Imperial troops and with the assistance of his companions won still more brilliant victories. But here the enemy, commanded by Tsao Tsao (the classical villain of Chinese history) who had the Sovereign in his power, was too strong and too treacherous to be openly defeated. Liu Pei tried to have him poisoned, but the plot was discovered and the attendant physician made to drink the potion. Thereafter, Homeric combats ensued in the western provinces where the Sworn Brothers combined every effort to strengthen the Han dynasty.

Thus, in brief, is the history of the beginning of the Three Kingdoms, a favourite plot of Chinese plays and a favourite theme of story-tellers. Tsao Tsao finally reigned in the provinces north of the Yangtze; a famous general, Sun Ch'üan, in those south of the great River, and Liu Pei in the west. Even this division of territory, however, brought no peace and the life-and-death struggle continued between those old-time Tuchüns, Liu Pei and Tsao. After a brief triumph of the Brotherhood, defeat followed. Kuan Ti was captured and executed in A.D. 219; Liu Pei died of grief, leaving his throne to his young son, and, finally, Chang Fei was assassinated. But before his death we have one last and striking picture of the latter weeping tears of blood for his sworn brother, Kuan Ti. Even this savage giant evidently had his softer moments. Strange mixture of frankness and ferocity, intrepid defender of his friends, soldier of fortune by choice (since he owned rich properties and might have lived out his allotted span in peace), hard drinker and never a harder fighter than when under the influence of wine, a cruel disciplinarian who had his soldiers and even his officers beaten when they disobeyed him, he finally was killed by two of his own captains whom he had ordered flogged.

Of these three blood-brothers Kuan Ti alone attained divine rank, as he was alone to die fighting for his liege lord. From ten to forty years after his death he received his first title: "Marquis of Martial Dignity," but thereafter he appears to have been neglected, till the Sung dynasty conferred further honours on him and ordered a temple erected to his memory. From that time he continued a posthumous triumphal progress down the ages.

He was regularly promoted to godship in the XVIth century, and received at the same time the embroidered robe and State-cap which were deposited at the little temple in his honour still standing outside the Ch'ien Mên at Peking.

The Manchu dynasty treated him with particular respect. In fact, he became, in a sense, the patron-saint of these Emperors. Tradition ascribes the origin of their veneration to the days when their ancestor Nurhachi, then ruling in Manchuria, begged the Ming Emperor to send him the statue of a god. The image sent was that of Kuan Ti, and Nurhachi, finding it resembled his own father, paid it special reverence. Later, Kuan Ti repaid this reverence by miraculous intervention on behalf of the Manchu dynasty. He fought for them against the "White Lotus" rebels who attacked their palace in 1813. He aided them again the following year to defeat rebels in Honan, and "again, in the middle of the last century, he is said to have aided the Imperial troops against the T'ai P'ing rebels, and on one memorable occasion, in 1856, to have turned what would have been defeat into a glorious victory—" for all of which miracles he received new honours, the most picturesque being "Winged Helper," an allusion to his timely appearance in the sky at the head of angel battalions.

Doubtless, he believed, as most gods do, that mortals who ask for heavenly help must also learn to help themselves and thereby deserve the aid they seek. This may explain why, with apparent ingratitude, Kuan Ti (the spiritual president of the famous Triad Society) withdrew his "support from the dynasty to which he was indebted for so many honours." Though it was under his standards that the Boxers started their campaign in 1900, he gave them no miraculous assistance, nor did he come to the aid of his erstwhile patrons at the time of the revolution of 1911.

The soldiers of the Republic have adopted him, and, twice a year—on the 13th of the fifth moon, and on the 15th of the second—defile through his sanctuary (the converted ancestral temple of Prince Ch'un, the last Regent, near the Têh Shêng Gate in Peking). There he sits, side by side with Yo Fei, together with his adopted son, Kuan

P'ing, his faithful follower Chin Tsang, and various tried companions in arms, and a model of his famous charger, "Red Hare,"—the charger capable of galloping ten thousand miles a day on which this Chinese Bayard was ever to ride off in all directions at once, so to speak, to succour the distressed.[9] His mighty sword, "Blue Dragon," —the Chinese "Excalibur"—is also at hand, that sword that no modern soldier could lift.

Now, everyone loves a hero, everyone admires a knight of abnormal size and strength, but peaceful peasants love Kuan Ti best because of his less martial characteristics. They believe he can cure disease, and often does "appear at the bedside of sufferers and lay a healing hand upon their bodies, giving them golden drugs and other mysterious medicaments that speedily bring them back to health." He is also to some degree a rain-god, and these two features are quite sufficient to install him in the usual village temple, where he is portrayed wrapped in a long tiger-tawny robe which was given to him once for stopping a flood. In addition, he is a secondary god of Wealth, a patron-saint of several trades and professions, and a determined champion of sound morals, protecting those who live rightly according to the Chinese theory that "all loyal and high-principled people become active spiritual agencies when they die, and the deadly enemies of those who fail to fulfil their duties to society and their families."

---

[9] Gray, in *China*, says: "Even armour-bearer and horse have their votaries," and mentions having seen, in Kuangtung, "women worshipping these images, and binding small bags, or purses, to the bridle-rein of the charger."

As a sideline, Kuan Ti has taken on the functions of a patron of literature, though his claims to this title are of the slightest, based simply on the tradition that he was a devoted student of the "Spring and Autumn Annals" of Confucius, studying them at night without taking off his armour. We suspect that at one time in China the "complete hero" must have been a combination of soldier and scholar, since Confucius himself, "pre-eminently the literary sage, is, or was supposed, also to exercise the functions of patron saint of soldiers." At the same time, precedence in worship and sacrifice was always carefully given to gods and sages whose chief attributes were civil rather than military, to show that the national ideals were peaceful, not warlike.

Johnston [10] says that, "if Kuan Ti's reputation as a patron of literature seems based on a rather slender foundation, still slighter are his claims to be regarded as a good Buddhist." Actually, the Buddhists took him into their pantheon for no better reason than that they could not afford to ignore so powerful and popular a figure. Although he "did nothing whatever in his life to promote the cause of Buddhism," they invented his conversion by a famous monk who died in A.D. 597, and gave him a place among their *kalan*, or "tutelary guardians."

Doubtless, he is quite willing to fight the enemies of the faith in company with Wei T'o

---

[10] Cf. "The Cult of Military Heroes in China," *New China Review.* 1921.

and the Four Heavenly Kings who stand at the entrance and protect the sanctuary proper. For Kuan Ti has ever proved obliging, ever ready to take a hand and help his friends. A humorous instance is his victorious fight against the Japanese—of whom in life he certainly never heard—when they attempted to land in Kiangsu and Chekiang during the XVIth century. This patriotic action gained him special temples all over China.

Last of all, we find him in favour with Chinese spiritualists "re-visiting the 'glimpses of the Moon,' at the behest of mediums," and prophesying, by means of the Chinese *planchette*, many historic events, including the capture of K'ai Fêng by the Golden Tartars, and the ignominious end of the Northern Sung Dynasty.

* * *

Towards the end of the Fifth Moon, clouds gather in the sky, and scatter again after occasional showers which bring relief from the dryness accentuated by the yellow winds.

In Peking, this is the beginning of the rainy season, when the Lord of Heaven "divides the Dragon Hosts," assigning to each dragon his task for the year.

The farmers try to guess which dragon shall dominate their district. In case it is a "lazy dragon," the rainfall on their fields may be scanty, and prayers must be offered early and often to the Lung Wang, Commander of Dragons (*see* "The Sixth Moon"), to stimulate his subjects.

Sometimes, too, the dragons cluster all together in one place, when the Lung Wang is supplicated to divide them. It is their struggles in the sky which bring local showers, but the great battles of the true rainy season in the next month must be fought over the largest possible area, so that the needed moisture shall be fairly distributed.

The Dragon's nature, as an apparition, is best expressed in these myths.

CHAPTER X.

THE SIXTH MOON, OR "THE LOTUS MOON."

Dragon Cult

THE kingdoms of the world, so the Chinese believe, are controlled by dragons whose spheres of influence are redistributed each year at the beginning of summer. As the proverb says: "The dragons mount to the heavens in the spring, and return to the deeps of the ocean in the fall." The monsters sleep during the cold dry season. They begin to stretch and stir with the first heat (*see* "The Second Moon," the *Lung T'ai T'ou*). Soon they rise up to the clouds, gather in little groups and battle with their rivals. These combats cause local showers (*see* "The Fifth Moon").

But the real rains come only in the sixth moon, when the King of Dragons, the Lung Wang, celebrates his birthday on the 13th. Then he orders his scattered hosts to assemble and distribute water to the thirsty earth during the three *fu*, or three ten-day periods of greatest heat, corresponding to our dog days.

The dragon is not a conception of the Chinese mind alone—not even as a rain god. Dragons, or creatures very like them, are mentioned in the

Bible. Dragons ornamented the gates of the the Kings of Thebes and Delphi, as they did those of the Emperors of China. They were prominent in the palace of Sennacherib. They appear in many European myths and legends, to wit, the story of Perseus who rescued Andromeda from a dragon, the story of Siegfried who killed the dragon at Worms, the story of St. George and the Dragon—to quote only a few examples. In addition to these national dragons, we have records of similar monsters whose traditions are attached to cities or rivers, as, for example, the dragon of the Nile, the dragon of Naples, the dragon of Arles, of Marseilles, of Norwich, etc.[1]

The dragon in Western mythology is generally pictured as a cruel monster, an evil creature symbolic of sin. The enemy of God and man, saints and martyrs throughout history engaged him in mortal combat. Even pagan gods like Apollo held "the honourable office of dragon slayer," taking the succession of still older Greek heroes.

The Asiatic conception differs entirely from that of our forbears. The Chinese dragon is a beneficent being, giver of gifts to humanity. He rules the oceans. He dominates "every twelfth hour, day, month, and year of the lunar calendar." Above all, he regulates the rainfall. Hence, in an agricultural country like China, where men depend on good harvests for sufficient food, he is revered and worshipped.

---

[1] See *The Chinese Dragon*, by L. Newton Hayes.

In fact, the only link that connects the European dragon with his Oriental prototype, except for some similarity in appearance, is an association with water. We find a dragon guarding sacred springs in ancient Greece and causing the floods of the river Rhône.

From the most ancient times the dragon, in China, has been considered the "chief of the four spiritually endowed beings," of which the other three are: the *Ling*, or Unicorn, the *Fêng*, or Phœnix, and the Tortoise, *Kuei*.

The origin of the dragon is, of course, purely mythical. Some regard him as the primary image of creation. Others trace his descent from the Nagas, or Serpent-Kings of India. His kinship to the sea-serpent, the boa constrictor, or the alligator still to be found on the Yangtze valley, has been suggested. It is most likely, however, "that some antediluvian saurian was the true source from which the dragon idea has sprung."

In the reign of Huang Ti (supposed to have ruled 2698-2598 B.C.), there is a record "of the first appearance of a true dragon." By this time, Chinese imagination had begun to fix the type as a composite creature with the head of a camel, the horns of a stag, the ears of a cow, the neck of a serpent, the scales of a fish, the claws of an eagle, and the paws of a tiger.

As he became more prominent in symbolism and mythology, the original Chinese dragon was divided into various species—all belong to the genus *lung*. Some were several miles long, others no bigger than a silkworm. Some had wings and horns, others purple whiskers. All, however, were known for their remarkable eyesight. Indeed, their vision was so keen that they could distinguish a blade of grass at a hundred *li*.[2]

The eight great varieties include the dragons of the sky, of the marshes, and the "big five" who preside over the seasons and the divisions of the world. "In the East is the blue (or green) Shên Lung (*lung* means dragon, *shên* spirit) who is identified with spring; in the West, the white, identified with the autumn; in the North, the black, associated with winter; and in the South, both the red and yellow, who divide the duty of ruling the summer." The Sung dynasty in A.D. 1110 officially sanctioned these Protectors of the Empire and granted them princely titles. The minor varieties include little local water-spirits to whom lovely maidens used to be offered as brides, water-phantoms that are deified ghosts of the drowned.

Four dragons guard the capital: the dragon of the Summer Palace lake, the Black Dragon of Hei Lung T'an in the Western Hills, the dragon of the Jade Fountain, and the White Dragon of the lake of the district of Mi Yün. Under the Empire, princes were delegated to worship these guardians, whose cult is mysteriously connected with that of the Ma Tsŭ P'o, Goddess of Sailors (*see* "The Second Moon").

Undoubtedly, the most popular dragon in China

---

[2] In this connection, it is interesting to recall, says Newton Hayes, that the English word "dragon" is derived from the Greek *drakon* which means to gaze, or to see, and the classics have more than one reference to the animal as sharp-sighted.

is the one known simply as the "Lung Wang," or "Dragon King," who has come to represent the fecundating principle of nature and the essence of the *Yang*—male or creative element—symbolising prosperity and peace, because he controls the rain.

How dragons in general, and the Lung Wang in particular, first became associated with water is not easy to trace. Probably, when simple folk, with no explanation of natural phenomena, watched the evaporation of water to the sky where it took the form of clouds, they saw in those changing cloud-shapes the reflection of the monster already conceived from the model of the huge saurian. And when the sucked-up moisture descended in the form of rain, their fancy suggested that it was the gift of the writhing dragon. Is there not an ancient Chinese saying that "clouds come from dragons—or dragons from clouds—as wind from tigers"?

Because of his power to rise from earth to heaven, the Chinese dragon presently became the symbol of the Emperor who also rises above ordinary mortals. "The Sovereign's most reverential title was 'The True Dragon' and, in harmony with that ideal, the word 'dragon' in the adjectival sense was used in names of all that had to do with his life and position." Thus his throne was the "Dragon's Seat," his pen—the "Dragon's Brush," and the tablet which represented him in every big temple was called "The Dragon Tablet." Under the Mings, the Imperial crest was a red dragon, because red was then the national colour, but under the Manchus the yellow or golden dragon was proclaimed the official crest of the Empire.[3]

Emblem of masculine strength and power, controller of the destinies of the country, the dragon appeared on the Chinese flag and on official robes. Even nowadays, when Imperial symbolism is gone, nearly every phase of Chinese life still retains the influence of this unique creature. People believe in him just as they always did. He still appears, occasionally, to mortal eyes. He still figures prominently in legends and art-conceptions, and his name is constantly used in every-day talk. Thus, a male child is addressed as "little dragon." Betrothal certificates are called "dragon-phœnix papers," and locust blossoms "dragon's-claw flowers." Even modern things like fire-engines prove how he has coiled himself round the popular heart, for, though foreign importations, they have been christened "water dragons." Likewise the faucets of Western plumbing bear the name of "dragon's heads."

The Taoists, always so charitable to native myths, have given a special religious significance to dragons. They have invented, or adapted, the Four Golden Dragon Princes who live in the depths of the ocean and "feed on pearls and opals." They claim also "Old Mother Dragon" who, to judge from her clay images in the temples, "seems to be a quite ordinary and rather benevolent lady, the last person in the world one would expect to give birth to an uncanny son."

---

[3] Chinese dragons are of many colours in addition to the above-mentioned. They may be purple, blue, white, or black.

The Buddhists count their dragons "in numbers equal to the fish of the great deep." They, too, have their Dragon Kings of the Four Seas, originally conceived as serpent-gods but, when Buddhism reached China, superimposed on earlier Chinese beliefs connected with dragons, and later merged with them. The Lotus of the Surprising Law quotes Eight Dragon Kings, such as the Poisonous Dragon, the Princely Dragon of the Anavadata Lake (north of the Himalayas), and the Dragon of the Blue Lotus Flower. While some of the Buddhist dragons and, especially, the Mountain Nagas imported directly from India are, by virtue of their stranger origin, mischievous spirits, the Buddhist Lung Shên inhabiting lakes and rivers are invariably helpful and friendly—a necessary concession in a dragon-loving land like China.

Meanwhile, both Buddhists and Taoists appear to have adopted the Lung Wang, the Divine Dispenser of Rain, sometimes called the Chinese Neptune, who lives in a beautiful palace under the Eastern Sea surrounded by the treasures of the deepest deep—a place drenched with ghostly twilight, whose loveliness cannot be described in words.

Rain processions

A well-beloved god, his birthday is universally observed. Dragon-processions are held in his honour at various times, but especially in the fifth moon, when the soil of the fields and even the mud bricks of the farm-houses are split and cracked with the dry heat, and again towards the end of the sixth, when the rains have broken, and people give thanks for benefits received.

THE SIXTH MOON, OR "THE LOTUS MOON" 341

In both cases a paper effigy of the rain dragon is paraded through the streets, a dozen coolies acting as living vertebræ under a blue cloth skin painted to look like scales and illuminated by lanterns. Such a monster, undulating in life-like fashion, is most impressive, even terrifying, especially as pandemonium attends his passage. Gongs clang, crackers pop, and bonfires of paper money blaze, while people sprinkle him with water to show what is expected. As the dragon zigzags through the grey old city gate, children scream with terror and delight, and women desirous of sons crowd to pick up the burned-out candle-ends that drop from his body, believing them talismans.

The gaudy show does not last long. An hour after dark and all is over, the last light extinguished, the gongs silent, the crowd dispersed. The tattered dragon-skin on its bent bamboo framework is stored away in some dusty guild-house, neglected till next year when it will be patched up, freshly painted and paraded again.

Details of these rain processions vary. The dragon himself sometimes gives place to an effigy of the Dragon King, who has no feet, but the head and face of a monster and the body of a man. Strange servitors, creatures of the sea and of the storm, are carried round him, and coloured banners, yellow and white symbolising wind and water, black and green to represent clouds. A human attendant bears two buckets of water suspended from his shoulders and a branch of bamboo with which he sprinkles passers by, saying: "Come, O Rain! come, O Rain!" Now and then a halt is made *en route*, and a feast offered to the Lung Wang. Instead of a menu, an official list of contributors to the banquet is placed upon the table.

A rare variation of the rain-procession, sometimes seen in remote villages, includes a paper boat instead of a dragon. In this case, a man disguised as a woman appears to be seated on the deck. The illusion is complete until the boatman, standing alongside and brandishing a pole, pretends to punt rapidly. Then the false lady-passenger, serene to the waist, but with her feet upon the ground, is obliged to run as fast as she can while attempting to preserve her dignity and smile calmly, even if out of breath. Meanwhile, the merriment of the yokels standing by may be imagined.

From the Emperor down to the humblest mandarin prayers have been addressed to the rain gods in times of drought since remote antiquity. Sometimes the Sovereign called upon the Taoist Pope (*see* "The Fifth Moon") in his retreat in the Dragon Tiger Mountains of Kiangsi to intercede and, if his petitions proved unavailing, his salary, paid by the Imperial Treasury, was withheld. Sometimes His Imperial Majesty made his official demands in person, or by princely deputy, but always in solemn stately fashion with colourful paraphernalia, for the Chinese were adepts at taking the Kingdom of Heaven by storm when they wanted favours. A feature of the ceremony was the continuous beating of drums. Sinologues suggest that the large and symbolic part which the drum plays in Chinese life is due to its early use in these rain processions. This may

PEASANTS PRAYING FOR RAIN BESIDE A SPRING.

FESTIVAL PROCESSION, KANSU.

well be. Legend ascribes the origin of thunder to a drum a thousand *li* long buried in the earth. The growl of drums certainly resembles the distant rumble of thunder. Therefore, a slight stretch of popular imagination suffices to identify them with the downpour that accompanies the storm.

Even now provincial prefects hold intercessory services lasting several days, during which the people often fast till a favourable answer comes. Formerly, when the Lung Wang would not listen to the junior official, the Viceroy petitioned personally. Cases have been known in Kuangtung province—and probably in others—where this high dignitary dressed in mourning robes, with chains round his neck and his ankles fettered as a sign of humility, proceeded to the Lung Wang's temple escorted by his fellow citizens. There he *k'o t'oued*, burned incense, offered up a written prayer to the god, and placed upon his altar banners with the characters for Rain, Thunder, Lightning, and Wind upon them in honour of the Lung Wang's satellites—members of the Ministry of Thunder which we shall presently describe.

Still more amusing expedients to obtain rain are recorded. The figure of the Lung Wang has been taken out to look for himself at the parched countryside, or a certain city gate closed. Once, in a great emergency, frogs were collected in tubs, and a cunning soothsayer engaged to tickle them and make them croak—an unfailing device to bring the rain. Iron tablets are also used as rain charms since, for some unknown reason, dragons fear iron.[4] The iron tablet brought to the capital for the personal prayers of the ex-Emperor Hsüan T'ung after the very dry season of 1925 was an excellent example of methods of intimidation used against the Lung Wang. Bouillard, in his *Usages et Coutumes à Pékin,* tells a similar tale of a high official sent to fetch an iron cash which was to provoke the needed rain. The first special train on the Peking-Hankow line was reserved for this great personage in order that he might go to Southern Chihli and get the fetish.

Old superstitions die hard among all classes. The farmers in the villages still march to their local shrine with willow-wreaths on their heads. A few years ago, Republican soldiers were ordered to present arms to a "dragon snake" carried in a palanquin at Lo Yang. The spring rains in 1926 were directly attributed to the prayers of the Panchen Lama, and one of the Tuchüns has lately threatened to punish the rain god for overdoing his gifts of water.

In Shansi, Chihli, and Honan, men and boys in seasons of drought go about wearing wreaths made of fresh willow-branches, the willow being, as we have seen, a water-loving tree and, therefore, considered a rain-charm (*see* "The Fifth Moon"). In Kansu a recent spring-drought brought out old Chinese methods of praying for rain, curious enough to describe in full:

"From the 1st of the moon, the country-folk coming from different directions paraded the

---
[4] "Iron is feared by demons and is, therefore, used in making swords and knives. Often something made of iron is put in a baby's cradle to ward off evil spirits."—Plopper, *Chinese Religion Seen Through the Proverb.*

main thoroughfare of Hochow to the noise of gongs, each wearing a willow-branch on his head and carrying an incense-stick and a piece of paper. These companies of rain-makers proceeded to the dragon temple, and also to the city-god's temple, where the antique rites were gone through, while all classes of the male population of the city turned out to make their petitions known to the rain-god. *K'angs,* or earthen jars filled with water, and with willow-branches inserted therein, were placed in front of shops; while ropes with slips of paper containing rain-prayers were stretched across the principal streets. The school-boys in the procession sung a song as follows:

> ' May the big rain come continuously,
> And the small rain drop by drop.
> Great Heaven, Great Heaven!
> Pearly Emperor, Pearly Emperor!
> City God! Earth God!
> Have pity on all things!
> In my hand I hold two willow-branches.
> Scatter the rain under all heaven,' etc.

"In Chinese, this rain-making process is called *ch'iu yü* —' beseeching for rain.' If the drought is persistent, recourse is had to what is called *ch'ing yü*—' inviting rain,' in which performance the prominent actors are geomancers, self-appointed men who choose a leader among themselves. He occupies the rôle of Shang Ti, or God. During the 'inviting' ceremony, this leader ascends the second storey of a temporarily constructed platform on the street, and his thirty or forty satellites take up positions on the first floor of the platform. A table with a tablet is placed on the upper floor. Six posts are planted in front of the platform, and four *k'angs* (two on each side), with one in the centre, are placed near them, each containing water and willow-branches. Live frogs, fish, and tadpoles in these *k'angs* are regarded as necessary concomitants. Papers bearing appropriate Chinese characters for the occasion are pasted on the posts, and even in the lime spread on the ground hieroglyphics are inscribed.

"The orthodox time-limit for securing results, *i.e.* rain-fall—is seven days, but there can be an extension to nine days. This, then, finishes the programme for inducing the Huang Ti to give rain.

"The ordinary custom of prohibiting the butchering of animals was indulged in for a considerable time, and even the south gate of the city was ordered to be shut, but the latter proved to be a farce—only for an inconsiderable time was the gate partially closed. A ludicrous cast was given to this dramatic exhibition in that a certain renowned goddess, styled 'the golden flower goddess,' esconced in a temple 110 *li* from Hochow, is supposed to have acted the part of suppliant on behalf of the people for rain before the Pearly Emperor. The latter not only refused to grant her petition, but actually resolved to kill her . . . A prayer was then written out and distributed, so that the populace could lend a helping hand in sending up petitions to rescue her . . . "—*North China Herald,* August, 1926.

A unique religious ceremony took place at the

station of Changchun (in Manchuria) in June, 1926, when the Chinese population asked the Russian Orthodox priest to hold a church service for rain:

"For about a month the neighbouring region had not seen a single drop of rain, and the farmers were anxiously searching the sky for the sight of a cloud which would save the parched fields. In despair, the people tried to mollify the 'God of Rains' by prayers and offerings, but the angered Rain God remained deaf.

"Finally, the local Merchants' Guild organised a mass prayer meeting of farmers and merchants in front of the Buddhist temple. At every shop, incense and prayer-candles were burning, all butcher stalls were closed, and a general fast was ordered until the first rainfall. The religious ceremony was conducted by several Buddhist priests, and when this was ended, all those present were led by the priests and representatives of the Merchants' Guild toward the Russian Orthodox church. Acceding to the demand of the Chinese, the Russian priest held an impressive church service imploring God to grant rain.

"The church was surrounded by a huge crowd of Chinese, bare-headed and kneeling down at the sign of the priest. Many Russians joined in these prayers. The religious ecstasy was tremendous and the Chinese, although not understanding the Orthodox rites, fervently prayed to the Christian God in their own fashion.

"Later the whole procession, headed by the Russian and Buddhist priests, proceeded to the settlement near Changchun.

"Several young Communists who were met on the way and did not take off their hats were obliged to do so by the Chinese police. The ceremony, which started at 10 a.m., ended only about 4 p.m.

"It is curious to note that three days later the first rain of the season fell, starting the periodical rainy season of Manchuria—this being about two weeks ahead of the usual date."—*The Peking Leader*, July 1st, 1926.

On cannot help remarking sometimes that either the Rain God is a stupid god, or else the prayers of his devotees are too fervent. Where showers are needed, he sends torrents that melt the dry clods of the rice-fields to liquid mud in a few minutes. Every dry river-bed straightaway becomes a swift stream, and the terrors of drought give place to the dangers of floods, so that the last state of the farmers is worse than the first.

In former times, when this happened, armies were called out, and we read of a certain Han general who deployed his troops in battle order for three days along a river bank. Here they beat drums and shot arrows, in a word—fought the river like an enemy until the flood-waters subsided.

No longer ago than 1872 a magistrate, after making unsuccessful efforts to combat a flood which menaced the city of Tientsin, drowned himself. His sacrifice caused the waters to subside, and he himself became a water spirit, taking the form of a small snake. This little creature, found by a peasant and immediately recognised as an incarnation of the Rain God, was carried in state to Tientsin and placed in the Lung Wang's temple. The whole civil administration, headed by the famous Viceroy, Li Hung-chang, came to do him honour. Later, Li memorialised the Throne to grant the serpent a special honorific title according to historical precedents—a request duly complied with by the Sovereign. No one thought the procedure ridiculous, because it has long been the habit in China to give honours and titles to little river snakes discovered in local waters.

In China snakes may become dragons, though the change requires hundreds of years. This power of metamorphosis explains why a humble water snake often receives honours as a water divinity; also why the souls of the drowned, automatically deified by the manner of their death, sometimes take the form of water snakes, acting as emissaries of their lord and master, the Lung Wang.

So long as the gods behave themselves, the Throne and the people have always been generous to them. They get high-sounding names, silk robes, jewelled head-dresses, feasts, outings, incense-honours and amusements. But if they withhold expected blessings, they are punished like mortals who fail in their duties. A local dragon, for example, will be torn to pieces during his progress through a city if he does not grant what the people ask, and even His Godship, the mighty Lung Wang, will be punished by exposing his image to the rays of the hot sun. That discomfort will sometimes bring him to reason where everything else fails, is proved by the legend of the Emperor who went to Hei Lung T'an, in the Western Hills near Peking, to pray for rain. The dragon remained deaf to his supplications. Furious, His Majesty there and then issued a decree banishing him to the desolate wastes of Heilungkiang in Northern Manchuria. A start was made for this disagreeable place of exile. Now, it was just in the "Period of Greatest Heat", and the farther his tablet travelled, the hotter and thirstier the dragon grew. So, after a few miles, he repented and sent the rain. Thereupon, Ch'ien Lung gave the order that he be taken back to his pool, and the peasant-carriers of his palanquin splashed gratefully along the rain-rutted cart-roads and through the rain-blurred empty streets of the villages. Not even the passage of the Imperial procession could bring the farmers out of their houses, because there is an old superstition in China, and especially in the neighbourhood of Peking, that it is unlucky to be out of doors while it rains. When the sky is fertilising the earth, decency and prudence forbid men to assist at the mystery. Official work relaxes, and military operations subside while the Lung Wang waters the world. Even now a foreign cinematograph advertises on its Chinese posters that, as a special feature, "neither rain nor wind will stop the performance". Nor is it considered proper to ask a Chinese if he thinks it is going to rain, for that is to accuse him of presumption. How shall mortals dare to predict the will of the gods? Or else it is equivalent to comparing him with the despised turtle whose damp neck foretells the coming downpour.

* * *

Nowadays, any tourist may bathe in the dragon-pool at Hei Lung T'an without fear of *lèse majesté* and then, if he pleases, climb the hill above to the Lung Wang's sanctuary. The steps are steep, yet it is well worth while making an effort to see a typical and well preserved shrine to a god whose images we shall constantly meet in Chinese towns and villages. The feature of Hei Lung T'an, absent in smaller temples, is the group of weather gods,—let us call them "Departmental Kings of Nature,"—supposed to rule the elements under the direction of the Lung Wang. All are acolytes of His Dragon Majesty, officially grouped together as the "Ministry of Thunder," because the Chinese believe, as we have already noted, that the affairs of the Universe are managed by Bureaux or Ministries, the counterpart of those on earth.

"Ministry of Thunder"

This Ministry of Thunder has over eighty officials connected with it, including the little goblin-imp called the "Cloud Pusher." Only six of its members, however, appear in most temples and receive popular worship. Their cult in human form is of comparatively recent origin—dates, in fact, from the Han dynasty—but that does not prevent them all having pedigrees, biographies, wives, concubines, and birthdays.

It is not astonishing that such gods were invented. The whole world had them—spontaneous creations of the infancy of man's intelligence when the simplest scientific knowledge was non-existent. Thor and Lei Kung, the Chinese Thunder God, had common roots in the mental habits of primeval humanity. Both represent the earliest recorded conceptions of men concerning the visible phenomena of the world into which they were born. What more natural for primitive peoples than to deify overwhelming physical forces they could not understand? Lightning to them was a death-struggle between the day-god and the demons of darkness. Thunder was the beating of celestial drums. Massed clouds, a gathering of dragon-hosts; rain, the spittle of these hissing monsters in mortal combat. When the blackness of the storm, like a heavy curtain pulled across the sky, seemed to portend the end of the world; when the wind with fierce cries tore at the trees, and heaven's artillery crashed as though tons of metal were being dropped from a great height, skin-clad savages fell upon their knees praying for deliverance to they knew not what spirits from they knew not what calamities.

The God of Thunder

A quaint legend explains how the President of the Ministry of Thunder (chief festival—24th of the sixth moon) got his bird's head and feet in popular representations. Once, a mother-in-law was ill-treating her son's wife. Lei Kung came to punish the old lady, but she outwitted him by throwing a dirty cloth over his head. He was helpless till the rain came and washed it off, allowing him to rise to heaven. As he ascended, the neighbours saw with surprise that their mysterious divinity had a bird's head and claws, and ever since he is easily recognisable by these peculiar features. Lei Kung as a rule holds a hammer and a drum, or string of drums, or a hammer and a chisel with a bandolier of drums around his shoulders. Mothers sometimes say to naughty children: "Lei Kung will split you," and the babies behave at once, frightened by the threat of the sharp chisel being driven through their heads. When youths indulge in mock fights, as they love to do, and inadvertently grow over-rough and hurt one another, it is quite sufficient for a bystander to shout: "Don't you see the sky is clouding over?" This reference to the thunder-hammer which may bump them in revenge is enough to stop the horseplay, for "thunder is Heaven's anger."

Although he is such an ugly "black bat-winged demon," Lei Kung is also supposed to be a benevolent divinity who protects human beings from evil and, through his connection with the rain gods, assists good harvests. A person who throws his rice on the floor, or treads it under his feet, however, thus showing lack of appreciation of the gifts of nature, risks being struck by Lei Kung.

Some Chinese think that people are killed not by the lightning but by the actual noise of the thunder. Others believe that meteorites fall on the unjust, and these meteorites are often called the "Hammers of Lei Kung."

Next in line to Lei Kung often stands Yü Shih, the Master of Rain, an imposing figure in yellow armour, supposed to be the incarnation of a mortal who, during the reign of Shên Nung

The "Master of Rain"

(2838—2698 B.C.), relieved a terrible drought. He holds a measure called the *"ch'i,"* hence the expression *"liang t'ien ch'i,"* meaning "the measure with which you can measure even the sky." The land he measures out with his yard-stick he sprinkles with his watering-can. His peculiar gifts include the power to go through water without getting wet, through fire without getting burned, and to float through space. Stranger still, he can metamorphose himself into a silkworm-chrysalis. An equally weird concubine is attributed to him, a lady with a black face, a serpent in each hand, and red and green serpents reposing on her ears. Finally, his household is completed by "a mysterious bird with only one leg, who can change its height at will and drink the seas dry." His attendant and assistant who stand beside him is the Sheat-Fish Goblin who helps to pour out the rain-waters.

The "Mother of Lightning"

On the right of the Lung Wang in the weather-pantheon generally stands the Goddess, or Mother, of Lightning, a female figure in parti-coloured robes who holds two mirrors. When crashed together or drawn apart they cause the flash. The idea is that they represent the male and female elements which, when combined, produce sparks, on the principle of the tinder and the flint. The Mother of Lightning, or "golden snakes," assists "the God of Thunder by revealing the hearts of men" (Plopper, *Chinese Religion*, etc.).

The God of Wind

Next to this Goddess is Fêng P'o, God of Wind, the Chinese Aeolus, represented as an old gentleman with a long white beard, or, sometimes, as a woman riding a tiger, the animal associated with

THE SIXTH MOON, OR "THE LOTUS MOON"

wind. In whichever form, the figure holds a wind-bag, and looses the wind which jumps like a drunken man. The Shrimp-Goblin is in attendance, a quaint imp holding a vase out of which rain is poured.

As we study this interesting group of grotesques carefully, we learn that to the peasant they are simply masters of the weather, playing with the weather as they please. The little group of blue-clad farmers, that smothers the altar-pedestal of the thunder-god with artificial flowers on his birthday, hopes for direct assistance and a gathering of clouds in his cloudless valley. The student sees behind the images the *Yang* and *Yin*, personified forces of Nature, and the astrologer-diviner who comes up from his shelter in the village, in an idle hour when his humble clients are away working in the fields, to burn an incense-stick before his patrons, sees them as manifestations of stars and constellations. All three are right, inasmuch as they represent vital, if divergent, Chinese points of view on divinities they still believe in.

* * *

Birthday of the Lotus

When the people's prayers are answered, and the Lung Wang and Company send moisture to the thirsty earth, then the lotus, loveliest daughter of the Rains, is born. Indeed, poets call the sixth moon the Lotus Moon, because the 24th is the birthday of Buddha's flower, just at the season when the summer rains break. As in Japan people come out in hundreds to see the cherry-blossoms, so in Peking a fashionable and popular amusement is lotus-viewing. The finest setting for the blooms is Central Park, with its frame of pink walls, colour of ashes of roses, dominated by the noble wings of the Wu Mên. Here we can overlook the moats of the Forbidden City now wrested wholly from their more proper occupant—the water. A dense growth of leather-like leaves, above which rise in majestic isolation the solitary flowers, encircles the outer rampart, shutting the Palace in as if it were the palace of the Sleeping Beauty.

Sightseers install themselves in open tea pavilions to enjoy what a Chinese friend quaintly calls the "sceneries," or stroll about, admiring, commenting, or lost in delightful dreaminess. Since all classes flock to view the blossoming, the study of types is interesting.

The students are there in gauze gowns and incongruous broad-brimmed foreign hats. Schoolgirls with bobbed hair walk hand in hand—quaint figures dressed in the extremest of the extreme new fashion with tight trousers, very short jackets, and velvet tam'o-shanters. More conservative maidens wander about in pale silks, blue, pink, mauve, or maize—any colour but white, which is still reserved for mourning. Ladies of high degree with painted eyebrows and shoes of pink brocade sway to and fro on tiny feet, supported by soberly clad maid-servants, and little children with gay caps trimmed with gilt figurines of the gods to protect them against evil spirits copy the dignity of their elders.

Contemplation, not action, is the Far Oriental's ideal of life. Hence, we find many a man, who has not lost the art of dreaming, sipping tea, or calmly poetising with a group of friends, oblivious of the passing crowd. Or else an old-school artist stands silently wrapped in effective isolation, studying a subject for a picture—a bud just breaking into bloom against the sunset sky, or a leaf-goblet pouring out its pearls of dew. It is against his tradition to paint from the model. Visual training has taught him to absorb nature, and become one with the spirit of Nature. Then, and then only, he will return home and bring forth an impression whose intensity, expressed without detail but suggested with consummate technique, is "directly proportioned to the singleness with which it possessed his thoughts."

Rival crowds go lotus-viewing to the Northern Lakes near the Drum and Bell Towers, wandering about under the old willow-trees whose branches, played with lazily by the west wind, toss and ripple like green spray. When their leaves turn over they gleam silvery like little white fish. One unforgettable summer evening, we joined the throng at the mystic twilight hour when the perfume of the flowers was sweetest, when the bordering reeds clattered their long lances faintly together, and fire-flies sailed in the gathering gloom. A Buddhist monk from one of the neighbouring temples joined us at the water's edge and we spoke together. "How beautiful the flowers are to-night," said he; "it is a sign that men's hearts are pure. Otherwise, they would wither! Perhaps you do not know, you foreigners, all that the lotus means to us—all that it has meant." "Pray tell us good sir. Though we are ignorant, we desire to know."

"I shall not speak of the religious significance of the lotus—that is too vast a subject—except to say that it is a symbol of our Buddhist Paradise, land of serenest calm and eternally fragrant beatitude. Hermits have fashioned their garments from its leaves. Its roots, prepared in fresh milk, will cool the heart in heat. Its blossoms adorned the thrones of the T'ang and Sung Emperors. Have you not heard the story of the poet who fell asleep in one of the Palace pavilions when the Empress invited him to take tea with her, how the party lasted far into the night and she had him conducted to his apartments by the light of torches in the shape of golden lotuses detached from her throne?"

Sao Ch'ing Niang Niang

As we stood chatting, a few raindrops fell, and the ladies, of whom many, both Manchus and Chinese, were in the crowd, hastened to shelter. Conveyances were called. "They will be hurrying home," said the monk, "to cut out a paper figure of the Sao Ch'ing Niang Niang. Lately it has rained for days on end. Therefore, they will hang her figure in their apartments, behind the Orchid Door! She has a broom in her hand and can sweep the heavens of clouds and bring fine weather."

"But who is this quaint lady of the skies? We have never heard of her."

"She is the spirit of the star San Shou, the Broom. And now," said he, bowing and putting his hands together, "I too must leave you. Our meeting has been agreeable but, as the Buddhist proverb says: ' meeting is only the beginning of separation.' "

We learned later that the Sao Ch'ing Niang Niang was probably invented as a check on the heavy rains of the sixth moon which may become dangerous for the crops. As this is the season when the harvest is gathered and must be beaten and winnowed, it is important that there be a few sunshiny days. Indeed, the farmers have a proverb: "No matter how much money one has, no man can buy dry weather during the sixth moon." Or else they say: "If the skies continue dark throughout the sixth moon, no one can afford to eat heartily."

Too much dampness will spoil the grain unless it is well aired before being stored away in mat baskets. Hence the expression: "Wheat is not food till it is safely stored in baskets." This is only an Oriental variation of our own adage: "Never count your chickens till they are hatched."

* * *

A special day—the sixth—of this wet month is set aside for "airing the Classics" in the temples. Buddhist and Taoist monks take this occasion to place their sacred books on tables in a sunny courtyard where they are opened, dusted, and interleaved with herbs to protect them against the insects which swarm with the heat. The custom originated in the time of the T'ang dynasty, when a sick Emperor sent a priest to the Western Heaven in search of a religious manual. While crossing a river, the holy book got wet and an old woman passing by advised the messenger to spread it out to dry on the ground. But a breeze sprang up and blew away some of the leaves, and though he sought diligently the priest could never find them, and he sadly collected those that were left and delivered them to the Emperor. All this happened on the sixth of the sixth moon, wherefore it is kept as a day of remembrance. From being a purely religious feast, known as the *hui* or "union," because it takes place in all the temples simultaneously, it spread to civilian life. The Imperial archives were aired on this same day, and even ordinary folk took an opportunity, sanctioned by custom, to air their clothing. Also, women washed their hair, and cat and dog-owners gave their pets a bath. Likewise, the Imperial elephants were ceremoniously taken from their stables near the Shun Chih Mên under the charge of an official, and given a swim in the moat outside the City Wall to the great delight of the populace.

These elephants, a tribute from Indo-China, disappeared after 1900. They had been used in ceremonial processions since the T'ang dynasty. The site of the special palace built under the Mings to house them is now occupied by the Republican Houses of Parliament.

* * *

The Patron of Horses

During the sixth moon several lesser divinities are marked for worship in the Chinese calendar. One is Ma Wang (whose birthday falls on the 23rd), Protecting Spirit of Horses and Patron of Carters, often seen sharing a shrine with Niu Wang, Guardian of Cattle—a shrine sometimes without images but simply with an uncarved stone. In olden times, the Horse God seems to

THE SIXTH MOON, OR "THE LOTUS MOON" 361

have had a triple incarnation. The Ma Tsŭ, or Ancestor of Horses, as he was then called, received sacrifice in spring, the First Breeder in summer, and the Celestial Horse-Breaker, a spirit who sends equine diseases, in winter. The three are combined in the pictures of the Ma Wang when he is shown with three faces. On the *chih ma*, sacred posters (*see* "The Twelfth Moon") printed in his honour, we sometimes find him a king seated on a throne with three attendants round him; sometimes represented as the Celestial Charger who conveys the dragon or the star *Fang*. His cult resembles that of the Kitchen God (*see* "The Twelfth Moon"), and used to be part of the official ritual. The example of sacrifice was set in the Imperial stables with an offering-table prepared in his honour between eight and ten in the morning, and followed in humble farmyards where each peasant, owner of a lean Rosinante that drew the plough, burned a paper figure of a horse in the Ma Wang's honour. None neglected him throughout the northern provinces where horses are so much used, especially carters, *mafoos*, and stable-owners who depended on his favour.

* * *

The Fire God

The Fire God, whose birthday falls on the 22nd, received, on the contrary, his greatest reverence in the South. Gray says that for a whole month rejoicings were held in honour of the God of Fire in Kuangtung province. Speaking of Canton—in the old days, of course—he describes the streets of that once picturesque city "illuminated not by lanterns, but by crystal chandeliers suspended at frequent intervals from beams extended from house to house. The lights are protected from rain by sheets of canvas stretched across the streets which are very narrow. . . During this month, groups of figures in wax-work, attired in silk dresses, are carried in procession. They represent certain episodes in the ancient history of the empire. The figures are very well executed and would do credit to Madame Tussaud's well-known exhibition. In the principal streets, altars are erected to the Fire God, and Taoist priests are engaged all night long in chanting prayers. The idols of this deity are occasionally borne in procession through the city, accompanied by numbers of boys on horseback and ladies in triumphal chairs."

The Fire God is also worshipped "during and after a conflagration, when his 'red crow' is supposed to be flying about, ready to ignite unburned buildings." His temple is specially visited by shop-owners. The Chinese consider the sun, the "Fiery Wheel," to be the source of fire and, formerly, on the day of the spring equinox, new fire was obtained from the sun by means of a concave metallic mirror, after the old fire had been put out three days previously. Or it might be obtained by rubbing two pieces of willow-wood together (*see* "The Third Moon," the *Han Shih*).

Exactly which fire god is meant is not clear, for there is a Ministry of Fire, as of Water, with a President and several subordinate Ministers, (most of whom are Star Gods) associated, like so many other Chinese deities, with officials under the tyrant Chou Wang (XIIth century B.C.)

THE SIXTH MOON, OR "THE LOTUS MOON" 363

Most probably Gray refers to Huo Shêng, "Stellar Sovereign of the Five Virtues," who in his own person seems to arrogate worship and sacrifice for the whole group. Originally a Taoist priest who changed himself into a giant with three heads and six arms, this worthy now appears dressed in Imperial robes, or else with an animal body and human face, the colour of the "ripe fruit of the jujube tree." He has a third eye in the middle of his forehead, red hair, a red beard and a red cloak, a peppery disposition, and a fiery charger on which he travels from place to place, punishing people for their evil deeds. No friend or neighbour cares to give hospitality to persons or property saved from the flames sent as a visitation by the Fire God, "lest they call down his wrath on their own heads." Sufferers from fire promptly go to his temple and worship, begging him to leave their burning homes quickly. When he takes his departure, they thank him for the punishment, and once a year, on his festival (the 15th of the fourth moon), people worship him as a preventive measure, parading his image through the streets. Many associate this fearsome figure with the Red Emperor of the South, Lord of the holy mountain Hêng Shan (*see* "The Tenth Moon"), who taught the people the art of forging metals and, after reigning two hundred years, became an Immortal. His chief characteristic is extreme propriety. To outwit this puritanical god, people paste indecent pictures on their walls, especially in the kitchen. Shocked and scandalised, he avoids the vicinity of such a house, and thus the inhabitants are safe from fire.

In his temples, this goody-goody personage sits with an unctuous expression surrounded by four acolytes corresponding to four stars. One is Hui Lu, a celebrated magician of mythical times who holds a fire-pigeon that carries fire. Another has a fire-wheel for the same purpose. A third has a pig on his head (corresponding to the name of the star which he represents) and holds a gourd from which flames issue, and the fourth grasps a serpent with a fiery tongue.

The Patrons of Artillery

Vaguely akin to these Fiery Stars are the Wu Hu Shên, or "Five Tiger Spirits," patrons of artillery, worshipped on the 25th. Their cult was connected with the inventor of firearms (Huo P'ao Ta Chiang Chün), and it is they, with their powerful cannons, who blasted the Nankou Pass near Peking through the mountains. Who knows if, in the battles recently fought in this strong defile, some humble artilleryman has remembered his long-neglected patrons and set alight an incense-stick for them among the rocks?

* * *

Lu Pan— Patron of Carpenters

Such warlike figures contrast markedly with a peaceful personage likewise feasted in the sixth moon on varying dates in various places, Lu Pan, the Chinese Archimedes, whose legends seem to veil the authentic biography of a great inventor. In life, Lu Pan was a youth with the family name of Pan who lived in the ancient kingdom of Lu in modern Shantung. Born in 606 B.C., a contemporary of Confucius, he has gathered many traditions about his name. One stamps him as a skilled carpenter and is the basis of his elevation to the godship of his craft. At the age of forty he left the work-bench for a hermit's retreat and learned the secrets of sorcery, and, henceforth, we find him performing marvellous feats. He built a palace for the Hsi Wang Mu. He repaired the Pillars of Heaven when they threatened to collapse. He made a miraculous wooden carriage for his mother with a wooden coachman and a clever mechanism that took it up hill and down dale. The Chinese might doubtless claim this as the first taxi-cab. He fashioned a wooden magpie that flew away as if alive and remained three days in the air, and a wooden vulture on whose outspread wings his father travelled to the Kingdom of Wu (Su Chou, in modern Kiangsu), as he might to-day have done on an aeroplane. There the local people, believing the old man to be a devil—and who can blame them?—murdered him. In revenge, Lu Pan constructed a wooden genie whose outstretched hand pointed towards the land of Wu, and cursed it with drought for three years. It was only after the stricken people placated Lu Pan with numerous presents that he chopped the avenging hand away, whereupon a copious rainfall came to Wu. Lu Pan finally returned to the Li Shan mountains, near Tsi Nan Fu, where he completed his mastery of magic, and ultimately vanished skywards, leaving behind him his axe and his saw.

Whenever a new house is built, a feast is spread for Lu Pan. Spirit-money is burned in his honour, and incense and fire-crackers let off, all on a "fortunate day." As it is always dangerous to disturb the earth in China because the spirits may wreak vengeance, workmen usually tie bunches of leaves to the scaffolding-poles they erect, hoping that evil spirits will take the building for a grove of trees and pass it by. Sometimes a basket-work sieve is also hung up, and a mirror fastened to the centre, because good influences can easily pass through the sieve, and the mirror is supposed to have power to transform evil influences into good fortune. Often Taoist priests are hired to sprinkle the ground with holy water, to bless the ridge-beam, carefully selected without knots or cracks, and to chant prayers to Lu Pan all the time the beam is being raised into position.

The carpenters of Canton are famous for their splendid processions in honour of Lu Pan, who is the general patron of the proletariat, though by no means the only Chinese Craft God. In Peking, on his local birthday, a fair is held at the Pai T'a Ssŭ, near the P'ing Tsê Mên. During the Ming dynasty, this famous pagoda cracked and threatened to collapse. Legend says that at this critical moment somebody dressed like a workman walked round and round shouting the words: "Mend the Big Thing! Mend the Big Thing!" A few days later it was observed that the crack had been filled up and solidity restored. "That must be the work of Lu Pan himself," said the people, "no one but he could have done it so well." The episode is commemorated in a street rhyme:

> "At the temple of Pai T'a Ssŭ
> There is a pagoda shining white.

THE SIXTH MOON, OR "THE LOTUS MOON"

On that pagoda there are not bricks, but tiles.
On its pedestal a gaping crack appeared
And Lu Pan himself came down to repair it."

A practical god is a precious thing anywhere—but especially in practical China.

CHAPTER XI.

THE SEVENTH MOON, OR "MOON OF HUNGRY GHOSTS."

IN the seventh moon falls the period of the *Li Ch'iu,* or "beginning of autumn." It is the loveliest season in North China. New vigour and strength come into the air. All the heavy storm-clouds of the rainy season have rolled away. A brilliant sunshine, still hot at noon, bathes the world in golden glory.

Fruits are ripening in the orchards round Peking, aromatic wild dates, rosy crab-apples, and gilded persimmons. The sweet-scented cassia (*kuei hua*) sheds its perfume over temple courtyards. In the fields, the *kao liang,* ready to be harvested, stands in serried ranks waving brown tassels high above a man's head. The orange maize, set out to dry on farm-house roofs, makes brilliant splashes of colour in villages, showing as little brown islands in a sea of crops. Sunflowers stand like sentinels in every garden, turning their solemn faces to the lord of light.

Among the rocks of the hill-sides, brown crickets chirp cheerfully. The commoner varieties are left in peace to sing out their tiny lives. But there are certain kinds, like the much prized "Little Golden Bells" (found in the district of the

Hsi Ling, Western Imperial Tombs), which are hunted and caged.

At almost every fair during this moon, cricket-peddlers sell such singing insects in miniature cells made of bamboo or rushes, and passers-by pause, listen to the orchestra of wee musicians, and choose the particular singer they prefer. It is considered very *chic* for a scholar to keep one of the little creatures, whose cry is like the whirr of a small bronze gong, or the dry rapping of Chinese castanets, either in the folds of his gown or on his study table. "Not only do we like the cricket's song," our old teacher explains patiently, because rude foreigners find it so difficult to understand the refinement of Oriental pleasures, "but it reminds us of many things. The characteristic note of autumn, it brings a suggestion of parting that comes with the death of summer, and a hint of the colour-changings and leaf-whirrings which typify the impermanency of earthly desires. And our women-folk, hearing the song of this autumn singer, know that it is time to repair the winter garments of the household before the coming of the cold weather."

Subtly connected with this idea of the approaching sewing season is the seventh of the seventh moon, a day of omens when girls lay a needle on a bowl of pure water set out at daybreak in the courtyard. The shadow cast by the needle at the bottom of the bowl tells whether or not she will be skilful at her embroidery. If it takes the form of a leaf or a flower, that is a sign of proficiency. But if it looks like a stick, thin at one end and thick at the other, she will be incompetent, and her companions make such fun of her that she is obliged to hide to escape their teasing.

Various other devices are resorted to by little maidens eager to know whether or not they will be skilful needlewomen. On the evening of the sixth, some catch a spider and put it in a box set among offerings of fruit and flowers, cups of tea and wine, making ceremonial bows meanwhile. Then, next morning, after her hair is carefully dressed, each girl opens her box to see whether or not her spider has spun a web during the night. If the web takes the form of a cloud, a blossom, or a lucky character, this is interpreted to mean success.

Even the Court ladies, in the days of the Emperor Ming Huang (A.D. 713–756), used to catch spiders and put them in lacquer boxes for purposes of divination. When the boxes were opened, if the spiders had spun thick webs, the omens were good but, if they had remained idle, the omens were bad.

* * *

The Cowherd and the Weaver

These quaint rites were inspired by the Festival of the Weaving Lady and her lover, the Cowherd, who meet once a year on the seventh night of the seventh moon. It is easy to trace the link connecting the spider with this Chinese Arachne. Indeed, she is often called "the Spider Princess" in Japan, where her myth has been borrowed, and most beautifully borrowed (see *The Romance of the Milky Way and Other Stories*, by Lafcadio Hearn).

Herself a weaver, therefore patroness of the loom and the needle, herself once a mortal woman, therefore friend of women, herself a tender lover, therefore compassionate towards maidens seeking love, this Star Goddess is one of the prettiest and most poetic figures in Chinese mythology, and of her appealing story there are many versions.[1]

In all of these, the lovers are identified with two stars, one on either side of the Milky Way, a luminous river called "the Silver Stream of Heaven" by the Chinese. The Herdsman is supposed to be Aquila, and Lyra his beloved Weaving Lady (one of the seven daughters of the Kitchen God, all of whom are collectively honoured on her festival.)[2]

Once upon a time, the Weaving Lady, who passed her time weaving garments for the gods, left her loom and with her sisters descended to earth to bathe in a stream. Near by, a poor Cowherd watched his cow at pasture. She was a magical cow, with hair like spun gold, and suddenly she began to speak in a human voice saying: "Master! yonder are seven maidens, daughters of Heaven. The seventh is beautiful and wise beyond all measure. She spins the cloud-silk for the gods and presides over the weaving of earthly maidens. If you go and take away her clothes

---

[1] The tale of the Cowherd and the Weaver is a popular theme for plays. Quite recently the well known actor, Mei Lan-fang, gave a new and charming drama founded on the old legend.

[2] The inclusion of these two star-spirits into the Chinese pantheon appears to be the logical sequence of domestic events that occurred, of old, at the approach of winter:—the return of cattle to the barns from their summer grazing grounds, and the preparation of warm clothing by the women. The date on which the Cowherd and the Weaving Lady, respectively patron of cattle and patroness of needlework, are honoured has been changed in the course of the ages (see "The Eleventh Moon").

while she bathes, you may become her husband and gain immortality."

The Herd-boy hastened to find the girls and hide the red robe of the loveliest among them. Thus, when her sisters put on their fair garments again to fly back to heaven, one found herself earth-bound, and willingly, since in the youth she recognised her long desired lover and lord.

The cow agreed to obtain permission from the gods for their marriage. They became man and wife, and lived together happily for three years, while two star-children were born to them. But the loom was silent, and the sound of the shuttle was no more heard, till the gods, growing angry at the idleness of the Weaving Lady, interrupted the course of true love by ordering her back to work.

The Herd-boy grieved so bitterly at the loss of his wife that the devoted cow said to him one day: "Do not be sad, I will gladly lay down my life for you. When I am dead, wrap yourself in my skin. Then you can follow your beloved to Heaven." So it came to pass. But when the Herd-boy reached the sky, his Celestial Mother-in-law, fearing that further tender dalliance would mean idleness on her daughter's part again, traced a line across the Heavens which became the Celestial River,[3] and the two lovers, transformed into stars,

---

[3] Chinese legend says that the Heavenly River, or Milky Way, is the source of the Yellow River. A mighty rebel in the days of the mythical Emperors butted his head against the north-western pillar of the sky and broke it down. This allowed the River of Heaven to enter the Earth. Equally, in ancient days, there was a mystical connection in the popular mind between the Heavenly River and the Yellow River, since along the course of the latter the gradual expansion of the early Chinese settlers took place.

found themselves separated by the stream, doomed to live apart forever.

Greatly they grieved, so greatly that their pitiful case was submitted to the Jade Emperor and he, in his mercy, decreed that they be permitted to meet once a year on the seventh night of the seventh moon. On that night, provided the skies be clear, the crows and the magpies make a bridge with their wings and bodies—a bridge across which the Weaving Lady goes to meet her husband. You will not see a single magpie in the trees on that day, after the hour of noon, because all have gone up to the skies to help the lovers. When this festival actually fell in the eleventh moon, the "bridge of birds" was, perhaps, an allusion to their yearly migration.

But if there be rain on the seventh night of the seventh moon, the River of Heaven rises and the bridge cannot be formed. Such rain is called the "Rain of Tears"—the tears that the true lovers shed at their parting. So, after all, the Cowherd and the Weaver may not always meet even on the day of their festival. Indeed, it may happen that, by reason of bad weather, they may not meet for three or four years at a time. But their love remains eternally young and eternally patient, and they continue to fulfil their duties each day without fault, happy in the hope that the seventh night of the next seventh moon may be propitious and bring them together.

Since the Weaving Lady shares with the Mao Ku Ku the title of Patroness of Women and Patroness of Needlework, it is but natural that her festival is primarily a women's festival. In fact, only women take part in the services appropriate to the day, and no males are present, except little boys still under their mother's wing. The ladies, formerly, went in for needlework contests with prizes for those who excelled in embroidery, or amused themselves, in gay rivalry, with needle-threading competitions. Humbler maidens still try to thread needles either by moonlight or beside a glowing incense-stick. In Kiangsu province, they thread a seven-eyed needle with red silk, almost without the aid of a light—under a table, for example. Such as succeed are sure to be good needlewomen.

In olden days, a most elaborate offering-table was spread for the Star Goddess with clay figures of men, animals, and fruit. Nowadays, we still find water-melons, cakes, and feminine toilet articles, imitated in paper, prepared in her honour. These gifts are tastefully laid out on tables in the halls of houses which, as the festival is celebrated by night, are brilliantly illuminated with lanterns. After a series of bows, the flowers, powder-puffs, rouge-pots, etc., are thrown up on the roofs, so the goddess may use them, and the cakes are divided among the children.

Gray, describing this festival as it was held in Canton forty or more years ago, says: "Women, especially unmarried women, embroider shoes and silk garments as offerings for this anniversary. Flowers, sweetmeats, and preserved fruits, are also presented to the Weaving Lady, and amongst these are placed basins containing shoots of the young rice-plant; they are so arranged as to appear as if they were growing, and in the centre of each cluster a little lamp is placed whose spark of light reminds one of a glow-worm or a firefly. Miniature bridges, formed of flower garlands, grains of boiled rice, and almonds cemented by gum, connect the various offering-tables." Some ladies make a habit of taking a few stitches just after making their gifts, to consolidate, as it were, the benediction which they have asked on their work. Others lay a threaded needle among the flowers.

"At midnight, the young girls with their female attendants go out to draw water, most Chinese houses being provided with one or more wells. The water is poured into large earthenware vases, arranged in order around the mouth of the well. After the Goddess and her Sisters have been invoked to give the water medicinal properties, the jars are hermetically sealed and put in a place of safety, to be opened only when a member of the family requires a draught of the disease-dispelling beverage. . . . The festival ends with the burning of richly embroidered garments, which thus ascend in smoke for the use of the Heavenly Sisters."

The Weaving Lady is invoked also for happiness in married life and for the gift of sons. She shows special kindness to orphaned girls, protecting them, and sympathising with their loneliness. They, in their turn, steal out on the seventh evening, hide behind a vine, and listen to the lamentations of the lovers of the skies. Peeping between the leaves, in the silence of a transparent night, the heavens seem to these little maids very near and warm and human, and they dream of a love "unchanging, immortal, forever yearning and forever young, forever left unsatisfied by the paternal wisdom of the gods."

*   *   *

All Soul's Day, Second Feast of the Dead

Many old things in China are doomed to die, are dying fast. The gorgeous ceremonies of ancient times in honour of the Star Lovers are already relegated to the limbo of forgotten things. Little attention is paid nowadays to the pretty festival, save in old-fashioned homes.

The seventh moon feast that does *not* die, however, is the Feast of the Dead, All Souls' Day, or, more properly speaking, "the Festival of Hungry Ghosts," intended especially for the unhappy spirits who no longer have human descendants to care for them. Such spirits suffer from hunger and thirst, and must so suffer uninterruptedly, unless someone attends to their needs. Otherwise, they meddle maliciously with human affairs. Policy, therefore, no less than charity, requires the faithful to satisfy their wants. "Sometimes, the spirit-tablets of childless ghosts are collected in special temples where they are placed in a room apart, and a caretaker is hired to burn incense before them. Such a shrine of the neglected is deeply pathetic. The tablets—perhaps several hundreds of them—standing side by side, old, discoloured, crumbling, and covered with dust—make us feel, in their silence, as if spirits actually hovered round them, for these objects signify their viewless presence."

The feast begins on the 15th of the seventh moon, and lasts until the 30th. Throughout this period, families visit and repair their graves, as they did at the Ch'ing Ming, but on odd calendar days only. Offerings of food, money, and paper reproductions of useful articles like clothing, conveyances, and furniture, are made to their personal dead, and to departed spirits in general. These offerings being burned, their ethereal counterparts reach the needy ghosts.

The roots of this festival, entwined with primitive spirit-worship, reach down to remote antiquity. To-day, it has become identified with the popular Buddhist festival of the *Yü lan p'ên*, when for one whole month the souls are released from hell and permitted to enjoy the feasts prepared for them. On the last night of the sixth moon, the "mouth of the pit" opens, to close again on the last night of the seventh—in much the same way as the official seals were opened and closed under the Empire at New Year.

The primitive Chinese had no notion of hell at all (*see* "The Second Moon"), though they did have a firm belief in the life beyond the grave, imagining spirits attached to and wandering in this world. This idea, endorsed by Confucius and other philosophers, exercised a supreme influence on Ancestor Worship. When the Buddhists came, they were faced with the problem of how to graft the doctrine that men must work out their salvation by passing through a series of Infernos upon the old and vague belief. As usual, they compromised. "Accept our hells," said their teachers to the people, "and our priests will show you how to escape them." Special liturgies, composed to mitigate torture and stress the sentiment of compassion for the dead in general, and forlorn souls in particular, ingeniously turned the antique festival into an instrument for promoting the influence of Buddhism in China.

At the *Yü lan p'ên,* or *Yü lan hui,* the "Magnolia Festival," Buddhist priests everywhere hold masses for the dead. In temples, services are said all day and all night for lonely souls, or such spirits as have suffered injustice, and especially for beggars who have died within the twelvemonth.[4] Thanks to these prayers, and the accompanying gifts of food and raiment, their destitute spirits, soured by calamity and misfortune, may attain a happier condition, and refrain from haunting the living. Only the souls of condemned criminals can not, or will not, be satisfied but remain malevolent and constitute no small danger to the community. Therefore, to avert their evil influences, propitiatory incantations are said by the monks who, in return, receive food from the superstitious.

The services, called "Bringing Help to the Needy Spirits," are very impressive. Priests and acolytes assemble in vestments of great ceremony, the chief officiant facing the altar—a field of flames with all the candles and votive lights, and the gilded Buddha looking down from above, so wise and quiet. Assistants seat themselves to right and left, so as to form two ranks facing each other. The ceremony begins with a Buddhist hymn chanted to the accompaniment of two bells,

---

[4] These masses remind us not a little of the Catholic masses for the dead celebrated on the *Jour des Trépassés*—our November 2nd—to assuage the souls in Purgatory. Between the two, there is a distinct connection—in sentiment at least.

SEVENTH MOON, OR "MOON OF HUNGRY GHOSTS" 379

a large deep-sounding bell and a little bell with a sweet small voice which a monk holds in his hand, while a slender youth beats rythmically upon a queer wooden drum, shaped like the head of a fish and lacquered red. The effect of the chant, sung in unison, is admirably suited to a service of incantation, and each strophe is accompanied by special gestures of the hands and fingers of the monks.

Weird words, talismanic words, summon from the Ten Directions of Space the Hungry Ghosts freed by the "Breaking of the Gates of Hell." Next are repeated the verses announcing the "Bestowal of Food-offerings," and the prayers by which these are transformed into heavenly nectar and ambrosia.

"We, devoutly presenting these vessels of pure food, do offer the same to all, without exception, of the Spirits dwelling in the Ten Directions of Space, . . . and in every part of the Earth, not excepting the smallest atom of dust within this temple. We offer to the spirits of those newly dead, and to those long since passed away, and likewise to the Lord-Spirits of mountains and rivers, and soil and waste places. To all, out of our pity and compassion, we desire to give nourishment. We wish that each and every one of you may enjoy this, our food gift. And, moreover, we pray that all dwellers within the Zones of Formlessness, and every being still tortured by desire, may be enabled to find contentment, . . . may reach the higher knowledge, and, at last, be freed from every pain and attain re-birth in the Zone Celestial where is everlasting bliss. . . .

And we beseech you to watch over us and guard us by night and day, never to harm us, and even now to help us obtain our desire in bestowing this food upon you. Let the merit of this action be extended to all suffering souls, and let the power of this merit spread the truth through the Dark World, and assist all beings there towards the Supreme Enlightenment. It is our desire that you may quickly become Buddhas, and thus be fully and finally delivered out of the Hell of Hungry Ghosts."

Rich families will invite priests to say masses in their own homes—usually in the evening, for the spirits are twilight creatures,—and square lanterns, called "lanterns of the way," are hung out to light the dead back to their old dwelling-places. People who are far from their ancestral tombs—too far to make a personal visit—prepare paper bags filled with mock-money. On each bag, a strip of red paper with the name and date of death of the person for whom it is intended is written. These are laid on an improvised altar and, while the priests chant *Sutras,* members of the family in turn make deep *k'o t'ous* to the spirits of their forefathers—even the little children who can not understand the meaning of their filial obeisance. They find it so hard to be serious when, after the mass is over, the bags are taken into the courtyard and set alight. "Oh, the pretty bonfire!" a small boy exclaims. "Hush! Little Dragon," whispers his mother, drawing him aside into the shadows to tell him the reason for this beautiful and touching custom. "To-day," she says softly, "all the dead leave their tombs and come back to us. The sky is thronged with an invisible procession."

"Why do they come back, mother?" he murmurs.

"Because, my treasure, they love us and expect us to love and serve them. Therefore, irreverence is very wrong and cruel." Unwise too, since naughty spirits are also abroad these days, ready to harm little boys and girls who, for this reason, are forbidden to go out after nightfall during this festival.

"Little Dragon," thoroughly sobered now, bobs his head in a jerky *k'o t'ou*. It is his attempt at an apology to spirits, bad and good. Thus, very tenderly, children in China are given their first lesson in politeness to the dead.

In the southern provinces, and notably at Amoy, tables with food-offerings are placed in every doorway. Incense is burned, so that the perfume may make the festal meats more appetising, and candles lit to guide the ghostly guests to the banquet where they are supposed to eat, drink, and make merry, as if they were alive. The effect, in narrow alleyways filled with suffocating smoke and heat, is very weird.[5] Now and again, we hear a whispered prayer, as a mother lights an extra taper for her dead child: "Candle, little

---

[5] From the great quantities of food left over comes the popular custom of giving dinner parties in the seventh moon, parties at which much wine-drinking and hilarious gaiety are not considered bad form, with exciting games of "morra," the finger game, to enliven the proceedings. This is the *micare digitis* of the ancient Romans, and still exists in Italy. In China, two persons play, both opening certain fingers of the right hand and naming the number they are supposed to have spread out. If both are wrong in their guesses, the game recommences. If one is right, the loser must drink a cup of wine. As the game is played very fast, it requires good and quick psychology.

candle, burn brightly for my eldest born, bring him close to me to-night!" Her faith and love, she believes, can draw him to the make-shift hearth, as, alive, they drew him into her arms.

Presently, a group of priests in bright embroidered robes comes slowly down the street chanting the invocation: "*O mi to Fo*"—"Buddha Amida," uttered as a prayer, and especially as a prayer for the dead. They stop at each table, repeating formulas that magically increase the viands so that they may nourish an infinite number of spirits. The procession ends at a local temple. Here a scaffolding is erected high above the usual altar, and decorated with flags, coloured streamers, and lanterns. The subscription-gifts are lifted up on to this platform by men detailed to do so. Also, whatever private persons may bring, such as cakes or meats arranged in the form of pyramids.

According to the rites, the abbot of the temple places a doll on top of the scaffold. The effigy represents a deity, either Yama, the Chief God of Hell (*see* "The Second Moon"), or an incarnation of Kuan Yin who has, as we have seen, miraculous power to deliver souls from Hades, and is able to keep order among the Hungry Spirits.

In addition to food, the sacrifices include paper imitations of houses, gardens, furniture, boots, clothes, money, also paper boats for crossing the canals and lakes which may exist between Hell and the place of festival; means of land transport too, horses, sedan-chairs, etc., according to the people's imagination. Now the abbot raises his hand, giving the order to set this high pyre alight.

SEVENTH MOON, OR "MOON OF HUNGRY GHOSTS" 383

Everything must burn, and, burning, miraculously multiply, in order to satisfy the millions of hungry and naked spirits. Even the half-consumed incense-sticks are thrown into the flames to serve as carrying-poles for the ghosts who, after eating their fill, are thus able to take away with them what food is left. The ceremony ends in a scene of fire and pillage. The paper images blaze up, and flames attack the figure of the giant god on his pedestal. Their red tongues first lick his lower limbs, then rise and gnaw at his vitals, till he topples over and crumbles into a heap of smoke and ashes.

Before the assembly scatters, strips of red paper, with the names of subscribers and contributors to the festival, are also burned, so that the ghosts may know who have remembered them. Then follows a struggle for the food-offerings, the people clutching greedily, pushing, jostling, surging to and fro, in a wild stampede where children are sometimes trampled under foot. After the uncanny solemnity of the service, this human outburst, called "pillaging the scaffold of homeless ghosts," is startling.

We find, as usual, local variations in the celebrations of this Feast of the Dead. Sometimes, children collect money to help the souls of their little comrades, and make them miniature donations. Sometimes—in Kiangsu—lights called the "Lamps of Hades" are lighted in the homes, and the women perform a very old rite, called "taking off the skirt," beside these flickering tapers, to facilitate the birth of sons. Sometimes, theatricals are organised for the pleasure of dead souls.

More general is the custom of burning paper boats, called "boats of the law," supposed to convey the saintly souls of Buddhist and Taoist monks across the Heavenly River on errands of salvation to those suffering in hell.[6] Such boats may be twenty or thirty feet long with a whole population of divinities on board: the Prince of Demons (*Kuei Wang*), easily recognisable by his two tufts of hair, his tiger-skin and trident; the "Horse-Headed Devil" and the "Ox-Headed Devil", guarding the cabin in which sit the Ten Kings of Hell (*see* "The Second Moon"); the Ti Tsang Wang, and the quaint figure of the "Fisherwoman," whose presence in this company is due to a pun on the words *yü lan*, which can mean fish-basket as well as magnolia.

These paper boats, with their paper crews, are carried in picturesque procession either to a temple or to an open space on the banks of a lake, a river, or even a stagnant canal. Here, at rude altars rigged-up on trestles, masses are chanted, while lanterns curtsey to their reflections in the water, and fire-balloons float skywards. The burning of the boats concludes the open-air service. In a surprisingly short time every trace of the Gods of Hell and their ghostly argosy has disappeared, leaving only a ring of ashes. Then

---

[6] A replica of a Yangtze River steamer was recently burned at a memorial mass for Chinese soldiers killed in the civil war. An armoured train in paper likewise ascended in smoke, dedicated to the memory of the defenders of the Nankow Pass. At the same ceremony cart-loads of imitation silver dollars and notes payable at the "Heavenly Exchange Bank" (signed with a foreign signature) were offered to dead warriors whose pay was in arrears.

the crowd disperses and the priests depart in colourful procession, their acolytes trudging behind with bells and praying-stands.

On this same evening, the fifteenth of the seventh moon, children parade the streets with lanterns in the shape of lotus flowers, or real lotus-leaves with candles fixed in them, singing couplets intended to express the impermanency of all things: "Lotus-flower lamps! O lotus-flower lamps! to-day we light your tapers, to-morrow you will be gone!"

Similar lanterns, resembling lotus-buds, or little boats, or even squares of plain unpainted wood, are launched upon the waters for those who have been drowned. Pious fancy imagines that the *kuei* (spirits), forever tossing and shivering in the waves and tides, seize these votive lamps and, with them, the possibility of re-birth in human form, or else by their light are guided to the dole cast for them upon the waters.

Launching the lanterns, so unexpectedly poetic in materialistic China, is a widespread custom. It is common in all the coast provinces, among all river-dwellers. It extends from Canton to Peking, where the Boatmen's Guild, on the T'ung Chou Canal and on the Pei Hai, place thousands of these little lamps on the water, while large barges with shrines attended by Buddhist priests float to and fro. But nowhere is the touching ceremony more lovely than in the Wind Box Gorge above Ichang, on the Yangtze; nowhere the belief, so much older than Buddhism, stronger that the waters are thick with souls. At dusk, the peasants assemble on the shore and launch their little fleet of twinkling lights. Lazily, the frail glowing shapes drift out into the current. The impulse of the gentle breeze and flowing stream scatters them widely apart, then drifts them together again in the narrow gorge beyond. Each, with its quiver of flame, seems a soul afraid, trembling on a blind current which bears it into outer darkness. Soon the lights themselves burn out; "then the poor frames, and all that is left of their once fair colours must melt forever into the colourless void." Because it is not good to meddle with the lamps of the dead, not a single boat ventures upon the river, and fishermen catch no fish on that day. After a whispered farewell to the frail barques, launched, it may be, in loving memory of their own kith and kin, they return home as soon as the last floating candle is extinguished, hoping and believing it has been caught by some drowned soul desirous of a return to earth.

* * *

Ti Tsang   The Festival of Hungry Ghosts ends on the last day of the seventh moon, when the gates of Hell close once more on the spirits who return to torment for another year. The thirtieth is also the birthday of the Ti Tsang Pusa, whom foreigners think of as the King of Hell, confusing him with the Yen Lo Wang, or Yama, that very antique Aryan divinity (*see* "The Second Moon"), whose cult Buddhism transplanted on Chinese soil. The error arises through one of Ti Tsang's popular titles, "Lord of the Underworld, or "King of the Dark City." But this does not mean that he is the pitiless judge of the Infernal Regions, or holds the position of the merciless sovereign of the damned. How different, in fact, is his true rôle! Ti Tsang is really the "Vanquisher of Hell," the Redeemer who descends into hell for the purpose of releasing those in pain.

In Japan his gentle and gracious figure is known as Jizo. At many a wayside shrine we find his statue decorated with cotton bibs, offerings of mothers who invoke him as protector and playmate of dead children (*see* Lafcadio Hearn's *Glimpses of Unfamiliar Japan,* for a beautiful description of the Jizo cult).

In China, he is often represented with eighteen hands, the all-embracing arms of mercy, or else with the gifts of Buddha: a Staff—to open the gates of Hell, a Jewel—to shed the light of salvation on souls in anguish.

One of the most popular incarnations of the Ti Tsang Pusa was a certain monk, named Mu Lien. Like Buddha, he was the son of a king who ruled over a kingdom, long since extinct, in south-eastern Korea. Like Buddha, too, he forsook his luxurious home to become a priest. In the VIIIth century A.D. he travelled to China, where he took up his abode in the Chiu Hua mountains in Anhui province. Here he lived for more than half a century in a white stone hut built on a plot of land as large as his coat could cover, which he begged from the local magistrate.

There is a tradition that the poet Li T'ai-po—who named the Chiu Hua: "Nine Flower Peaks," because, when he first caught sight of them from his boat on the Yangtze, they reminded him of the up-turned petals of the lotus—visited the Sage. Whether true or not, the idea of the recluse wandering over the hills, deep in intellectual converse with the witty, but none too saintly, poet is interesting. The fame of the saint certainly brought many visitors to his distant retreat, if not actually Li T'ai-po. Some became his disciples. One, a certain Sheng Yü, built for him a beautiful monastic dwelling, and all sincerely mourned his loss at the great age of ninety-nine—even the spirits of the streams and peaks joining in the lamentations at the hour of his death, when a crashing of rocks and a sound of moaning was heard in the hills.

To this day Chiu Hua remains one of the most popular places of pilgrimage in China. From September to November, the pleasantest time for travel in the Yangtze valley, over one hundred thousand of the faithful visit the mountain, chanting their prayer: "Saviour of the Unseen World, save us from its sins and sufferings! Show thy love and pity towards our beloved dead." Many old women, wearing hair-pins shaped like the golden crozier Buddha gave Ti Tsang to open the gates of hell, tread the stone-flagged pilgrim-path which winds up the mountain side. Many a youth, wearing a tinselled head-band—sign that he represents a mother too aged and infirm to travel—treads the Holy Way, hoping that when she dies, and this same fillet is bound around her brows, the Compassionate Redeemer will recognise it and believe she came herself to worship him. Is the kindly god deceived? If not, surely he forgives a deception rooted in love—for the sake of his own mother who also sinned.

That mothers, often so nearly divine, are human too Mu Lien himself realised, when once, in contemplation, he saw the soul of his parent in hell—a prey to hunger. Filial piety bade him follow her with a mess of pottage, but the famished ghosts took the food from him, so that he was unable to help. Again he descended to the Underworld and looked for her, this time without success. She had been re-born as a dog. After a long search, he recognised her in this miserable re-incarnation, obtained possession of the animal, and tended it devotedly.

Since that time, Mu Lien, as an embodiment of the Ti Tsang Pusa, has continued his interest in all suffering souls. His vow before the throne of Buddha confirms his purpose to labour until he has finally brought every living being to the haven of peace and happiness. "The hosts of evil spirits can not daunt him, and from sorrow, danger, and pain, he will not shrink. He will take on himself the burden of the woes of all who trust him, and he will never regard his work as finished as long as a single soul languishes in sorrow or pain."

The Buddhist idea of hell is not eternal damnation. Therefore, Ti Tsang's task is not impossible of accomplishment. All must at last reach salvation. But while the humbler pilgrims climb to his shrine in the literal belief that Ti Tsang is a powerful deity and protector of suffering humanity, especially in the dark ways of death, that his staff literally opens the gates of hell and his jewel illumines the dark world, the wiser pilgrims know "that it is not in images of clay or bronze that they will find the real Ti Tsang, not in a garnished temple or curtained shrine, but in the secret places of their own hearts. Each human being is himself the bearer of the staff that will break open the gates of hell, and the possessor of a jewel that will illumine the darkness through which his own soul is groping. So long as he is sunk in sensuous delusion, or in sin, or in selfishness, or is led astray by the false glare of wealth or worldly honours, he will be encompassed by all the dangers that beset a blind man, wandering guideless in a strange land; but deep in his inmost nature (*ti*) there is stored a treasure (*tsang*) which, if he will only clear away the dust and rubbish under which it lies concealed, will assuredly prove to be everlasting and incorruptible. Similarly, the only hell that man need fear is the hell that he creates for himself, out of his own evil thoughts and deeds. Purity of thought is Ti Tsang's jewel, strength of character is his staff, and these are the weapons against which the gates of hell shall not prevail."

CHAPTER XII.

THE EIGHTH MOON, OR "THE HARVEST MOON."

HE eighth moon is the Harvest Moon *par excellence*. Though Chinese farmers gather several crops a year, this is the time when they cut the giant *kao liang*, green as jade and red as burgundy, when they carry home the golden maize. The simple rhythm of the flail resounds from every threshing-floor in North China, and the creak of the stone mill turned by blindfolded donkeys is heard in every village.

Their heavy labours over, the peasants again worship their earth-gods with the same old-fashioned rites used in the second moon.[1] How antique such harvest thanksgivings are, the special hymn quoted in the Book of Odes (VIIIth century B.C.) proves: "With offerings of millet and of sheep, we sacrifice to the Gods of the Soil and of the Cardinal Directions. Our fields have been reaped and the people rejoice. Let us play upon the lute and upon the lyre, let us beat upon the drums to invoke the Father of Husbandry (probably Shên Nung is meant). Let us humbly beg of Him the blessing of the soft, warm rain

---

[1] The Emperor also again made a solemn sacrifice at the Shê Chi T'an. *See*: "Imperial Ceremonies."

that our grain shall multiply during the coming year, and the land be blessed with abundance." The second half of the petition refers to the new crop of winter-wheat, sown late in autumn to be harvested in the next fifth or sixth moon.

Village theatricals Present-day Chinese farmers no longer play upon the lute and the lyre in honour of their divinities. We may, however, trace a remembrance of this custom in the theatricals held in almost every village. Although a precious amusement to the local communities, their real motive is to entertain the gods whose reactions in China are always very human. For this reason, the stage of the open-air theatre, generally attached to the temple although outside the compound, faces the main shrine so that the "Invisible Ones" may have a good view of the play over the heads of the audience.

Attempts have been made to compare these temple-dramas with the old miracle-plays performed in the porches of European cathedrals. But there is no real kinship between the two. In China, such plays have no religious significance, and are not usually connected with religion, even if held on holy ground. Their plots are taken from legend, or history, or episodes from famous novels.

Failing a temple-theatre, a matshed will be erected for the performance—put up the night before the play begins, and taken down immediately afterwards. Indeed, when actors come to town, a secondary mushroom-village grows up with restaurants, cook-shops, and peddlers' booths. The stage paraphernalia may be brought by the

VILLAGE THEATRICALS.

THE EIGHTH MOON, OR "THE HARVEST MOON"   393

dramatic company, or transportation may be provided by the village hiring it. Such arrangements are made by the local elders with the *pao tan ti*, or "programme-bearer" who corresponds to our advance agent.

The performers are professionals,—travelling troupes who wander about the country-side seeking engagements. They may be financed by some rich town-dweller who puts up the capital for costumes and properties, and is known as the "Master of the Chest." Their season begins in spring and lasts, with an interruption for the heavy summer rains, till autumn. It ends with the grand *galas* of the harvest moon.

Luckily, peasant audiences are not hard to please. They require no seating accommodation but patiently stand, squat on the ground, or, if there be room, unharness their mules and watch the performance from their carts ranged in a row facing the stage. They want no dramatic novelties, no elaborate properties. The stage is a simple affair, entirely open to inspection, but imagination supplies all deficiencies of scenery. Finally, they ask no all-star companies, but are content with even a rude troupe of mummers.

In their eyes, the coarse but gaudy costumes make a brave show. The hero strutting across the stage in his soiled "dragon robes" and turning a somersault at the critical moment, the "flower-faced" (*hua lien*) clown poking fun at the villain, the heroine whose shrill falsetto shrieks accentuate moments of intense feeling, the supers flashing gilded spears in the sunshine, are wonderful enough. And the orchestra is a never ending delight to them, though terrifying to our unaccustomed ears. Gongs and cymbals bang loudest at the most dramatic crises; long-necked guitars, bellied into round bodies covered with snake-skin, twang in a furious crescendo; squat instruments with goitered necks squeak as if in pain, and gourd-like mandolins wail.

To the ignorant foreign spectator, the village crowd is as good as the play, and better. These wide-eyed farmers, whose natural love of amusement is generally starved and suppressed by hard work, know delights the brighter for the usual homespun tints of their lives. Their narrow perspective enhances the significance of any amusement, however trivial, which is not a portion of the daily round. We, who thrill only to "first nights" that cost a fortune, wonder if their capacity for enjoyment of simple things is not a sign of the superiority of Oriental civilisation—in this respect at least. Watching, there comes to mind the eternal question: is it better to be backward according to the Western definition of the word, and content; or progressive, and restless? The answer may be our responsibility —and a heavy one.

Sometimes the farmers themselves form amateur theatrical companies after the harvest, and give "little theatre" performances. They enjoy the change and excitement of the boards, in the most literal sense of the word, and, though their earnings are meagre, they get their food free and their fun into the bargain.

Lion Dancers Akin to the strolling players are the "Lion Dancers" who wander from village to village.

LION DANCERS.

Photo by Yamamoto, Peking.

Photo by Yamamoto, Peking.

LION DANCERS.

Each troupe is composed of two or three mountebanks with rude but picturesque properties. Their entertainment has been handed down from the Indian jugglers and itinerant animal trainers who first appeared in China under the T'ang dynasty. As live lions are not obtainable in this country, a cloth lion serves their purpose, being manipulated by two men under a skin, one carrying the cardboard head, the other the hind quarters. The dances, called *Shua Shih Tzŭ,* or "Exercising the Lions," were originally supposed to have the power of a demon-expelling ceremony, because the lion was an emblem of Buddha and protector of religion. "In Peking, companies of acrobats have been organised to cultivate this specialty. The blue and yellow lions perform a contra-dance, displaying astounding skill and agility, the eyeballs, tongues, jaws, ears and tails in rapid motion, while the bells of neck-collars tinkle to the accompaniment of gongs. Because lions are credited with a fondness for playing with a ball, the main feature of the performance is their pursuit of an enormous globe thrown in front of them or across their path. They will even leap after it on to the roof of a one-storeyed house, or jump down from there into the courtyard."

As popular as the "Lion Dancers" are the stilt-walkers, with their carnival spirit. These groups of masqueraders, with false beards, painted faces, and humorous make-ups, sing as they parade through the villages. Men dress as women, with costumes of exaggerated gaudiness, and mimic the mincing steps of small-footed ladies. To amuse the rural population, they impersonate familiar

Stiltwalkers

types—the Fisherman, the Begging Priest, the Old Woman, the Wood-Cutter, etc. A living parody on life and manners, the performers stride along in single file, capering and cavorting, the more proficient turning somersaults in a crude spirit of merry-making, every man competing with his neighbour and trying to out-do him in fantasy and weirdness. Sometimes an equilibrist follows in the wake of the party, a man astride a pole carried on the shoulders of two companions. His attempts to maintain his poise, with a dead crow in one hand and a fan in the other, are very funny. In distant Kansu province the stilt-walkers' procession has been modified in a curious way. Here little girls, standing upon a small wooden board fixed to a high pole, are carried about by men whose dancing movements remind one of a hieratic cake-walk. The custom is called *yang ko*, *yang* meaning "to raise." It is a survival of an old *"No"* procession intended to drive away pestilential diseases. Legend has it that Confucius himself once "came from his house in his court robes" to watch a similar spectacle, as "an indication of his desire to conform to the habits of the country-side." The religious character belonging to the ceremony in his age has died out, and nowadays stilt-walkers perform to get and give amusement, nothing more.

* * *

After they have had their fun, it is usual for Chinese peasants, should they be fortunate enough to possess a spring in their neighbourhood, to give thanks for the water so precious to their crops.

THE EIGHTH MOON, OR "THE HARVEST MOON" 397

We have ourselves seen this thanksgiving service at the "Black Dragon Pool, *Hei Lung T'an*, near Peking. Half a dozen farmers, delegates from the nearest village, gather beside the water. One acts as leader of the humble band. He kneels, all kneel. He bows his forehead to the flagstones, once, twice, three times, and his companions follow his example. Then, in the name of the countryside, he burns a prayer written on yellow paper and lights a bunch of fragrant incense-sticks. As the smoke drifts slowly across the pool, a fish swims by in a shining coat-of-mail and a kingfisher with metallic blue plumage swoops down over the water.

* * *

The Moon Festival, Third Feast of the Living

The Harvest Festival coincides with the Moon's Birthday, so that the fifteenth of the eighth month is a double feast, one of the most important in the Chinese calendar. The Queen of Night represents the fluid element of the universe, because the Chinese early discovered her connection with the ocean tides. "The Moon," they say, "consists of the *Yin* fluid, or water." Now nature, according to their theory, is controlled, as we have seen, by two great principles, the *Yang* and the *Yin*, the *Yang* being male and the *Yin* female. The Sun personifies the *Yang*, source of virile energy, light and heat, the Moon the *Yin*, source of moisture, especially in the form of clouds and rain. The sun's cool companion of the night, she early came to be regarded as a feminine deity typifying darkness, water, cold, and womanly submissiveness—as opposed to light, heat, and masculine domination,—symbol of the Empress, and of officials whose allegiance to the throne was like a wife's allegiance to her husband.

The two form a heavenly couple who have charge of the world and its affairs—an idea not confined to China. The marriage of the Sun and Moon is reflected in the myths of most peoples. We find distant echoes of this belief in the stories of Zeus and Europa, of Minos and Britomartis. The fancy of our forefathers pictured the moon as a coy or wanton maiden who either fled from, or pursued, the sun every month, till she was overtaken at the moment when the luminaries are in conjunction; that is to say, in the interval between the old and new moon.[2] It is supposed that the Olympic Games originally celebrated their mystical marriage though, later, these sports came to be held in honour of the dead.[3]

The Chinese fixed the date of the Moon Festival in the eighth moon—the season when the female principle began to take the upper hand in Nature; that is to say, when summer heat gave place to autumn coolness, and summer brightness to winter darkness. Its fifteenth night is the moon's apogee. At no other time is she so bright or brilliant. Then, and then only, the Chinese say "she is

---

[2] On the principle of sympathetic magic, this interval was considered a favourable time for human marriages. The ancients likewise chose this interlunar period for the sacred wedding-feasts of their gods and goddesses.

[3] The relation between the Emperor of China and his consort has been likened to that of the Sun and Moon, or the *Yang* and the *Yin*, because "the Ruler of Men regards Heaven as his father, Earth as his mother, the Sun as his brother, and the Moon as his sister." Children of Heaven, the principal planets are, *ipso facto*, kindred of the Sovereign Son of Heaven.

perfectly round."

Feminine herself, the night lantern is, not unnaturally, the patroness of women, and in every family it is the duty of the women to worship her. There is a saying: "Men must not worship the Moon, women must not sacrifice to the Kitchen God." The single exception was the Emperor who honoured his Celestial Sister.

Once upon a time, the *Wu Ko,* or "Song Posturing Dance," was an essential part of the ceremony. Sometimes one girl, sometimes two, took part, one dancing while the other sang verses in praise of the mid-autumn moon. "It is as if polished," so ran the antique chant; "it is also entirely spotless. Nothing dims its splendour. Here is the symbol of conjugal love as conjugal love should be."

The Moon has lost her Sovereign Brother. The vestal virgins dance no longer. Present-day rites are simplified, yet, nevertheless, still picturesque.

Under the exacting eye of the "Mother," daughters and daughters-in-law dress the open altar. Five plates are set out filled with round fruits, apples, peaches, pomegranates (hinting at numerous posterity), grapes, and melons like little green balls. Their shape not only symbolises the moon but stands for family unity. But mooncakes (*yüeh ping*) are the distinctive offering of the feast. They are made of greyish (moon-coloured) flour—and piled thirteen in a pyramid, because thirteen represents the months of a complete Chinese year (*see* "The Chinese Calendar") and, likewise, a complete "circle of happiness."

In the XIVth century these cakes were the means of conveying secret instructions to Chinese patriots suffering from Mongol oppression. The conquerors had billeted one of their own men on every household. These spies proved overbearing, "taking to themselves the powers of rulers in the homes, and causing all to bow to their wills . . . The women, especially, were as slaves under their yoke." Great indignation underlay their helplessness, till someone hit upon the idea of writing a secret message on the little paper squares stuck on the moon-cakes. When sent, as they still are, from neighbour to neighbour and friend to friend, the pasties carried the order for a rising *en masse* at midnight. Though the oppressed people were without weapons save their kitchen choppers, hatred strengthened their arms. The surprise attack succeeded and the revolt ultimately led to the complete overthrow of the stranger dynasty.

Some women bake their moon-cakes themselves as an act of piety, in remembrance of the deliverance of their forbears from the oppressors, and stuff them extravagantly, if they can afford it, with bits of lard, spices, melon-seeds, almonds, orange-peel and sugar. In cities, confectioners present them to the poor—for, says the proverb, "even to dream of a moon-cake foretells riches," —and in villages "moon-cake societies" are often formed. A skilled baker acts as treasurer, and all the members contribute a few coppers monthly so that, when the festival comes round, every family is supplied with the luscious *yüeh ping*, decorated with rude pictures of the Moon Hare or the Moon Toad.

The Moon Hare appears also on every altar, represented by a special tablet, or a little clay

THE EIGHTH MOON, OR "THE HARVEST MOON" 401

statue, and near by is placed a bunch of beanstalks, his favourite food.

Just at the hour when the moon is clear of the tree-tops and sails like a full-rigged ship into the high heavens, the service begins. The courtyards in millions of small, poor homes are changed to fairyland because the Goddess touches them with her silver fingers. She hides the poverty and ugliness of everyday things. She smoothes away the wrinkles from tired faces, and lends a grace to awkward silhouettes as they approach her table. One after another the women go forward and make their bows. Two candles are lighted because it is customary to offer them in pairs. Bundles of incense-sticks are stuck flaming in the family urn, but their light glimmers faintly in the floods of moonshine. The whole service lasts but a few moments and concludes with the pasting of a brilliant poster against the wall of the house—a poster showing the Moon Rabbit under the Sacred Cassia Tree, pounding the Pill of Immortality in his Mortar. Ceremonious salutations are addressed to this quaint little animal figure. Then his picture is taken down and burned. Thus end the religious rites proper to the Moon Festival.

* * *

Now come the social pleasures, longer or shorter, more or less elaborate, according to the means of the celebrants. The inevitable feast is usually at midnight—the hour when the moon illumines the highest palaces. General festivities continue for three days. The evenings are devoted to moon-viewing parties which date from the time of the Emperor Wu Ti (about 100 B.C.) who had a special terrace, called the Toad Terrace, constructed to look at the toad in the moon, and gave elaborate banquets on it. The ladies adjourned to a verandah or a flat roof by themselves, out of sight of their men folk, as in old-fashioned Chinese society the sexes did not mingle. Blind musicians were called to sing for them the famous poem of Li T'ai-po, who made a cult of the Queen of Night and addressed his last ode to her.

The story goes that one evening, when the moon was full and he himself had drunk too much of the wine he loved too well, Li T'ai-po fell into the water and was drowned. His parting words were verses in honour of his idol.

> "How long, O moon, hast thou honoured heaven with thy presence?
> "I lift my wine cup and question the blue night sky . . .
> "On the wings of the wind I yearn to visit thy paradise,
> "Yet I fear the chill of thy jade and crystal palaces . . .
> "Better remain on earth and watch thee entering our halls
> "By painted doors and latticed windows,
> "Taking the sleepless by surprise . . .
> "Just as we mortals suffer joy and sorrow,
> "Thou, Goddess, suffereth light and darkness.
> "Now thou art round and full and brilliant,
> "Now broken like the finger-nail of my beloved.
> "Here, then, is proof that perfection exists neither for gods nor men.
> "The best wish we can wish is that you and I may long enjoy thy light together."[4]

---

[4] The above is in no sense a translation of Li T'ai Po's poem, but the freest possible adaptation in an attempt to render the Chinese attitude with regard to the Moon-Goddess, as expressed by her most famous poet.

The Moon's birthday was an occasion to consult the future. A lady would slip away from the company for this purpose. Secretly, as if half-ashamed, she lit three incense-sticks, whispered her question and awaited the answer, hidden behind the gate. The reply came through the phrases, lucky or unlucky, pronounced by the first passers-by. Much the same superstition existed in Europe, proving once again how the minds of all simple folk think and imagine along similar lines.

During the three days dedicated to the moon, it was customary to send presents to doctors, especially to those who practised electro-biology, a kind of hypnotism especially popular in Kuang-tung. "The person willing to be operated upon was placed directly in the moon rays. He had to stand leaning with his forehead on the top of a pole which he grasped with one hand while the other rested upon the ground. Burning incense-sticks were then waved over his head and about his body, and the operators, of which there were usually two or three, repeated prayers in a low tone to the moon. In the course of half an hour the mesmerised person fell down to be raised again, placed upon his feet, and made to go through various movements. As a result, the disease, or devil, disappeared."

It was customary also at this festival to invite friends to view the family curios which were unpacked and arranged on tables and in cabinets for the temporary exhibition. Children showed their toys to one another and, especially, the presents they had received. Toy-fairs were in full swing, and most charming gifts for the little ones could be purchased for a few cents. There was a whole group of clay statuettes to choose from, genii, fairies and godlets a few inches high. There were variegated pagodas and coloured clay Moon Rabbits dressed up as officials, —some very quaint in civilian robes, others as old-style warriors carrying flags. A certain kind of doll was popular in some provinces, notably Fukien. Parents who had a child born during the twelvemonth bought one of these and wrote the name of their baby on it. Henceforth it was regarded as the child's double, used to represent it in various home ceremonies, and, in case of death before maturity (sixteen years of age), buried in the same grave.

* * *

The Moon Rabbit   The Rabbit and the Toad are two recognised inhabitants of the Chinese Moon, which has a picturesque population. Indeed, the moon-folk, both human and animal, are among the most interesting figures in Chinese mythology.

How did they get there? What is the explanation of the actual mainsprings of the miracle? It seems probable that the world's moon-myths, including those of the Chinese, were originally derived from the same source—shepherds that watched their flocks by night, or primitive nomads who lived under the open sky and, when darkness came, lay down with eyes turned heavenwards. Darkness stimulates imagination. As they gazed up at the moon, they saw, or thought they saw, figures outlined on her silver disk.

Photo by the Asiatic Photo Publishing Co.

MOON RABBITS ON TOY-STALL.

Photo by the Asiatic Photo Publishing Co.

OPEN-AIR ALTAR FOR MOON FESTIVAL.

Soon they spoke of these visions to one another, and thus the myths were born to be carried down to our own days by the flowing stream of tradition.

Among all nations, the lunar spots have afforded a rich subject for the play of human fancy, but it is curious how often, and among what widely diversified peoples, we find the Hare an inhabitant of the pale planet, or associated with her. We read in Pausanias that the Goddess of the Moon, having been consulted by the soothsayers as to where they should build a city, gave the cryptic reply: "Where a hare makes its burrow." In Russia, there is a superstition that if a hare runs between the wheels of a carriage containing a betrothed couple, their married life will be unhappy. The idea is that, as the hare represents the moon —patroness of conjugal life— she sends him to show her disapproval. Should she herself deliberately cross the path of lovers, they are doomed to misery.

The Hottentots, and many other African tribes, say the Moon chose the Hare as her messenger to inform men that, as she faded and came to life again, so mortals should die but achieve resurrection. The Hare started off in a great hurry just to show how quick he was. More haste, less speed! Arrived, all out of breath and flustered, he got the message wrong: "The Moon says," he stammered, "as she dying will be born again, so you dying, shall be forever dead." Sorrowfully they believed what the Hare told them, until a second messenger, the Tortoise, who had been overtaken on the way, delivered a message of consoling immortality. Thereupon, those who had listened to the Hare grew angry, and one among them seized a stone in an attempt to kill him. But the missile only cut his lip open, and that is why the hare has a cleft lip to this day, and we still speak of a "hare lip." In some versions of the myth, the Hare, maddened by rage and pain, is believed to have flown at the moon and scratched her. She still bears the marks of his claws on her face.[5]

In China, the connection between the Hare and the Moon is obviously very old, since the ancient *Li Chi,* or Book of Rites, mentions the hare as one of the moon-sacrifices and calls this animal: "The One Who Looks At the Moon." Hence, the superstition common to both Japan and China that the hare (whose eyes never close) gives birth to her young with her eyes fixed on the moon, and the belief that, according to the brilliancy of the Lady of Night on her Festival, there will be few or many hares during the next year. An old proverb recalls the same connection: "Two only sleep with their eyes open, the hare and the moon."

But Buddhism is really responsible for the general popularity of the Moon Hare in China, because Buddhism brought from India the explanation of how and why he reached the moon. Hear this pretty and pathetic parable designed to teach the beauty of sacrifice, the virtue of un-

---

[5] It is significant that most primitive peoples attribute human mortality to the perverted message of some creature—a duck, a sheep, even a lizard—and believe that, owing to the blunder or wilful deceit of the messenger, God's beneficent scheme to make man immortal miscarried. See *Folk Lore in the Old Testament,* by Frazer.

selfishness, this legend soft and tender with the patina of ages!

Once upon a time, there was a forest glade where holy men came to meditate, a beautiful natural garden filled with fruits and flowers, carpeted with tender grass and refreshed by the waters of a sparkling stream, blue as lapis-lazuli. Now in this little paradise there lived a hare, a creature whose many virtues gave him ascendancy over all the other animals. By precept and example he taught his companions to perform their religious duties in a manner approved by the pious, until "their renown reached even the world of the Devas."

One evening the Buddha came to this garden. Certain of his disciples accompanied the Holy One, sitting reverently at his feet and listening while he preached the Law. All night long he discoursed and until the next day at high noon, when the sun darts his sharpest beam; when the horizon, enclosed in a net of trembling rays of light and veiled with radiant heat, does not suffer itself to be looked upon, when the cicadæ shriek their loudest; when no living creature leaves the shelter of the shade, and the vigour of travellers is exhausted by heat and fatigue.

In that time of the day, the Buddha chose to assume the figure of a Brahman, crying out like one who has lost his road and is consumed with weariness and sorrow: "Alone and astray, having lost my companions, I am a-hunger and a-thirst! Help me, ye pious!" All the little forest-dwellers heard the cry of distress, and one by one they hastened to the Holy Man, begging him wander no further but remain with them and accept their hospitality. And each, according to his means, brought food for him. The otter brought seven fishes, saying: "Accept these, and remain with us." The jackal brought his kill, saying: "Honour us with thy presence and grant us thy instruction." When it came to be the turn of the Hare, he approached empty-handed, and said humbly: "Master! I, who have grown up in the forest nourished by the grass and herbs, have nothing to offer thee but my own body. Grant us the boon of resting Thy Holiness in this place, and vouchsafe to me the favour of feeding thee with my own flesh, since I have nothing else to give thee!" Even as he spoke, the Hare perceived a heap of magic charcoal burning without smoke near by. He was about to leap into the flames when he paused and began gently picking out the little creatures lodging in his fur. "My body I may sacrifice for the Holy One," he murmured, "but your lives I have no right to take." He placed the tiny insects safely on the ground and then, "with the utmost gladness, like one desirous of wealth on beholding a treasure, threw himself into that blazing fire."

Resuming his own form, the Buddha praised the loftiness of the sacrifice: "He who forgets self, be he the humblest of earthly creatures, will reach the Ocean of Eternal Peace. Let all men learn from this example and be persuaded to deeds of compassion and mercy." Moreover, to reward the Hare, Buddha commanded that his image should adorn the disc of the moon, a shining example for all time. As for the other animals of the forest, they were translated to the world of the Devas, thanks to their holy friend. Ever since these happenings in the forest, the moon has been known to the Buddhist world as the "Hare-Marked."

With his irresistible appeal to the popular imagination, the Beloved Hare, looking down from the moon, and establishing himself on family altars, was a sharp pin-prick to the Taoists. They could not hope to oust him. Consequently, as usual in such cases, they adopted him and renamed him "the Jade Rabbit."[6] According to their fancy, he is pictured with very short front legs, very long hind legs and a white tail curled over like a feather. They too have set him—or left him—in the moon where he pounds the Pill of Immortality, sometimes called the Elixir of Jade (hence his title), that confers life everlasting and has all the qualities of our own "philosopher's stone."

To stand eternally pestle in hand is a dull duty. But, at least, Taoist imagination has given the Moon Rabbit an enchanted setting. Overhead droop the branches of the Sacred Cassia Tree, of miraculous powers and supernatural beauty. Poets praise its fragrant flowers that open for the moon's birthday, and physicians declare its aromatic bark a cure for all disease. In their musty treatises, dating from the IVth century A.D., we find a quaint prescription: "Thoroughly mix cassia-bark with bamboo-juice and frog's brains.

---

[6] Or rather "the Jade Hare"—the true Rabbit does not exist in China.

Then drink this potion, and seven years later you shall walk upon the waters,"—in plainer words, become an Immortal.

* * *

Companion to the Moon Hare is the Wood-Cutter, engaged in the task of attempting to cut the Cassia Tree down. He chops at it and at it, this poor man, in life a scholar and condemned to play the rôle of a Chinese Sisyphus for some misdemeanour. But no matter how hard he works, his task is hopeless. The tree miraculously repairs its own injuries. Giver of life to others, it is itself immortal. A noble tree, it bears no grudge but has, among its many virtues, the virtue of forgiveness. A scholar seeks to destroy it, yet the cassia is a friend of scholars, and the elegant euphemism for taking the second literary degree[7] is to "pluck a cassia flower from the topmost branch by the Pavilions of the Moon."

These Moon Pavilions are palaces of exquisite enchantment, whose towers and pinnacles are indicated on the posters burned in honour of the Jade Rabbit. But such crude drawings give scarcely a hint of their ethereal loveliness. Imagine a surging glitter of all the colours of earth and heaven, pagodas of crystal reflecting the shadings of the rainbow, storeyed silver pavilions, one shining tier rising above another, enclosed in walls of jade and perfumed wood, staircases of agate leading from terrace to terrace in fairy gardens with mysterious flowers and dark trees,

---

[7] The examinations for this degree were held every three years in the eighth moon.

with birds white as camphor nesting in their boughs!—All this, and more, is found in that enchanted place.

Yet, only once have mortal eyes looked upon its loveliness. The fortunate man was the Emperor Ming Huang who, on the fifteenth night of the eighth moon in the first year of his reign (A.D. 713), while walking in his Palace enclosure with a learned priest, inquired of what material the moon was made. "Would Your Majesty care to go and see for yourself?" the Sage replied. Of course, the Emperor said he would. The holy man threw his staff—some say his girdle—into the air, and it became a bridge by which Sovereign and Sage reached the Moon. They saw everything there was to be seen, the Moon Hare making his pills under the Cassia Tree; they saw Hêng O, the lovely châtelaine, whose legend we shall presently recount; they saw her attendant fairies, singing and dancing in the gardens of delight and, finally, they saw a terrifying white tiger, whereupon Ming Huang decided it was time to leave.

On the return journey His Majesty, who was renowned for his skill as a musician and had taken a lute with him, enlivened the road by playing sweet airs. This celestial music brought the people of his capital running from their houses into the streets where they fell upon their knees. The Sage advised the Emperor to shower cash upon the city. It was an astute suggestion—confirming evidence of an adventure the Sovereign, not unnaturally, later thought must be a dream, until an official report arrived from the

Governor, describing the music and the rain of copper coins.[8]

* * *

The Moon Lady Ming Huang's hostess, the Lady of the Moon, is metamorphosed into a three-legged toad, according to a myth of Indian origin like that of the Moon Hare. The legend fell on fruitful soil since, long before Buddhism arrived, the Chinese already had established a connection between the Moon and the waters, and believed that amphibious animals belonged by their essence to this element.

Now Hêng O, the Moon Lady, was not incarnated in an ordinary rain-calling frog of the rice-fields, as she doubtless would have been had her myth been invented by the Chinese themselves, but in the miraculous *Ch'an*, or three-legged toad. In his throat the character for "eight" is always traced, so he belongs by right to the eighth moon. He secrets water in his feet, so that identifies him with moisture. He has power to deflect arrows with his mail-coat skin, hence his connection with the Divine Archer, husband of Hêng O. Finally, he lives forever—the faculty of toads to live long is well known—which fits in very well with the Pill of Immortality swallowed by the Moon Lady, and bears out the idea of the Eternal or Indestructible.

Hêng O and her husband Hou Yi lived in the days of the Perfect Emperor Yao, about 2000 B.C. Hou Yi was a fairy-kind of person who trod on

---

[8] The origin of the modern Chinese theatre is traced to this visit. Ming Huang was so impressed by the singing and dancing of the moon fairies that, on his return to earth, he taught their songs and postures to some of the palace youths, who were called the "Disciples of the Pear Garden," a name that sticks to Chinese actors even now.

side-walks of air and fed on the essence of flowers. At the same time, he was an officer of the Imperial Guard and a skilled archer, but his bow was not an ordinary bow; it was enchanted. One day, ten suns rose in the sky. They shone so brightly that earth-dwellers could not endure the fierce heat. The Emperor called for his lieutenant: "Many times already," said His Majesty, "hast thou served thy country. We recall when thy arrows, shot into an engulfing flood, caused the waters to subside. Shoot now at these false suns, and deliver my people from their miseries." Hou Yi did as he was bid and achieved great fame. Even the Hsi Wang Mu (*see* "The Third Moon") in the distant K'un Lun Mountains took notice of this promising young man, ordered him to build her a palace of multi-coloured jades,—evidently he had a streak of Jack-Of-All-Trades in him,— and for reward gave him the Pill of Immortality. Like a wise woman who knew men and their greedy instincts, she gave advice with her gift, saying: "Make no haste to swallow this pill, but first prepare thyself with prayer and fasting for a twelvemonth." Like a wise man, he listened, feeling instinctively that Immortality must not be gulped, and, on reaching home, hid his treasure in the roof-thatch under a rafter while he began his moral dieting. Unfortunately, in the midst of it, he was called away to pursue a strange criminal called "Chisel Tooth." While absent, his wife Hêng O saw a beam of white light issuing from the roof, smelt a delicious perfume, discovered the pill, and ate it. Immediately, the laws of gravity ceased to affect her. She found she could fly, and when she heard her husband coming home ready to scold, fly she did right out of the window. Quickly he sped after her, bow in hand, and the pursuit continued half across the heavens. But he was turned back by the force of the wind, whereas she reached the moon and there, being breathless, coughed up the outer covering of the pill, which immediately turned into the Jade Rabbit, while she herself was metamorphosed into the *Ch'an*, or three-legged toad, probably for deceiving her husband. At any rate, since that eventful night she has lived in the Moon, while Hou Yi has built himself a palace in the sun. As *Yang* and *Yin* they direct the affairs of 'the Universe, meeting once a month on the fifteenth of every moon when he comes to visit his wife. Moon-gazers will tell you this is the reason why the Queen of Night is most beautiful at this particular date.

* * *

The Old Man of the Moon

Another one of the moon-divinities in mortal shape is the Chinese Man in the Moon, *Yüeh Lao Yeh*, the Matchmaker, an old greybeard who presides over all marriages made on earth. His duty is to attach betrothed couples with a red cord which shall bind them for life. The proverb says: "Marriages are made in Heaven, but prepared in the Moon," and there is no escaping destiny, as we may judge from the following T'ang dynasty legend. A certain youth, named Wei Ku, when travelling, saw an old man sitting in the moonlight by the roadside with a book in his hand. The traveller, advancing, inquired politely what the book was about. "It contains," said the Sage, "the marriage fates of those on earth. Look!" he added, producing a red cord from his sleeve: "with this I tie the feet of men and maidens whom fate decrees to be joined together. And joined they are, no matter if they belong to widely separated provinces, or clans at enmity with one another. Now, if you wish, I will show you your future bride. She is the daughter of that old vegetable-seller yonder." The maiden, alas! was a mere infant of low descent, and, desirous of confounding the unfortunate prophecy, Wei attempted to have her done away with. But the man hired to kill her only succeeded in scratching her over the eyebrow. Fourteen years later and in another province, when Wei had quite forgotten about the Greybeard and his Cassandra talk, he married a handsome girl. As the bridal veil was raised, a scar showed over her eyebrow, and inquiries proved her to be the same maiden decreed for him by the Old Man of the Moon.

All the myths concerning the *Yüeh Lao Yeh* are reflections of the mystical marriage of the sun and moon, and mystical allusions to the pale planet as emblem of marriage, because each month she throws herself into the arms of Her Radiant Lover, is absorbed by him, and leaves him only when he has rekindled new light in her. Wherefore Chinese maidens burn candles to the Matchmaker and to the Moon, seeking to know the handsome stranger whose home they shall enter in the red bridal chair. Scotch maidens, and English maidens too, bow to the brilliant Lady set in the soft black onyx of the sky, begging her to reveal in dreams the names of their future husbands. In Holland, as in China, lovers sighed to the moon twenty-five centuries ago. The expression "honeymoon" is rooted in moon beliefs. In fact, from India to the Orkney Islands the moon is thought to influence marriage.

* * *

Marriage Gods   Most peoples have special marriage gods, and the Chinese are no exception. Their *Yüeh Lao Yeh* divides his matrimonial responsibilities with various divinities, some helpful, some hindering. Among them is an extraordinary personage known as *Hsi Shên*, God of Joy. He is a disreputable character, incarnation of the infamous Chou Wang, last ruler of the Shang Dynasty—a human vivisectionist sometimes called the Chinese Nero. In life, his fondness for women seems to have been exaggerated. He did not even respect the goddesses. One day, when worshipping at the temple of Nü Kua, the sister of the Perfect Emperor Fu Hsi—she of the female head and serpent body—he dared to praise her beauty by writing on the wall of her shrine, and express his desire for a woman as lovely to take to wife. To punish his impiety, Nü Kua devised a plan whereby not one, but three fairy women were sent him. The first of these ladies was a pheasant turned woman, all but the feet. Her attempts to hide her origin are supposed to have set the fashion for foot-binding among Chinese ladies. The second wife was a stone guitar miraculously transformed. The third—the spirit of a fox

A SOOTHSAYER.

Photo by the Asiatic Photo Publishing Co.

WAYSIDE SHRINE TO A FOX-SPIRIT.

with nine tails, proverbial synonym for a flatterer.[9] It was she who became Empress, with a unique reputation for cleverness and cruelty. Why the husband of this malicious trio, this libertine, whose amorous adventures were by no means confined to what one might think was an adequate domestic love-nest, should typify the sane, safe, sensible joys of marriage, is a mystery. But he does, and it is to this gentleman of more than doubtful reputation—to whom no temples are erected and no sacrifices made—that every Chinese bride turns her eyes on her wedding day, looking towards him in the direction of the planet Venus where he now lives. She does it shyly, not daring even a whispered prayer, because he is not a fit person to be prayed to. Yet she does it—perhaps with the formless wish that her husband may have the ardour of the *bon vivant* without, of course, his fickleness. All women are like that, wanting the impossible.

Very different from this rake is the sedate goddess, the Tzû Sun Niang Niang (*see* "The Fourth Moon"), a popular Chinese female divinity, who has got her cult mixed with his. Our Lady

---

[9] The fox, in China, is a fairy beast with wonderful powers of transformation, and the fear of these animals, who are often malicious, is wide-spread. People especially dread were-foxes who take the forms of beautiful young women. It is always wise to be polite to a fox lest he become your sister-in-law, or even your wife. Fox-worship in China, though less universal than in Japan, is fairly prevalent and often carried on at very small shrines. Father Mullins, in *Cheerful China*, speaks of a shrine in Shantung of peculiar structure, with an opening so narrow that worshippers were obliged to crawl in and out on their hands and knees. Tiny women's shoes were given as offerings, and fir-branches covered with paper flowers fixed into a ball of clay. This shrine was built to *San Lao T'ai Yeh*, the fox-spirit, over a spot where foxes were supposed formerly to have had their dens.

of Many Children lived in the reign of the Husband of Many Wives. She was a high official's consort, very virtuous—so virtuous that she committed suicide to escape the attention of this too ardent sovereign. Mother of a complete family—for she had the classical number of children, five boys and two girls—and a model of wifely chastity, she was deified as a shining example, and became a patroness of marriage. Her presence, spiritually speaking, at the wedding feast is recognised by the young couple when they eat the special cakes called *tzû sun po po* in her honour, as they sit side by side on the *k'ang* after the ceremony.

A very curious family is also connected with marriage in China. We remember Cheng Wu (*see* "The First Moon"), canonised as the Ching Tu Fo, or "the Buddha without Vitals." The knife with which he disemboweled himself and the scabbard from which he drew the blade were transformed into a youth and maiden. This maiden, vowing chastity, has become an enemy of marriage, whereas her brother, with brotherly contrariety, attempts to counteract her influence by assisting lovers. In the larger temples father, son, and daughter sometimes appear in a group—little figures under glass tucked away in a side shrine. One runs across them, as it were, by accident—perhaps by following a group of women who have already burned incense to the more important gods, and have one fragrant stick left for the Little Man and his Little Family. Much rarer is the image of the wicked spirit Hsiung Shên, whom Cheng Wu's daughter sometimes uses to disfigure and "spoil" brides; he is so ugly that,

THE EIGHTH MOON, OR "THE HARVEST MOON" 419

when he sees himself in a mirror, he runs away ten thousand *li*. In his destructive activities against the human race Hsiung Shên is assisted by two other evil spirits, the "Tall Jinx" and the "Short Jinx." Tall Jinx helps people into the noose when they hang themselves. Short Jinx is useful for errands where you have to climb through small holes,—secret errands, we suspect rather caddish errands.

* * *

Gods of Luck and Longevity

More sympathetic figures at the marriage-feast are the God of Happiness, or Luck, Fu Shên, dressed in blue official robes, and the God of Longevity, the old man with the high forehead painted or embroidered on birthday scrolls.

Fu Shên is rather young as the gods count youth,—a Taoist conception which owes its origin to the Emperor Wu Ti (A.D. 502-550) of the Liang Dynasty. This sovereign had a fancy for dwarf servants in his palace, dwarf comedians in his theatre. The number of these unfortunate Tom-Thumbs, levied from a certain district in Hunan, seriously affected family ties. Thereupon, a famous judge in the grand and fearless manner reminded His Majesty that, according to law, dwarfs were subjects like other men, not slaves. The Emperor saw the point, and ordered the levies stopped. How delighted the people were at being liberated from the hardship of sending their tiny husbands and fathers to be shut up in the Palace, instead of tilling their tiny fields! So delighted, they canonised the judge, and worshipped him as the Spirit of Happiness.

What could be fairer than that? Many men have become gods, but none have found a simpler, more human way to popularity than Fu Shên whose portraits and images are spread all over the country, whose worship is as universal as that of the powerful Tsai Shên, God of Riches.

His colleague, Longevity, is unmistakable, wherever we find him, on account of that so high top to his head which always seems in danger of boiling over.[10] Legend identifies him with the star *Shou Hsing,* or Canopus, also known as the Southern Bear, or the "Old Man of the Southern Pole." Ch'in Shih Huang Ti, after the unification of China, was induced by Taoist influence to build a temple to this star. The T'ang sovereigns recognised the stellar figure with sacrifices on the day of the Autumn Equinox. The Mings, on the contrary, neglected him. Cynical, the ups and downs of the Gods in China, faded and dejected under one regime, glittering and brilliant under another, dropped by one Sovereign, hotly taken up by the next! Only the people are faithful— sometimes — and they remain constant to "The Old Man of the South," or "Southern Measure," supposed to control life, and to his confrère, another old gentleman with a long white beard who personifies the North and the "Northern Measure," supposed to control death.

The worship of the "Measure," Northern or Southern according to the season, is very general, especially in South China. The Moon's birthday, or the anniversaries of the children of the house,

---

[10] The proverb says: "The God of Longevity takes arsenic," meaning he cannot be killed.

are favourite occasions for this little home ceremony. The standard measure in the north is the amount of grain that could be contained in a square stone box. One of these boxes stands on a marble terrace in front of the Emperor's Palace, and we know of another in the courtyard of Fa Yüan Ssŭ, in Peking.

Mrs. Ayscough, in *A Chinese Mirror*, describes the worship of the "Measure" in Kiangsu province as she saw it. She says that every family has its own rice-measure, "made in an infinite variety of sizes and each filled (for the festival) with large pieces of sandal wood, and embellished with a gay paper Moon-Palace which rises from the centre around a pillar of incense-sticks. Early in the morning, the tip of this pillar is lighted, and throughout the day it smoulders slowly till, by midnight, the whole erection is consumed. A figure of the Patron Saint of Literature is placed in the Palace, because he is supposed to inhabit one of the stars of the Northern Measure, the Great Bear Constellation." How, we naturally ask, does the Patron of Authors get into this *galère*? The whole cult is mixed and tangled, and hitched to various stars (*see* "The Third Moon"). Old Man Longevity has a share in it too. His picture is on the *chih ma* (*see* "The Twelfth Moon") burned before the altar, for no better reason, apparently, than the flimsy pretext of a legend. Long, long ago, a lad met a famous fortune-teller, all on a summer's day when the white clouds flew up above the hills like a flock of enormous birds and spread over the sky. "You are a fine lad," said the old man.

"What a pity your life is so short!"—"Tell me, tell me, how long?" he begged. "Well, it is best not to speak of such things, but if you urge me, I must. Nineteen years is the span of your life." The boy ran home crying to his mother and she, crying also, sent him back post-haste to the soothsayer to ask what could be done about it. "My advice to you," said he, "is this. Take a haunch of venison and a jug of wine, mount yonder hill, and there you will find two old gentlemen playing chess. Lay down your provisions without a word and await the end of their game. Then, and then only, make your request,"—and he whispered in his ear what the young man should say.

Immediately the youth set out, found the chess players, waited while they finished their game, waited while they enjoyed the food and drink. Then, with tears in his eyes, he besought them to save him from such an early death. So well and so politely did he strike the pathetic note that they took out their records, and there and then changed the one in nineteen to a nine, making the entry in the Book of Fate read ninety nine. "Now young man," both said together, "return home and warn that fortune-teller never again to divulge to any mortal his appointed life-span. We can't be changing our ledgers all the time. Besides, it is not good for human beings to know the future. It makes them uppish, or it makes them sad, and it's none of their business anyway."

We can picture the joy of the mother when her beloved son came back with a new lease of life. Probably, it was her gratitude,—though tradition does not fix the point,—which instituted the first sacrifices in honour of the Old Man of the Northern and the Old Man of the Southern Measure for, as you can guess, they were the chess players. Parents whose children are ill, or delicate, still burn incense to them for longevity and honours. Also to the Bushel Mother, a vague figure identified with another of the seven stars in the Great Bear constellation. A pretty custom, unfortunately dying out, made the host of an inn or tea-house fly a flag during the festival, with these seven stars arranged in proper order. It was a polite way of wishing long life to his guests.

* * *

The God of Thieves

As the "measure," naturally associated with the harvest, suggests the season of plenty, we find the birthday of the God of Thieves fixed about the same date, on the seventeenth. Practical as always, the Chinese consider it suitable to burn incense to this doubtful divinity while the stealing is good. Not only his followers, busy with their nefarious trade, worship him, but many who are not professed thieves. From his title, "Midway in the Heavens," it appears he is unwilling to come down to earth, unwilling to accept the shelter of a temple roof. That is why men make no images of him. Their prayers are said under the open sky. This horror of confined spaces is, doubtless, a characteristic acquired when he was on earth, a clever thief himself and not anxious to be caught in a trap. Robbers like room to run. Villainy implies liberty.

Now it would be suitable, and proper, to condemn the God of Thieves, to point him out as an example. But the blackest gods are not all black, and he had a redeeming feature. Filial piety off-set his crimes. True—he stole, but he stole food for his mother. The motive makes him seem more lovable. Sympathy dilutes our condemnation, as pure water crude wine.

CHAPTER XIII.

THE NINTH MOON, OR "CHRYSANTHEMUM MOON."

"SPRING," says the Oriental poet, "is the season of the eyes, autumn the season of the ears," meaning that in spring the pleasures of nature are visual,—the tender green of budding trees, the magic of morning haze, and the pink mist of blossoms on the hill-sides,—whereas in autumn the appeal is rather to our hearing, through the music of chirping crickets, the whirring of dragon-fly wings, the soft click of leaves in the breeze—gold striking against gold—the sharp whip-crackle of dry branches as high winds rattle them, and the musical cries of the pines.

Nowhere is this truer than in North China where, in the twilight of the year, the bare brown hills and bare brown fields leave us utterly dependent on the song of the lingering meadow-lark, the quivering of insect voices, and the deep notes of the frog-chorus.

The only touch of colour that relieves the drab landscape is the gold of the persimmon orchards. In the old Chinese books there is a pretty story concerning a persimmon tree, which peasant-mothers tell their children while their fathers gather the golden fruit and pack it into the paniers on their little grey donkeys. Long ago, when Chu Yüan-ch'ang, the first Ming Emperor, known as the "Beggar King," was still a poor man and had had nothing to eat for two days, he came to a deserted garden ravaged by the disturbances in the land. Now the wall of this garden was broken, and most of the trees ruined. But the future emperor, led by the shrill voice of a cricket to a distant corner, found a branch covered with luscious fruit. He satisfied his hunger and went his way. Years later, the Beggar King passed this same garden at the head of his victorious troops, leading a successful rebellion against the reigning Mongol dynasty; he stopped and dismounted. Turning to the assembled officers, he remarked: "Observe yonder tree! When I was weary, it offered me shade. Gratitude is a princely virtue!" So saying, he took a crimson robe and covered the tree, begging it to accept the title of *Liang Shun Hou,* or "Marquis of Ice and the White Dew."[1]

As if to compensate lovers of beauty for the denuded country-side, the chrysanthemum, the flower of the ninth moon, lends a glory to Chinese gardens. The chrysanthemum typifies the fulness of the year, for it marks, as it were, "the golden wedding of the spring, the reminiscence in November of the nuptials of May." "When these flowers are gone," says the Chinese proverb, "there are no others left in the world." Therefore, the pleasure of their gorgeous

---

[1] The title is a reference to the period of the *Han Lu,* or "Cold Dew," which generally falls in the ninth moon.

NINTH MOON, OR "CHRYSANTHEMUM MOON"

colourings and infinite varieties is the greater because it is tinged with melancholy and the shadow of parting. Rich men arrange scaffoldings of varying heights, called "chrysanthemum hills," to show off their finest specimens,—flaming chariot-wheels with petals for spokes, balls of fire with lambent tongues, variegated pin-wheels in nature's day-fireworks, jewelled stars of her own Legion of Honour,—each plant in a porcelain pot, itself a work of art, but fading in comparison with the loveliness of flowers painted by the Great Artist. Other connoisseurs build "flower-towers," gold, or mauve, or copper-red, in the corners of their reception rooms.

* * *

Such impromptu displays are appropriate to the ninth of the ninth moon because that is the festival known as "mounting the heights" (*Têng Kao*, or *Chung Yang Chieh*), whose origin is connected with the *Yang* or Male principle, symbolised, in general, by odd numbers and, especially, by the numeral nine. Once upon a time there may have been also a suggestion of sky-worship connected with the day, but this has long since faded from recollection.

"Mounting the Heights"

The popular legend regarding the holiday is no older than the Han dynasty, a late invention to account for the primitive myth. It tells of a soothsayer who warned a virtuous scholar of a terrible calamity impending. "Hasten!" said he to his friend, "with all your kith and kin, climb to the shelter of the mountains, till there is nothing between you and the sky, and take with you food and drink." The scholar thanked his counsellor and followed his advice, carrying with him a paper bag containing food, and a jug of chrysanthemum-wine.[2] Returning at the end of the day, he found his cattle and poultry had died a violent death. "That," said he to his family "would probably have been our fate, but for the warning."

History does not relate what plague they escaped, nor why it was sent. People still picnic on the hills, however, in memory of the miraculous deliverance; not the peasants, who ignore the custom, but well-to-do folk and, especially, scholars who compose poems and discuss classical texts in picturesque spots. Besides chrysanthemum-wine, excursionists carry with them wheat-meal cakes called *têng kao*. There is a play on these Chinese words, which mean "cake," and also "promotion." He who eats these cakes supposedly secures his own advancement in official life, just as he who climbs the heights metaphorically climbs the ladder of scholarship.

Fashionable dinner-parties used to be given, with a delicacy appropriate to the season—crabs cooked in wine from Liang Hsiang Hsien (on the Peking–Hankow Railway near the capital), and

---

[2] Chrysanthemum-wine is believed to prolong life. The plant has supposedly magical properties because of the resemblance of the flower to the radiant sun-disc. Hence its old name of *yeh ching*, or "soul-energy of the sun." Chinese artists make frequent use of the chrysanthemum as a symbol of longevity, for almost all their designs, however decorative, have a hidden meaning. A good example is a panel painted on the Imperial Palace eaves where a cat crouches, intently watching butterflies fluttering above chrysanthemums. This appears to us a very informal, even a trivial, decoration in such a stately setting, until we realise that everything represented, flower, animal, and insect, owing to a sound-similarity of name, typifies continuity and length of days.

Chinese connoisseurs were as particular about the temperature of this wine as Frenchmen are about their best Burgundies. Outdoor luncheons were served on the old Mongol Walls, the site of the first city of Peking; at Wang Hai Lou, the quaint old fishing-palace; at the T'ien Ning Ssû Pagoda; on the hillocks in the Chinese City behind the Temple of Agriculture; and at the monasteries of the Western Hills.

In olden days, this "festival of the *literati*" was marked also by elegant festivities at Court, and chrysanthemum parties were held to view the exquisite purple, or "autumn-coloured," blooms brought to perfection for these occasions with infinite pains. Guests in graceful, flowing robes walked through the fairyland of Palace gardens, paused by some dreamy pond overhung by trees, took paper, ink, and brush from their silken girdles, and composed a sonnet on the view and the feelings it called up. Or they adjourned to some open pavilion with a distant glimpse of the hills, framed picture-wise between the pines, and there sipped the "wine of longevity" and poetised, passing their couplets, as they did their cups, in honour of one another.

De Groot, in his *Fêtes Annuellement Célébrées à Emoui*, puts forward some interesting and plausible theories with regard to the "Festival of Mounting the Heights." Tracing it back to prehistoric times, he suggests that, the harvest being in and marauding neighbours free to make war (*see* "The Eleventh Moon"), it is probable that detachments of men were sent as advance-posts into the mountains carrying provisions with them and that, for this reason, the conservative Chinese, (who will continue a custom long after the reason for it is lost) mounted the heights yearly at this season with food and wine. Thus, what had once been a warlike expedition with a serious purpose degenerated into a simple picnic excursion.

Confirmation of this idea may be found in the fact that the Chinese, of Fukien province at least, believe that in eating the cakes of the 9th of the ninth moon men become clairvoyant. Probably such offerings made to the gods in primeval days were expected to permit the soldiers to discover in good time, and therefore to prevent, the advance of their enemies.

* * *

Kite-flying

We might push the theory a step further, and guess that the need for giving distant signals was the reason for inventing the kite, which certainly was an engine of war long before it became a toy.

In many places, the ninth day is still largely given over to kite-flying. Around Foochow, thousands of people gather on the hills if the weather be fine, and special police are sent out to keep order, because rivalry among the kite-flyers sometimes leads to clan-fights. The festival kites are huge affairs. They require four or five grown men to manipulate them—men with the strength of arm, yet the lightness of wrist, of an expert fencer. Chinese gentlemen of leisure, averse to more violent sports, form societies and purchase a monster-kite to take the air in competition with their neighbours. Certain guilds make and fly kites specially for the occasion.

NINTH MOON, OR "CHRYSANTHEMUM MOON"

On the kite-festival at Black Rock Hill, near Foochow, the sky is populated with dragons, frogs, butterflies, centipedes, and a hundred cunningly devised figures which, by means of simple mechanisms worked by the wind, roll their eyes, move their paws, and flutter their wings. The ground is thick with people intent on sending up new and strange shapes to join those already floating overhead. Here stand a company of Weavers with a wonderful gold-fish eight feet long, complete with fins and multiple tails. It can only be made to rise from the ground with difficulty, and many a joke and many a curse speeds it on its way. Yonder the Tinkers have a jointed centipede difficult to manage. Men climb on stools, each holding a section, and, when the breeze comes steadily from the right quarter, shout directions as to how best the ungainly creature shall be made to catch the wind. It is an anxious moment. Air currents are fickle, and a false manœuvre may bring the kite crashing to the ground. But skilful hands seldom make mistakes, and the huge insect crawls upwards as if alive, amidst applause from the spectators. Derisive taunts stimulate the last competitors, a company of Brassworkers, still struggling with a dragon immensely long and very hard to fly. But it too leaves the ground at last, the monster-body floating in graceful curves, fiery emerald eyes blazing. Up, up they go, the rivals, trying to outfly each other, straining at their strings. Sometimes one is armed with a hook, or blade, to bring down an opponent captive.

The battle in the skies is accompanied by eerie sounds—shrieks and moans from the tiny Æolian harps attached to each competitor.[3] A pretty legend explains the origin of these harps. In the reign of the Emperor who founded the Han dynasty, a general still deeply attached to the deposed royal house determined to make one last effort to drive the usurper from his throne. He gave battle unsuccessfully. Then, finding his army caught in a trap and threatened with annihilation, he conceived, as a last desperate ruse, the ingenious idea of frightening the enemy with kites fitted with singing metallic strings. He waited for a favourable wind, he waited till the dead of night. As the opposing army slept in their camp, wrapped in silence and the soft summer darkness, it was aroused by sounds from heaven, sounds that seemed to say: "Beware, Han! Beware, Han!"—"Hark!" cried the soldiers in a panic, "those must be the voices of our guardian-angels warning us of danger. Let us flee!" And they fled, hotly pursued by the loyal general and his army.

Private individuals will sometimes set kites deliberately adrift in the hope that, when they fall to earth, the evils which may be hanging over their families will likewise crash to destruction,—or the belief that, given its freedom, a kite will fly higher than its cord allows, thus prophesying great honours for their kith and kin.

---

[3] Old records mention a special kite-harp called the *yao ch'in*, a gourd-shaped frame of bamboo, across which several thin bamboos were fastened. When attached to a kite-string, the *yao ch'in* made a loud humming noise. It belonged to the group of "stringed instruments vibrated by air."

NINTH MOON, OR "CHRYSANTHEMUM MOON" 433

As the sun swings low behind the hills, a good part of the crowd streams back to the city of Foochow, but some enthusiasts wait on till dark, attach lanterns to their kites and fly them again. The effect is fairylike, as if tiny stars from the Milky Way had broken free and were floating lazily to earth. From the ground, lamplets, crimson, azure, emerald, and gold, start off into the fear of the night together. They rise steadily against a dark wall of sky. They catch the current of a breeze too high for earth-bound mortals to feel. They curtsey and bow to one another in greeting, and their tiny flames, like little cannibals, devour a patch of darkness till they themselves are devoured by the night.

* * *

Another seasonable amusement is cricket combats.[4] Towards the end of summer, men go out on the hills, especially in the southern provinces, and capture these insects, by enticing them from their nests with a fruit called "dragon's eyes." They say that fighting crickets are produced from feathers buried in the earth, and that a good cricket is often found guarded by two centipedes, or even by snakes.

Fighting Crickets

The best fighters are specially bred and trained, and they may cost many taels of silver a-piece. Those that chirp loudest command the highest price. Favourite varieties have such quaint names as *Pai ma t'ou,* "Bald White Head," *Huang ma t'ou,* "Yellow Bald Head," *Mei hua chih,* "Plum-

---

[4] For singing crickets, one of the æsthetic pleasures of early autumn, see "The Seventh Moon."

Blossom Wing," *Hsieh kai ch'ing*, "Crab Claw," *Pi pa chih* "Guitar Wing," and *Chu ch'ieh hsin*, "Beard of Bamboo-Knots."

Very fine specimens are as dear as race ponies, and young Chinese fanciers have been known to ruin themselves by investing in large studs of these insects, which are not only expensive to buy, but to keep. They require special attendants, as horses require grooms, and separate stalls— earthenware pots lined with fine mould and fitted with a microscopic cup for their drinking-water. They must be fed on a carefully varied diet,—two kinds of fish, a certain grub, a special meal, boiled chestnuts, boiled rice, and honey to strengthen them. If they are sick from over-eating, red insects, called *hun ch'ung*, are given them. If they catch cold, they get mosquitoes; if fever, shoots of the wild pea-plant; and in case of difficulty in breathing, "bamboo butterflies" are administered. Nor is their moral welfare neglected. No smoke is allowed in a room where they are kept, as it affects their temper, and every male is allowed to have a lady in his tiny earthen cage for two hours each evening.

On the day of the fight, champions are carried carefully down to the cricket-pit, where a mat-shed is erected and hundreds of spectators gather. The real arena is a circular tub or bowl with a flat bottom. Two combatants, matched in size, weight, strength, and colour, are placed in it and excited to fury by having their backs tickled with a rat's bristle fixed in the end of a pen-handle. The game little creatures rush at each other like fighting cocks, chirruping their tiny battle-cry, till one is left defeated—and probably minus several legs —on the field of honour.

Hundreds of dollars are sometimes wagered on these contests, and punters, before making their bets, consult a placard setting forth the prizes previously won by contesting crickets. Stakes are picturesque as, in addition to money, they include a roast pig, a piece of silk, and a gilt ornament like a bouquet of flowers.

During the struggle there is intense excitement. After the victory, a successful owner is congratulated like a Derby winner. His champion suddenly acquires great value, and is priced according to the further events it is likely to win. Such a hero of many fights gets the title of "conquering cricket, and when it dies it is buried in a small silver coffin with the hope that more good fighting crickets, attracted by the honourable funeral, will be found next year near its grave." The most notorious cricket-arenas used to be in the vicinity of Canton, but the majority of Chinese village lads indulge in these insect-battles. Their champions cost only the trouble of catching them. Their arena is the dusty street, their pit an old chipped rice-bowl. But they get as much excitement in gambling for a cash as richer town-dwellers who wager silver ingots.

* * *

Most of the amusements and anniversaries of the ninth moon reflect the "male principle," being connected with things and persons of special interest to men.

The warlike expeditions to come later in the autumn (*see* De Groot, *Fêtes Annuellement Célébrées à Emoui*, and "The Eleventh Moon") were preceded by martial pageants. The troops marched out in their picturesque uniforms—red jackets with black velvet ideographs, turbans twisted round their heads, ten-foot pikes in their hands, lacquered bows and arrow-cases shining in the sun, triangular flags embroidered with fierce beasts—and sacrificed to the Celestial Standard-Bearer. Stepping in time to the beat of their drums, officers burned incense to their patrons publicly on the parade-ground, choosing for the ceremony the day known as the "descending of frost."

* * *

Yen Tzû   Meanwhile, the scholars, who despised the soldiers for their ignorance, quietly honoured two famous men intimately connected with Confucius. On the 11th of the ninth moon they kept the anniversary of Yen Tzû, or Yen Hui, the favourite disciple of the Master, whose fame as a saintly man rings down the ages like the sound of a bell hung in the sky. While other and more brilliant pupils learned quicker and understood better than he, Yen Tzû practised what his Teacher preached. Thus he stands for the best type of disciple, unable or unwilling to originate, content to build his reputation on his devotion to the Master, and following him so closely in spiritual development that Ssû Ma-ch'ien, the great historian, likens him quaintly "to a fly which travels far and fast by clinging to the tail of a courser." Gray-haired

NINTH MOON, OR "CHRYSANTHEMUM MOON" 437

at twenty-nine, he died at thirty-one, bemoaned by his Great Teacher as the Beloved Disciple, the St. John of the Confucian family. His tablet occupies the place of honour at the left of the Sage in all his temples.

On the 15th of the ninth moon, incense is burned before the tablet of Chu Hsi, or Chu Fu Tzû, the prince of Chinese philosophers (A.D. 1130-1200). In youth a Buddhist priest, this remarkable man later became the implacable enemy of Buddhist —and Taoist—doctrines and was always advocating more schools and fewer temples. His was a checkered career. At various times he held high official posts including, twice, the governorship of a province. But, in the midst of a successful administration, he retired to the White Deer grotto near the Po Yang lake where stands one of the oldest universities of which we have any record—Salerno, the oldest European university, not excepted. Unfortunately, official life claimed him again, and though he proved that he could detach himself from philosophy when necessary and be a good and practical official, absurd charges were trumped up against him—breaches of loyalty, of filial piety, and sedition. In those days an accusation could ruin a man, and his enemies did finally get him out of office. His innocence was proved later, when he was too old to serve the State again. Then, for the last time, he went back to his favourite retreat and stayed there for the rest of his life. Legend pretends that the origin of his superhuman wisdom was a pearl of great value which a fox-fairy, in the likeness of a beautiful young woman who came to serve him

Chu Hsi

in his grotto, brought as a gift. Yielding to her entreaties, he swallowed the precious jewel, "and the pearl became in him the fountain of wisdom such as is not possessed by mortal man."[5] It inspired the writings that made him great. Indeed, Chu Hsi can never be forgotten so long as China lives, because he is the most famous commentator on the Confucian canon, and his instruction, practical as well as profound, rightly interpreted the doctrine to the people.

After his death, his coffin hung, like Mahomet's, suspended in the air. It only descended to the ground when his son-in-law fell upon his knees beside it, and reminded the "departed spirit of the high principles of which Chu Hsi had been such a brilliant exponent in life." This appeal to the corpse of a great man to be consistent, and not make a fool of itself by practising black magic, is another indication of the practical Chinese psychology.

* * *

City Gods The 25th of the ninth moon is the birthday of the most celebrated of the *Ch'eng Huangs,* he of Anhui, where the cult of these hero-gods originated, and where the first historical temple to one of them (dating from 240 B.C.) may still be seen.

The Ch'eng Huangs are sometimes compared to the *Urbani* of the Romans, and, in general, to the

---

[5] Chu Hsi was, in fact, served by two female fairies, a fox-fairy and a frog-fairy. They could not live together in peace, and the philosopher, like lesser men, was long troubled by the squabbles of his women-folk. Finally, one day, after a bitter altercation "a dead frog and a dead fox were found near his retreat, and buried with due ceremony in the university grounds where a little stone still marks their resting place."

THEATRICALS AT THE TEMPLE OF A CITY GOD.

Photo by the Asiatic Photo Publishing Co.

TEMPLE TO A LOCAL GOD.

"local protectors" of various peoples. There is, however, one important difference between them and their Western prototypes. They are of human origin, whereas the antique gods of Europe—as, for example, Pallas of Athens, Mars, patron of Rome, and Aphrodite of Cyprus—were not.

The Chinese Ch'eng Huangs, being ascending humanity, not descending divinity, serve as a convenient link between the higher gods of the official worship and the people's immediate needs. They act as spiritual go-betweens, accessible to popular prayers—qualified agents who attend to the petty affairs of men.

Legend traces the vague beginnings of their cult back to the time of the Perfect Emperor, Yao, when a deified "guardian of dykes and ramparts" is considered the vague ancestor of the group. His Majesty Yao, who ascended the throne in 2357 B.C., instituted the composite *Cha* sacrifices, consisting of eight separate ceremonies held as a thanksgiving after the autumn harvest, that is to say about the ninth moon. The seventh sacrifice honoured the so-called "Superintendents of Husbandry," ghostly overseers of farm-buildings and boundaries—gods of the land and the locality. It was called *Shui Yung, shui* meaning water, and *yung* a mud wall, or rampart. Ultimately, the *yung* altered to *ch'eng*, a city wall—though exactly when is unknown—and *shui* to *huang*, the moat left after a deep embankment is dug. From these far-off beginnings the Ch'eng Huangs developed. Later they became, and remain, identified exclusively with cities. One of them is still regarded as the spiritual official in charge of every walled town in China, and their cult, "with that of the *T'u Ti*, the local tutelary divinity, and Tsao Chün, 'Lord of the Kitchen Stove' (which it antedates by at least a millenium) is probably the most popular in China to-day, because all three are intimately concerned with the affairs of the Dark World, and with the well-being of the *Kuei Shên*—spirits of the Dead." [6]

The Ch'eng Huangs increased in authority under the T'ang dynasty, when he who guarded the capital, then at Hsi An Fu, was raised to the rank of Prince. "His worship spread throughout the Empire, but the seasons of its observance were still not indicated in the *Ssŭ Tien*, Book of Religious Rites, nor had he, as a rule, a temple of his own." Simple *t'an*, or earth-altars, were raised in his honour, or he had a niche in the sanctuary of some other god. . . . It was not until the dawn of the Ming dynasty—when Chu Yüan-ch'ang set his newly won empire in order, both materially and spiritually—that the worship of the Ch'eng Huang was fully regulated. "In the second year of this Emperor's reign, temples were generally established to the city-god, and the plan on which they should be built definitely laid down. It was decreed that they should be in the form of official residences, that the Spiritual Magistrates should be represented as officials judging wrong-doers, and that they should hold ranks varying in degree with the importance of the post they held."

---

[6] See "Cult of the Ch'eng Huang Lao Yeh," by Florence Ayscough, *Journal of the North China Branch of the Royal Asiatic Society*, Vol. LV, 1924, also *A Chinese Mirror*.

Henceforth, the usual official hierarchy was established. All the Ch'eng Huangs together formed a Celestial Ministry of Justice, presided over by a Ch'eng Huang in Chief, usually the Ch'eng Huang of the capital, and subordinate to him. These canonised heroes held posts that corresponded exactly to the mortal provincial administration. The Emperor came with, but after, Shang Ti, the Supreme Lord; the Earth-Gods corresponded to their central administration, and so on, in a careful precedence, down to the smallest Ch'eng Huang who, like a junior magistrate, was in charge of a townlet. The system of dual control required a *fu* city to have two Ch'eng Huangs, because it had a Prefect and a Sub-Prefect, whereas a *hsien* city, in charge of only one magistrate, needed but one guardian-god.

There was nothing ridiculous to the Chinese mind in transferring a Ch'eng Huang from one town to another like an ordinary member of their civil service. It was even admissible to depose "Prince-Protectors," should they fail to fulfil their functions. The Taoist "Pope"—the Chang T'ien Shih (*see* "The Fifth Moon")—used to unfrock a Ch'eng Huang now and then, electing the *shên*, or spirit, of a deceased official of outstanding virtue to fill the post. He made such decisions at the instigation of the Yü Huang, Taoist Lord of Heaven. The public thought it just that gods, like men, should pay the penalty for slackness in the discharge of duty. On the other hand, promotions were given for good service. The Ch'eng Huangs might rise in rank by being transferred to larger cities, just as their mortal colleagues, after death, might be deified, proving that official careers were not necessarily ended in this world.

Naturally, it was essential to have an efficient Ch'eng Huang in charge of an important town, because his power was enormous. His was the whole duty of protection against enemies and epidemics.[7] His, too, the responsibility for reports on the actions of the people under his jurisdiction. His the right to suggest to the Lord of Heaven rewards for the virtuous and, to the Lord of Hell, punishments for the wicked. His, above all, the care of the dead and the duty of sending his attendants to accompany departing souls to the Courts of Punishment.

As it is usual to establish good relations with such a powerful personage, the temple of the city-god is a much-frequented place. Regular worship is held there twice a month—on the first and the fifteenth—and offerings are made to him on New Year's day at sunrise. Each official, on taking up his post, went to *k'o t'ou* at the "Municipal Palace." At all times, when danger or disaster threatened, he used to return and burn incense, or pray for help and advice. On the Ch'eng Huang's chief festival, his natal anniversary, it was the custom for the Prefect to present him with a suit of silk clothes in the name of the Government. First, he washed the face of the image with his own hands, then put on the new robes. Great crowds gathered to witness the

---

[7] The Ch'eng Huang of the capital, who was specially venerated under the Manchus, did not always succeed as a protector—for example in 1900. His principal temple, on the Shun Chih Mên Street, in Peking, has been since turned into a police station.

ceremony, and many petitioners begged the official to stamp garments belonging to their sick relatives with the jade-stone seal of the god, always under his care and only produced at this festival. The impression entailed a costly "squeeze." Poor folk had to be content with a chop from the copper seal kept in the shrine. But both had curative powers—in the eyes of true believers.

The usual Ch'eng Huang temple in a big city is the exact counterpart of an official *yamên* with flagstaffs at the gate, and a hall containing a big figure of the god dressed in a long robe, a scholar's cap, and black satin boots with thick white soles. He wears the same insignia of rank as his corresponding mortal colleague. The deep shadows of his dais give a Rembrantesque effect to the image, seated on a large carved chair facing a table with the badges of his office. The numbered "arrows of command" are for his messengers, to signify that they carry his orders. The scarlet boards bear the inscription: "Receive the Decree to go forth and inspect." The stands of the archaic weapons, square-bladed halberts, curved scimitars, and pewter maces mounted on long red staves, are symbols of his authority. In two rows, to the right and left of their master, stand his satellites, note-book in hand to jot down men's virtues and vices, and, behind him, lictors armed with rods to mete out punishment. A curious pair who wait upon the Magistrate are "Instead of a Boy," and "Instead of a Girl." You will often see parents *k'o t'ou*-ing before them. They have a sick child at home, and suspect the Ch'eng Huang desires the spirit of the little one as a page, or a handmaiden.

The local soothsayer has advised them to make offerings here and pray that "Instead of a Boy," or "Instead of a Girl," will consent to act as a substitute. If the petition is granted and their baby recovers, they vow that it shall march in the next procession in honour of the city-god, dressed in scarlet.

Sometimes in this same hall we find a war-junk to protect the "Protector" when he travels; sometimes a group of virtuous citizens who have volunteered to attend upon him. Often, an abacus hangs over the door to compute the sum total of men's good and evil deeds.

In some Ch'eng Huang temples, the principal courtyard is surrounded by plaster-groups representing the tortures of the Ten Hells shown in revolting detail. They remind us how close is the connection of the city-god with the Underworld. Indeed, his authority to rule is derived from King Yen Lo (*see* "The Second Moon") whose agent and official spy he is.

That is the reason why, three times a year, on the three great Festivals of the Dead, the image of the Ch'eng Huang is carried through the city in the midst of an extraordinary procession, though he may be taken out in his gilded sedan chair at any time when his help is needed—during a drought, an epidemic, or any public calamity. But his regular spring tour is at the *Ch'ing Ming*, when his image is carried beyond the western gate of the town, to "let loose the spirits" who return from the Inferno on this Chinese All Souls' Day to visit their earthly homes (*see* "The Third Moon").

On the 15th of the seventh moon he is escorted to the same place for the purpose of "counting the spirits," again freed from Hades for their long vacation; he, being specially in charge of the "Hungry Ghosts" without descendants to worship them, must call the roll on this, their festival. Finally, once more on the first of the tenth moon (*see* "The Tenth Moon"), he performs the ceremony known as "gathering the spirits," collecting and sending them back to hell after their last furlough of the year. On this occasion, the Ch'eng Huang's image passes the night in a house within the walls, not in his temple. He may not go home to sleep because his work is not finished. Next morning he does a round of the southern suburbs, returning through the principal streets of the city escorted by the Prefect, and warmly greeted by the populace who bow before him, present petitions, and perform acts of penance and thanksgiving.

The image carried in these processions is not the heroic-size figure of the Ch'eng Huang's sanctuary. As in most Chinese temples, the divinity is represented by two statues, the one immovable and always behind the altar, the other, as it were a reflection of the first, but small and portable. Both are treated with equal reverence. Thus, the processions in honour of the miniature Ch'eng Huang represent all the majesty of a great personage and a high office—all the antique, intricate, endless ceremonial of a complicated religious machine.

Under the Ming dynasty, when the cult of City Guardians reached its zenith of pomp and splendour, these processions were miles long, with banners, bands of music, soldiers, flags, penitents and painted devils—the whole cortège gay with cruel colours of barbaric intensity, scarlet and magenta, burning orange and blue, and the brassy gold of the flames of hell.

Only a few years ago we have ourselves seen such a procession, less grand perhaps than in former days, but still impressive, even terrifying enough to remain graven in the memory. The small central figure of the Ch'eng Huang, whose beard and moustaches were renewed beforehand by workmen who bored new holes in his expressionless plaster face, and stuck in fresh bunches of horse hair, occupied the centre of the stage —and rightly so. He sat with dignity in his chair; he carried a wand of office almost too heavy to hold. Behind him, attendants led a horse saddled and bridled in case he should feel inclined to ride. Beside him walked his faithful servitors, the "Tall White Devil" and the "Short Black Devil." The former was a fearsome figure, eight or ten feet tall, carried by a man whose knees and feet were visible. No wonder the children cried for fright, seeing his huge body-framework of bamboo covered with white cotton cloth, his head and face of ghastly white pasteboard, his high hat with a red band, his fan and his staff waving in his hands, his protruding eyes, and the bright red tongue hanging from his mouth. "Don't fear, don't fear," the mothers cry soothingly. "Pai Lao Yeh, the Old White One, has cash. See them hanging on a string from his mouth! If we ask politely, he will give us one to hang round baby's neck to keep away bad spirits, and bring us riches at the same time."

Scarcely less awe-inspiring is the "Short Black Devil," a small and stubby caricature of a man with a dark face, a tall black hat with a red band, and a protruding tongue waggled by a string. He capers and cavorts along the street. He fraternizes with other devils in the rear. For there are a whole population of sprites, and imps, and gnomes, accompanying the Ch'eng Huang, figures grotesque as gargoyles, fantastic and horrible. There are one-horned and two-horned devils, spindle-shanked, rachitic and malformed, carrying cudgels that bristle with spikes; devils with tattooed faces, colossal noses and paper hats, headed by their chief, "Golden Mask"; devils with huge nails piercing their heads and faces, and clothes bathed in blood; mischievous devils, carrying wine-pots, who lead drinkers astray; devils of the hanged, with chains about their necks; devils of the drowned, with willow-branches in their hands; special satellites called the "Ox-Headed Assistant," the "Horse-Faced," the "Cock-Headed," and the "Duck-Mouthed," all immense giants, alike save for their queer faces; the dreaded devil of suicide, and, finally, a ghastly figure whose entrails are half out of his body, like a pig lately killed. One shudders at the general impression of bestiality and impure magic, at the wealth of distorted imagination which has so securely and so horribly fastened the shackles of superstition on the populace. Pity the poor penitents bringing up the rear! They walk in terror, dressed in the red garments of condemned criminals weighed down with chains. And those with wooden "cangues" about their necks. And those others with spirals of lighted incense attached by little hooks imbedded in the flesh of nose, ears and arms, or iron skewers stuck through their cheeks, or knives buried in their flesh. All proceed in solemn and impressive silence one behind the other—acknowledged sinners prepared to wash their sins away with the sacrifice of their blood. Even troupes of children march in this sad company acting the part of spies for the Ch'eng Huang, an official letter to him pasted on their backs, and flags in their hands. As already mentioned, these little ones are generally fulfilling a vow made for them by their parents—a vow of gratitude for recovery from a dangerous illness.

There is a symbolic meaning to the whole procession, kindly explained to us by a skeptical Chinese gentleman from whose doorway we watch it return late in the evening, with the lurid murkiness of torches lending an appropriate weirdness to the convoy. The demon figures incarnate evil passions, crimes, diseases, and all forms of wickedness. Creatures of Hell, the Ch'eng Huang controls them because of his affiliations with the Underworld. So that is why they appear in the company of an essentially virtuous figure like the City God.

* * *

The Loyal Yo Fei Virtue, indeed, is the primary attribute of every City Guardian. The first Ch'eng Huangs, as historical records prove, were the most upright citizens of their native towns, specially canonised to rule them, men with the highest moral record, magistrates with a stainless reputation during their earthly administration.

On this roll of honour we find many noble names, but none more noble than that of Yo Fei, once Spiritual Official of Hangchow, the rival in chivalry of Kuan Ti (*see* "The Fifth Moon"), a hero whose deeds shine brightly in an age of darkness, a high-souled patriot and example of the most exalted personal and civic virtues.[8] Yo Fei's birth was miraculously announced by a great bird which alighted upon the roof of the house in the troubled years of the decline of the Sung dynasty.[9] His family, natives of Hunan province, must have been exceptional folk. They went hungry themselves to feed the poor, and gave their land to those who encroached upon it. Surely, they deserved their model son, who grew into a clever and high-spirited youth, touched by a moral earnestness uncommon in his generation. Yet, he was no milksop. Quite the reverse. He became a noted archer, and, in course of time, a brilliant general fighting, and fighting gloriously, for his sovereign against the Golden Tartars, then invading the northern provinces. His crowning victory routed a hundred thousand men commanded by the heir

---

[8] Newton Hayes in the "Gods of the Chinese"—*Journal of the North China Branch of the Royal Asiatic Society*, Vol. LV, 1924—says: "Yüan Shih-k'ai, in his brief enjoyment of Imperial power, decreed that Yo Fei should be recognised as a god . . . to preside over war, to rule over the destiny of Chinese military affairs, and to be co-equal in rank with Kuan Ti, already an occupant of the war-god's throne. A magnificent temple to this newly created deity has just been completed on the north shore of Hangchow's beautiful lake . . . in a portion of the large enclosure which contains the grave of one of the most honoured patriots in Chinese history."

[9] Hence, the second character of his name, which means "to fly," the same ideograph we find in the Chinese transliteration of "aeroplane."

to the Tartar throne in person. But such successes aroused jealousies. To achieve greatness is to achieve enemies, for greatness excites envy, and envy is the most fruitful of all the seeds of hatred.

The Sung Prime Minister Ch'in Kuei and his wife, traitors in the pay of the Tartars, hated the loyal Yo Fei. False and mean, writhing under the consciousness of their own shortcomings, they vilified and persecuted a successful patriot, hoping thereby to lessen the gap between himself and them. History testifies again and again that the more conspicuously great the individual, the stronger the incentive to slander him, for the interest of the slander is commensurate with the eminence of the personage assailed.

Yo Fei was finally thrown into prison on false charges, but when his case came up for trial he dramatically rent his jacket and showed four large characters which his mother had tattooed upon his back when he was a boy. These characters read: *Chin Chung Pao Kuo,* "Loyal and True to Death." On this curious evidence he was released, but only to be re-arrested later at the instigation of his implacable enemies, and secretly murdered in prison (in A.D. 1141).

After his death, tardy justice brought Yo Fei complete moral rehabilitation, and even canonisation, while the treacherous minister and his no less treacherous wife were doomed to everlasting infamy. While Yo Fei got back, posthumously, all his titles, Ch'in Kuei lost his. Men only remember the latter as "the false and foul," and his very patronymic has become the synonym for a spittoon.

Yo Fei, too, got honourable sepulchre. His tomb may still be seen near the margin of the lake at Hangchow, within the very district where his family owned their land. Here he lies at peace after his many wars, amid shady groves of bamboos whose straight trunks symbolise the upright official. The simple inscription his mother wrote indelibly on his flesh forms an appropriate epitaph, testimonial of his loyalty. A group of attendants, carved in stone, proclaim his dignity. Ch'in Kuei and his lady are also present, but in no honourable position. Two iron statues, near the entrance to the tomb, represent those dastardly traitors, kneeling and bound as if awaiting execution. Passers-by kick or spit upon them as an act of merit. The wheel of fate, turning, turning through the centuries, rights all wrongs at last. Unpunished in life, this couple expiate their crimes in effigy, exposed eternally and inexorably to heat, and frost, and the four winds that lash them. Who were courted, they receive contempt; who were tyrants, they are crowned never but with crowns of falling leaves.

An old legend interprets the life and career of Yo Fei and his enemies. Once, the holy meditation of the Lord Buddha was interrupted by an unseemly squeak from the Antique White Bat, a spirit born of chaos. The Phœnix, guarding the throne of the Blessed One, tore the Bat to pieces in an excess of zeal for the comfort of his Master. Nevertheless, Buddha rebuked the Sacred Bird and, as a punishment, sent him to earth to expiate his sin, giving him the body of a white eagle. As the latter is a bird of prey, the well-meaning, but misguided, Phœnix had all the predatory instincts to fight whilst working out his salvation. Temptation proved too strong. Seeing a tortoise asleep on a river bank, he forgot the Buddha's command: "Thou shalt not take life,"—and made a meal of the tortoise. For this new crime it was decreed that he be born once more to expiate in another life the wrong he had done. Thus it came to pass that he entered the body of Yo Fei, while the spirit of the tortoise was incarnated in Ch'in Kuei. As for the White Bat, he took the form of a beautiful maiden called Wang Shih. Destiny married her to Ch'in Kuei, their daughter became Empress, and her father was given the post of Prime Minister. Here are all the motives for tragedy and revenge, and here, plainly shown, the opportunity given to sinners to work out their salvation. The legend adds many ghastly details concerning the sufferings of Yo Fei, the anguish of his death and the tortures he endured in the hells prepared for murderers. Yet, at last, by reason of his virtuous life, he was purged of sin, and his spirit entered the presence of its lord, where it remains in constant attendance upon him, bearing a vase of peonies that spread a heavenly perfume; whereas Ch'in Kuei and Wang Shih, who persisted in evil, were condemned to eighty thousand re-births as miserable human beings. They were even forbidden re-incarnation as animals, because an unhappy beast is, still, more fortunate than an unhappy man.

*   *   *

Stories of the "Lords of Towns and Moats" would fill whole volumes. Every city has its local legends. There is, for example, an amusing one concerning the Ch'eng Huang of Yen Chin in Kiangsu. Here His Lordship figuratively "lost face" by omitting to prove the innocence of a worshipper accused of theft. The irate client thereupon returned to the temple and upbraided the god. At the well-merited reproach the plaster fell off the countenance of the image, and to this day it remains literally "without face"—a terrible lack in China—for, whenever the local people attempt to patch up their God-Magistrate, the plaster drops from his cheeks as fast as they put it on.

There is a sterner tale told of Chou Hsin of Hangchow, a Ch'eng Huang so dreaded for his severity that young and old fled at his approach. Once, while trying a case, a storm blew some leaves on his table. "Search for the tree to which these leaves belong," said he in a voice of ice and iron. None could be found in the neighbourhood but, eventually, one was discovered in a Buddhist temple far distant. Then the Judge declared that the priests must be guilty of murder. By his order, the tree was felled and the body of a woman found concealed in it, thus proving to everyone's satisfaction that the monks had committed the crime. At any rate they were punished for it.

Johnston, in *Lion and Dragon in Northern China*, tells a still quainter and more touching anecdote of the city-god in Jung Ch'eng, in Shantung. "The Chinese, as we have seen, regard three days in the year as specially sacred to the spirits of the dead. . . . On each of these spirit-festivals, the Ch'eng Huang is supposed to hold a formal inspection of his city. . . . The story goes that, during one of these periodical excursions, a young girl, member of a well-known local family, was watching the procession with the keenest interest. As the god's palanquin passed the spot where she was standing, she saw the image—or believed she saw it—deliberately turn its face in her direction, and smile at her with a look of friendly interest. A few days passed, and the girl became seriously ill. . . . In less than a month after the meeting with the city-god she was dead. During the night following her death, her mother had a strange dream. She was visited by the spirit of her daughter who told her that she was now well and happy, for she had become the bride of the Ch'eng Huang. Needless to say, the dream soon became the common talk of their neighbours, through whom it reached the ears of the district magistrate. After evidence had been given and duly corroborated, it was officially decided that the Ch'eng Huang's will had manifested itself in an unmistakable manner, and that to thwart it would bring disaster on the city. The girl's body was, therefore, buried with much pomp and ceremony within the temple grounds, her image, robed in real silks, was installed in the central pavilion beside that of the God himself, and she received formal recognition as the Ch'eng Huang's consort."

Such wonder-tales are interesting because they give a better idea of the powers and personalities of the Ch'eng Huangs, and their place in the popular esteem, than any scientific or historical explanation can ever do. Men may forget their shadowy relationship with the ancient *Cha* sacrifices; they still ask of these "fathers and mothers of the people" to right their wrongs, to heal their sick, to intercede for the dead.

* * *

As every town in China has its Ch'eng Huang, so every village has its *T'u Ti*. These T'u Tis are smaller magistrates, represented by the "Spirits of the Spot" (*genii loci*), and may be patrons of districts, buildings, or compounds, as well as villages.[10] Even more than the Ch'eng Huangs, these local tutelar gods fill an actual popular need, and were invented, long before the philosophic religions, for the comfort and protection of the simple folk. Peasants might, or might not, accept Buddha or the Tao. But they needed, and therefore conceived, some kind of Spirit to whom they could pray for success in all communal undertakings; "for protection against sickness, for the triumph of their lord in war, for succour in seasons of famine or epidemic." Even to day, we know that it is not Buddha, Kuan Yin, or the Jade Emperor that the farmers beg for bountiful harvests; neither is it to them they give thanks for a plentiful rice-crop. Their offerings and their supplications are made to the Rulers of the Soil (*see* "The Second Moon"), and to the Protectors of the Place, the T'u Tis. The old idea persists that each parcel of territory is still ruled by its

Local Gods

---

[10] The Household Gods (*see* "The Twelfth Moon") are in reality T'u Tis. Their cult is intimately linked with that of the Ch'eng Huangs.

own ancient local divinities, whose cult represents the "moral experience of the community, its most cherished traditions and customs, its unwritten laws of conduct, its sentiment of duty." Furthermore, because customs are identified with morals, any offence against the customs of a Chinese village is considered an offence against the gods who protect it and, therefore, a menace to the public weal. The cruellest fate that can befall a man is exile for misconduct, beyond the reach of his native gods.

The T'u Tis, as a class, are hazy heroes, or mythical old gentlemen, possibly the deified ancestors of the first settlers. In proof of this, we have the persisting custom of announcing the death of any individual in the community over which he presides to the T'u Ti, much as a death in the family is ceremoniously announced to the ancestors. The local god, in fact, holds the position of registrar of births and deaths in his village. We have ourselves been touched more than once by the sight of country-folk, dressed in white, with sack-cloth bands around their hair, wending their way down the village-street to report at his shrine that so-and-so has died. The men bow, the women wail, someone whispers the sad announcement, someone lights an incense-stick. The simple ceremony takes place in the street, for the T'u Ti, unlike the Ch'eng Huang, has no grand temple. In fact, his sanctuary is generally so tiny that no one can enter it. The god himself, and his lady-wife, are represented by rough clay effigies a foot or two high. Sometimes a concubine is added to the family-party, in which event the latter's statue is smaller still, denoting her inferior rank, and placed on his right—the left seat, the place of honour, being occupied by his legal wife. Around this quaint group, red rags are fluttering in the breeze, either on little masts, miniatures of the staffs that flank the gates of official buildings, or hanging on a tree whose branches shade the shrine. These pennants are offered by persons cured of sickness, and Johnston suggests that they may represent the blending of three old beliefs, "the cult of the local tutelary god, faith in the magic expulsion of disease, and the worship of sacred trees." A similar custom prevails in Tibet, at the *obos* of Mongolia, and in other widely-separated countries, the underlying idea being that the votaries have transferred their illness to the bits of cloth.

Kipling's lines about "a rag, a bone, and a hank of hair" come to mind as we watch a new flag planted one sunny morning. Nevertheless, superstition has its consolations. It must be a great comfort to these peasants this belief that their bodily ills can be cured by such simple methods. It is, without doubt, a satisfaction to feel that the spirits animating the battered stone images of their local shrine take a personal interest in their small affairs.

The T'u Ti is the recognised shepherd of his flock. He alone has power over the immediate neighbourhood; hence the saying: "The T'u Ti at the east end of the village is helpless at the west end." Frowned upon officially, repressed under the Sung dynasty as gross superstition, then reluctantly included in the list of Taoist gods, his authority persists despite the law. Actually independent of any authorised religion, he has more power than many of the greater gods with splendid temples. If they should disappear, he would doubtless remain, the personal protector of the country-side, friend of the peasants, invisible guardian of their village territories.[11]

---

[11] In Japan, the T'u Ti appears as the *Uji* god. He is greatly reverenced, and his parishioners are called *Ujigo*, or "Children of the Tutelar God."—A counterpart of the Chinese T'u Ti also existed in ancient Greece and Rome, countries where the ancestral cult was highly developed. There, every family had its own altar, and every group a communal altar to the local protector, the latter being, supposedly, more powerful than the individual family god.

CHAPTER XIV.

THE TENTH MOON, OR "KINDLY MOON."

"INDLY," "Helpful," or "Benevolent," are popular names for the tenth moon. The "Small Snow" may be expected in North China during this period—a gift of heavenly charity to the dry plains. The farmers rejoice when the first white flakes fall, softly kissing their sleeping fields. Later, the "Big Snow" will cover them with a thick white quilt as if to say: "Rest, weary workers, rest till spring comes again."

Poor folk have a seasonable proverb: "Now is the time to embrace one's shoulders," *i.e.* to keep the arms folded on the breast, as Chinese do when they feel cold. On the first of this month, according to custom, people light their portable braziers made of baked white clay set in an iron frame. Such stoves are open, and when filled with coal-balls give out a fierce heat. As no pipes are used, the gas escapes directly into the room. Hence the number of asphyxiation cases each year in Peking among those who sleep in tiny cubby-holes with windows and doors tight shut. To protect themselves against the odour of carbonic acid gas, *sung hua*, the tiny flowers of the pine, are thrown into the fire, while a mixture of various candied fruits, called *t'ang hu lu*, skewered on bamboos, is eaten to counteract its deadly fumes. Why fruit should have power to do this, is not clear. Perhaps the superstition is simply an excuse for enjoying sugared dainties which the Pekingese love, at a season of the year when they can be most successfully prepared. Clear, cold sunshiny days make the best *t'ang hu lu*. Hot or damp weather affects the syrup mixture adversely. That is the reason Chinese confectioners make a special display in the tenth moon of fruits preserved in honey, and peddlers hawk brochettes of candied grapes, cherries and orange-slices. Grown men, as well as children, may be seen strolling along Peking lanes sucking these bamboos of hatpin length. Small red crab-apples are also strung on strings, worn as necklaces, and gradually eaten on the way home from a fair.

Roast chestnuts are another delicacy sold on the streets about this time. They are cooked in big cauldrons filled with black sand, on the sidewalk outside the smaller restaurants. Old reed-mats serve as fuel, while an apprentice armed with a long spade turns the nuts over and over to keep them from burning. There is no need for him to cry his wares, as the delicious smell from his cooking-pot automatically attracts purchasers.

* * *

Third Festival of the Dead

The first of the tenth moon is the last of the three yearly festivals of the dead. Families again visit their graves, though this duty is not considered so imperative as at the *Ch'ing Ming* and

CARRYING PAPER MONEY TO BURN AT THE TOMBS.

PILGRIMS AT A TEMPLE.

the seventh moon Feast of Hungry Ghosts. Those, however, who still strictly observe the customs of their forefathers will repeat before their tombs the same ceremonies we have already described. A special association, called the "Society of Neglected Bones," takes this opportunity, when the autumn harvest is gathered and men have comparative leisure, to inspect lonely graveyards, point out necessary repairs to the persons responsible for their upkeep, and provide coffins and grave-sites for the poor.[1] Each member makes a small money contribution—perhaps only a few coppers—for this charitable purpose, and, in addition, is expected to give one month's personal service in neglected cemeteries, whether public or private. He gets his reward on the birthday of the patron-saint of the association, Wên Wang, father of the first Emperor of the Chou Dynasty (about 1100 B.C.)—a prince renowned for his charity and respect for the dead,—when, after a feast has been served in honour of the lonely souls, the "Neglected Bones'" members enjoy the "funeral baked meats" in the twilight of the sheltering trees.

Whether or not they visit their tombs again in the tenth moon, the Chinese, in their own homes, observe the ceremony known as "burning the clothes."

---

[1] It is interesting to note that among the Celts a festival of the dead also took place at the beginning of winter, when the powers of growth were at their weakest, the season of earth's decay being, not unnaturally, associated with the death of mortals. This festival has left survivals in various modern folk-customs, especially in Ireland, where the dead are still supposed to come out of their graves and visit the homes, where a good fire is made to greet them.

As the dead are supposed to have the same needs as the living, warm garments and other household necessities are sent to them at the beginning of winter. Paper imitations of wadded robes are packed into parcels with paper-money for current expenses in the Shadowy World, and carefully addressed to the ghosts for whom they are intended. Medhurst says that "writings are drawn up and signed in the presence of witnesses to certify the conveyance of the property, stipulating that, on its arrival in Hades, it shall be duly made over to the individuals specified in the bond." This deed contains a list of what is being sent, whether money, clothes, or paper effigies of man-servants or maid-servants, and is burned together with the offerings, so that the worshippers "feel confident their friends obtain the benefit of what they send them. Thus they make a convenant with the grave, and with hell they are at agreement."

The parcels are usually laid out on the *k'ang*, or brick bed-platform, and the spirits invited to graciously receive them before they are set alight. Kindly folk prepare an extra parcel to propitiate any wandering and shivering spirits. This serves a double purpose, helping the helpless and preventing mischievous or destitute ghosts from trying to steal the clothing from the family ancestors.

Two singular customs connected with Fukien province, and observed at this festival, are worthy of mention. Here a married daughter, if one or both of her parents are dead, is expected to offer a gauze trunk with shelves spread with a variety of household utensils in miniature—little chop-sticks, very small rice-bowls, such as children use, and tiny beds, chairs and tables like doll's house furniture—either to her surviving parent or, if both her father and mother have entered the spirit world, to her brothers. The trunk and its contents must be burned on the premises where her parents lived.

In addition, she is expected to make a food-present to her family, the dishes to include a duck, half of which will be returned to her. This is called locally "dividing the duck". Thus a married daughter gives yearly proof of affection for her deceased parents. Not only do these offerings show how thoroughly filial piety has saturated the ranks of Chinese society, but also, what foreigners little suspect, that a Chinese woman, though she has "changed her family", still keeps in touch with her own kith and kin who, despite the fact that they have no longer any legal control over her, often retain considerable influence over her married life.

* * *

The fifth of the tenth moon is the anniversary of Bodhidharma, or Ta Mo, as the Chinese call him—the first Chinese patriarch of the Buddhist Church who reached Canton after a three years' voyage from India about A.D. 526, when Wu Ti of the Liang dynasty was on the throne. Son of an Indian king, Ta Mo became a monk, rose to be twenty-eighth patriarch of Buddhism by succession of the great Kâsyapa, and then left his country as a missionary. Unpopular at home

Bodhidharma

as a sectarian who aroused the enmity of his fellow Buddhists, the pious Teacher came abroad to found the "Ch'an", or "Contemplative", School, whose doctrine is called "The Thought Transmitted by the Thought," that is to say—transmitted without words, either spoken or written. By the tradition of this doctrine, believers are to see the so-called key to the thought of Buddha, or nature of Buddha, directly by their own meditation. Legend says: "When the Buddha was preaching upon the Vulture Peak, there suddenly appeared before him the great Brahma who presented a gold-coloured flower to the Blessed One, and therewith besought him to preach the Law. The Blessed One accepted the Heavenly Flower, and held it in his hand but spoke no word. Then the great Assembly wondered at the silence of the Blessed One. But the venerable Kâsyapa smiled. And the Blessed One said to the venerable Kâsyapa: 'I have the wonderful thought of Nirvana, the Eye of the True Law, which I now shall give you. . . .'" By thought alone the doctrine was transmitted to Kâsyapa, and by thought alone Kâsyapa transmitted it to Ananda, and thereafter by thought alone it was transmitted from patriarch to patriarch even to the time of Bodhidharma who communicated it to his successor, the second Chinese patriarch of the sect."[2]

It is on record that the Emperor of the day, Wu Ti, called Ta Mo to his Court and had a long

---

[2] Ta Mo has been sometimes confused with St. Thomas who came to convert India, but Johnston, in his *Buddhist China*, emphatically corrects this mistake which arose through a sound-similarity between the name "Ta Mo" and the Chinese transliteration for "Thomas."

discussion with him. "Since my accession," said His Majesty, "I have built many temples and written many holy inscriptions; what has it profited me?" "Nothing," replied the Sage, with more sincerity than tact, "because all such things are like little drops of water that have dripped into a room, or like the shadows of clouds that follow an object, symbols without reality, shadows without substance." With more patience than we might expect, the sovereign again inquired: "What then is real?" to which he received the answer: "To give away all that thou hast and to perceive the germ of the Holy One within thyself."

The Buddhist narrator naïvely remarks that the "Emperor remained unenlightened." As a matter of fact, Wu Ti, "who loved the formalities of worship, had little sympathy with a sage who condemned them and bade him look for Buddha neither in holy books nor ceremonies but in his own heart." Equally unsatisfied with the interview, Ta Mo left the Court, floated across the Yangtze on a reed, and departed for the Wei Kingdom, where he had reason to think his doctrine of the non-reality of material things might be more acceptable.

Here he entered a monastery at Lo Yang,[3] and sat for nine years in uninterrupted meditation

---

[3] The famous monastery of Shao Lin, at the base of the Shao Shih mountain in Honan province. Founded in the Vth century, it still houses a group of Buddhist monks. The beautiful buildings are nowadays half in ruins, but the greatest treasure of the temple remains intact—the stone before which Ta Mo meditated, "wrapped in thought and surrounded by vacancy and stillness." True believers still see on it an imprint of the saintly figure.

with his face to the wall, reminding us of Simon Stylites and other Christian saints who sought salvation through silent contemplation in extraordinary attitudes. During this vigil, legend says, the legs of the "Wall Gazing Brahman," as the people called him, fell off. Wherefore, in Japan, to this day images of Ta Mo, there known as Daruma, are made without legs. Such images, wearing a red gown with a hood reminiscent of Indian draperies, may be seen on the toy-stalls of Tokyo bazaars so made that, however the little figure be thrown down, it will always bob up again into a sitting posture. " 'The Getting Up Little Priest' was originally modelled, or remodelled, after a Chinese toy made on the same principle, and called the 'Not Falling Down Old Man.' "

Legends about Ta Mo's life are favourite themes of Chinese artists, who generally represent him with a dark face and a curly beard, both decidedly non-Chinese, a shock of hair and a single slipper. Tradition says that, when he lay in his coffin, a disciple who came to inspect his body found the dead man with one shoe in his hand. When asked whither he was going, the corpse replied: "To the Western Heaven," and, sure enough, a few days later, when the coffin was opened, it was empty save for the slipper which the Saint had dropped.

The Ch'an, or Contemplative, School of Buddhism that Bodhidharma founded still counts numerous followers in China (and, as the Zen sect, many devotees in Japan) all sincere followers of the venerable Indian Sage, searcher of hearts and scorner of books. His esoteric school has since

THE TENTH MOON, OR "KINDLY MOON" 467

split into five branches which differ less in principle than in their methods of attaining salvation and acquiring personal "Buddhahood." The rivalry among these five sects led to the break in the succession of Chinese patriarchs, of whom the sixth did not appoint a successor because Ta Mo said: "With five petals the flower is complete, and I myself, the first of the six, am the stem on which the others shall grow." "The Separation of the Priests" occurred at a historic spot five miles south of Kiukiang, at the foot of the Li mountains in Kiangsi, where it was decided, after a conference of monks, to break the line. There was little difficulty about doing so as the Chinese patriarchs had no ruling power and were simply Defenders, Teachers, and Examples of the Buddhist Life and Doctrine, albeit with certain magical powers like the Taoist Immortals, being able to fly through the air, cross rivers on boats of leaves, and enter into trances.

The Ch'an doctrines are considered unorthodox by many Buddhists. Father Wieger goes so far as to stamp them as entirely non-Buddhist in their essence, but representing a development of the monism of the Vedas and Upanishads of Brahmanism. He reproaches Ta Mo's teachings with the decay of learning in Chinese monasteries, where "monkish energy concentrated itself on ecstatic meditation" instead of the pursuit of knowledge. Taken too literally, Ta Mo's teaching certainly led in many cases to mental somnolence but, on the other hand, it undoubtedly tended to "save Chinese Buddhism from the evils of priestcraft and clericalism and from a slavish worship of images, relics, dogmas and sacred books" which have stifled Buddhist thought in Ceylon and its principal mission field, Indo-China.

* * *

Buddhist and Taoist administration

When the line of Chinese patriarchs terminated about A.D. 800, there was little change in the Buddhist administration. The great majority of priests still reside in monasteries which depend, through the local civil authorities, on the Ministry of the Interior. In each province or district there are two Spiritual Supervisors (*Sêng Lu Ssŭ*), to whom are subordinate the Abbots (*Fang Chang*) of the monasteries. The latter, elected by the community and confirmed by their spiritual superiors, are responsible for the internal control of the establishment committed to their charge, receiving and disbursing the revenues. But the actual running of it is in the hands of the Master, *Chang Kuei Ti* or *Tang Chia Ti*. He is assisted by monks who supervise the behaviour of their brethren at services, meals, etc., and, in large temples, by a man in charge of the buildings. One priest also is appointed to the office of the *Chih Ko Shih,* and it is his duty to provide for the needs of guests and even to amuse them.

The strict rule of Chinese Buddhist monasteries requires the monks to rise at 1 a.m., and attend a service till 3 a.m. with no food but a little bread and tea. Another mass is held from 5 to 7 a.m., and still another from 10 till noon, after which comes a vegetarian dinner, theoretically the only meal of the day. A last service takes place from 6 to 8 in the evening, whereupon the brethren retire for sleep to be up again shortly after midnight. In practice, we may add, this hard road to salvation is trodden at a much slower pace. As an old abbot once admitted to us: "There are black spots on the sun, and Buddhism is not without its faults."

The same system of civil control is applied to Taoist monasteries, only, instead of the *Sêng Lu Ssŭ*, we find *Tao Lu Ssŭ*, or Supervisors of the Faith, with various titles and, in addition, Teachers, Sacrificers, Exorcisers, Magi, etc., all subject spiritually to the Chang T'ien Shih, or Taoist pope (*see* "The Fifth Moon"). It is worthy of note that the control of monasteries in China is still much the same as it was in the last days of the Imperial regime, as the Republic took over, practically intact, a system that had proved satisfactory.

* * *

The Five Rulers

Though various dates are given in different provinces—as so often happens—for the festivals of the Gods, the sixth of the tenth moon is the usual anniversary of the "Five Rulers", *Wu Ti* or *Wu Yo Shên*, not to be confused with the "Five Saints" or *Wu Shêng* (*see* "The Second Moon"), nor yet with the "Five Perfect Emperors": Fu Hsi, Shên Nung, Huang Ti, Yao and Shun. These Five Rulers are identified with the Five Sacred Mountains, the Five Cardinal Directions, the Five Colours, the Five Elements and even the Five Seasons.[4] They form part of

---

[4] For this curious division of the Moon Year *see* "The Chinese Calendar."

a mysterious group of spirits who preside over space and time. Their personalities and functions are so vague that a full account of them would only muddle and weary the reader. The Five Rulers, however, are of interest, because their cult is still alive. Men still worship the Yellow Emperor (known as the Spirit of the Centre) incarnate in the Central Peak, or Sung Shan in Honan, and Ruler of lakes, rivers, canals and forests; the Red Emperor of the South, or the Spirit of the Hêng Shan, or Southern Peak in Hunan, Ruler of the stars and all creatures dwelling in the waters, including the mighty dragons; the White Emperor, Spirit of the Hua Shan, or Western Peak in Shensi, Ruler of the mineral kingdom and the birds of the air; the Black Emperor, Spirit of the Hêng Shan, or Northern Peak in Chihli (there is a difference in tone between this Hêng Shan and the same word used for the Southern Peak, impossible to convey in English), Ruler of the Four Great Rivers and the animal kingdom; and the Green Emperor, Spirit of the Eastern Peak in the sacred province of Shantung,—the mighty T'ai Shan, Ruler of the destinies of mankind both here and hereafter.

How these mythical Rulers actually originated is lost in the mists of antiquity. Subdivisions of the Supreme Being, originally worshipped in all the directions of space, their cult began, shortly before Christ, around open altars like those dedicated to Heaven and Earth. It continued in temples built at the foot of the sacred mountains with which these Five Rulers became identified (how and why we shall see presently) and finally spread throughout the land. There could be no better illustration of how the human mind first invents its gods, and then proceeds to make them human, for soon we find these vague spirits not only with shrines and all the paraphernalia of ordinary worship, but with first princely and, later, Imperial titles given to them (the latter under the Sung dynasty), and mythical wives made Empresses by an edict from the Throne.

The cult of the Five Rulers remains especially popular in South China. One of the largest temples dedicated to them is in Canton, where five stones, supposed to be the petrified remains of five rams upon which these gods rode into the city, may be seen. The Genii were dressed in white, yellow, green, black, and red, respectively, and each bore in his hand an ear of corn. During their progress, they paused in the market place, where the rams were turned to stone, while their Celestial Riders disappeared in the air after prophesying that famine should never visit the city. Henceforward, Canton was called in their honour the "City of Rams."

There seems also to be some connection between the Five Rulers and epidemics. During the summer, when heat lies like a heavy hand on the southern cities where men live "too thick" with no hope of scattering to the mountains or the sea shore, the Rainbow Emperors, if we may call them so, are taken from their temples and paraded through the streets in the hope that they will ward off diseases and, especially, the dread cholera. Such idol processions are extremely picturesque.

Each god has his own sedan chair with volunteer bearers. He moves through a forest of bright red banners whose brilliant colouring, "angrie and brave", lends him a strikingly bold appearance. Behind him and before walk men with gongs whose brassy cries affect one like bells jangled, out of tune and harsh. Paper effigies of the Tall White Devil and the Short Black Devil, that we have already met as attendants of the Ch'eng Huangs, or City Gods (*see* "The Ninth Moon"), are carried in the procession. Paper copies of full-rigged junks, called "disease junks," also are borne shoulder-high. Finally, in the rear of the cortège, we find a well dressed man carrying two buckets containing pig's blood, buffalo hair and chicken feathers, symbolical of the dirt which causes sickness. Once upon a time, beggars performed this service for the community, but we have seen a volunteer from a respectable family bearing his filthy burden with slow, heavy steps out of gratitude for the recovery from illness of a near relative.

The gaudy procession wends its way through the city, and those who contribute money for it have the right to ask that the gods pass their doors, if the width of the streets permit. Thus, with many twistings and turnings, the cortège reaches the river-bank or the seashore. Here, under the burning sun, the junks are set afloat, carrying with them, the Five Rulers being willing, the pestilence that threatens the town. If the boats float straight to sea, so much the better. If they drift ashore, they bear the disease to the spot at which they land. A less selfish method is to burn the boats at the water's edge while the gods watch from their chairs and the two Devils are made to kneel, as if joining their prayers to those of the anxious populace.

* * *

Mountain Cult

The association of the Five Sacred Ones, petitioned to avert plague and epidemics, with the Five Sacred Mountains is more obvious. It is geographical, since these peaks are so situated that they conveniently represent the Four Cardinal Points and the Centre. Nevertheless, mountain worship existed in China long before the Five Rulers were conceived, having been a part of the Chinese Nature Cult ever since time immemorial, and born of that sense of mystical union with the universe, overwhelming yet including mankind, that for most people has been the beginning and the core of religious experience. Many hundreds, perhaps thousands, of years before Taoism or Buddhism came into existence, the sky inspired men's worship, and primitive imagination enshrined the gods upon the heights. To climb, therefore, was to approach the divinities, and the nearer to the heavens, the more acceptable man's sacrifices. "Lift up your eyes unto the hills whence cometh your strength" is a command which finds a sympathetic response in every human heart, for the noble peaks are the natural goal towards which our spirits soar, and the mountain tops our vision of perfection.

In a land like China, which is so largely a flat agricultural plain, there must have been a special spiritual need to look up towards mountains comparatively rare and, therefore, the more admired and desired. This motive was strengthened in course of time, when the dead were buried on their sheltering slopes. The ancestral ghosts, according to Chinese ideas, hovered near their graves and, being ghosts, became gods in the sense of acquiring supernatural power. What more fitting than that they should become gods of the peaks where they had their tombs? From this conception grew the custom of giving living personalities to the mountains. It was but a step further to attribute wives and anniversaries to them—as in the world of men. Then, the usual Chinese hierarchic spirit stepped in and identified the higher hills with the greater spirits, the lower with the lesser—human officialdom repeated as it were. Such, in brief, is the explanation of the development of a primitive belief.

Five sacred mountains were soon too few to absorb the entire mountain cult as it grew and spread. While they sufficed for the original animistic faith, the Taoists and Buddhists, when they appeared in China, also desired their holy hills. The Taoists, first on the scene, associated themselves at once with mountain worship, as they did with all the religious ideas they found rooted in the country, and practically appropriated the "Big Five," populating them with their own divinities, though without interfering with the old State Cult—as witness the temple erected as late as 1714 on Mount T'ai Shan to Confucius, in honour of the Sage's visit.

When the Taoists established the priestly guardianship which they still retain over the

North, South, East, West and Centre Peaks, the Buddhists, no less anxious for peaceful and profitable places of retreat and pilgrimage on the hill slopes—since they came from a land of mountain lovers—appropriated four mountains, called them sacred, and made them serve their purposes. These are the Wu T'ai Shan in Shansi, where the wise Wên Shu rules (*see* "The Fourth Moon"), the O Mei Shan in Szechuan (*see* "The Fourth Moon"), sacred to P'u Hsien, the Chiu Hua Shan in Anhui, the shrine of Ti Tsang (*see* "The Seventh Moon"), and the P'u T'o Shan, under the patronage of the merciful Kuan Yin, on an island off the coast of Chekiang (*see* "The Second Moon").

The four "Famous Mountains of Buddhism" were chosen to correspond to the four earthly elements of Buddhism: Air (Wu T'ai), Fire (O Mei), Water (P'u T'o), and Earth (Chiu Hua), the elements being thus assigned because of the geographical peculiarities of each place. But, in addition to the above-mentioned peaks, both Buddhists and Taoists have canonised any number of lesser hills with local traditions, and studded them with temples.[5]

---

[5] The list of mountains adopted by both religions is much too long to give here, for the *Shan Chih*, or "Mountain Chronicles" of China, fill many volumes. Mention should be made, however, of the "Eight Small Famous Hills" which include the T'ien T'ai in Chekiang; Yün T'ai and Ta Mao in Kiangsu; Chih Tsu in Yünnan; Wu Chih in Kuangtung (Hainan Island), Chi Yün in Anhui; Wu Tang in Hupei; and Wu Yi (Bohea hills) in Fukien, all lovely sites and all visited each year by many pilgrims. Near Peking we have the Miao Fêng and Shang Fang. Finally, there are many mountain-sites appropriated for special saints, such as the Lung Hu Shan, or "Dragon-Tiger Mountain," in Kiangsi, the home of the Taoist wizard, Chang Tao Ling, and, till quite recently, the residence of his lineal descendant, Chang Tien Shih LXII, the Taoist Pope.

None, however, have ever been able to compete in holiness with the original "Wu Yo" or "Five Sacred Mountains," which still hold their pre-eminent position because they once formed part of the official cult, and still claim among their worshippers not only the simple superstitious folk but many highly educated Confucianists.

T'ai Shan Great as are all these peaks, the T'ai Shan is far and away the most powerful and revered among them—Ancestor of all the Sacred Mountains, which are supposed to obey him as grandson of the Taoist Jade Emperor, himself the principal presiding deity over every mountain summit (*see* "The First Moon"). None dispute the Tai Shan's spiritual power, not even the dead whom he judges in his capacity as Emperor of the Infernal Regions. Their souls return to his mountain slopes as to the Elysian Fields. "My life ebbs fast," says the poet, "the T'ai Shan has given me a rendezvous." Even devils and evil influences fear the Eastern Peak. They know that a single stone taken from Mount T'ai carries with it some of the great power of that mountain. Thus, in many parts of China, one notices a boulder or rough slab set up at cross-roads, near bridges or at other places where bad spirits are supposed to congregate, with an inscription: "This Stone from the T'ai Shan dares to oppose"—meaning: oppose evil in any form. It seems likely that some of these boulders never saw Mount T'ai, but "the devils do not know that". Chavannes says: "We believe that the T'ai Shan is still one of the most active gods in China, one of those to whom popular imagination still attributes the greatest importance in the affairs of this world —and the next." Indeed, no better proof can be obtained of the vitality of the mountain cult in China, and of the T'ai Shan in particular, than to visit the Holy Peak in early spring, when the greatest of all Chinese pilgrimages draws the faithful in thousands to honour the mighty Divinity of the Heights, elevated by a Sung Emperor to the "Equal of Heaven," and further honoured by such titles as "Great Universal Mountain" and "He Who Sits In The Seat Of Honour," *i.e.*, the East, whence the sun rises and all life springs.

From T'ai An, where the pilgrims gather, they get a distant glimpse of the Sacred Mountain whose majestic head is raised above the lesser peaks of Shantung, a symbol of strength and refuge to them as to their distant forefathers who, in the beginnings of history, took shelter on its slopes from the flooded swamps of the Yellow River, and made here "a centre of pristine culture, nucleus of what we may call the Holy Land of China."

With each succeeding age the halo of sanctity around the T'ai Shan has deepened and brightened till innumerable historical associations cluster round the holy peak, associations confirmed by numberless inscriptions carved on the rocks. Now the entire mountain has become one vast book of records written upon stone. Every cliff surprises us with new evidence of how age and tradition have added to the spiritual power of the god of the peak. No wonder the humble pilgrim of to-day is awed as he treads the self-same road along which so many of the famous Emperors of China toiled to report their accession to the throne, as the Japanese Emperors still report a change of sovereign to their ancestors at the shrines of Ise.

It was the Royal policy, says the Book of Rites of the Chou Dynasty, to make a tour of inspection with sacrifices regularly to all Five Peaks and, usually, to the T'ai Shan in the second moon. Sometimes these Imperial pilgrimages included an immense retinue of officials, both civil and military, besides ambassadors from foreign nations, with their wives and concubines. Finally, these processions became so costly to the people that, after the extravagant royal progress of the year A.D. 1008, they were never repeated in full splendour.[6]

Let us follow in the footsteps of the mighty and join the crowd of modern worshippers who gather at T'ai An,—Buddhists, Taoists, and Confucianists, who alike expect the T'ai Shan to interrupt the machinery of the universe for their provincial little petitions.

Before starting for the mountain, it is customary to visit the T'ai Miao within the city walls. This vast enclosure, which occupies a quarter of the entire town, contains many shrines. There

---

[6] Seventy-two legendary Emperors and seventeen historic Emperors are believed to have climbed the T'ai Shan. Ch'in Shih Huang, builder of the Great Wall, sacrificed here. Kao Tsung of the T'ang dynasty was one of the best known Imperial visitors. K'ang Hsi, the great Manchu, came twice. Yung Cheng, the ardent Buddhist, ordered repairs done to many of the temples, and Ch'ien Lung arrived in 1770, preceded by hundreds of brilliant banners, escorted by nobles on horseback and accompanied by a white elephant, as is duly recorded on the frescoes of the T'ai Miao.

is a "Vestibule Pavilion" where the faithful prostrate themselves in a "salute from afar" to the Sacred Peak. There is a throne-room of magnificent proportions with an image of the God dressed in Imperial yellow robes and, beyond, in a grove of old evergreens, a "forest of tablets," —tall, gray stones recording the petitions and thanksgivings of the Ming and Ch'ing Emperors.

In a rear courtyard we find the private or living apartments set apart for the Spirit of the T'ai Shan—another curious proof of how the Chinese humanise their gods. Within the ruined bed-chamber stands an image of the Emperor of the Peak, his principal Wife and his two secondary wives, the Eastern Duchess and the Western Duchess, all much the worse for wear. The family party is completed by a figure of the T'ai Shan's son in a deserted shrine near by. He is the father of the beloved "Princess of the Coloured Clouds" whose miraculous image was discovered on the summit of the Eastern Peak.[7]

When their preliminary prayers are said to the holy family, the pilgrims pass through the city gate and cross the plain to the white stone

---

[7] This is the same goddess, whose Chinese title is the Pi Hsia Yüan Chün, that we saw enshrined at Miao Fêng Shan (see "The Fourth Moon"). She is also sometimes called the "Dawn Goddess," owing to her relationship with the T'ai Shan, or Eastern Peak: *ex oriente lux*. The pink tinted clouds that attend the rising sun seem to the poetic fancy of the Chinese to personify the ethereal fairy divinity, and gain her the title of Princess of the Coloured Clouds. She is much reverenced, while her parent, though supposed to be Regent of the Eighteen Provinces and surrounded by images representing them, is tucked away in a closed and roofless sanctuary fast falling to decay.

Doré, in his *Récherches sur les Superstitions en Chine*, II & IX, 990, gives an elaborate biography of the whole family of the T'ai Shan, as also of the Emperors of the other Sacred Peaks.

archway which marks the beginning of the ascent of the mountain. All along the well-worn stone road beyond it we find numerous temples; spacious buildings kept in good repair by the contributions of the devout; tiny shrines where "quiet drips from the shadows like rain from leafy boughs;" and neglected sanctuaries where the stone is peeling in great flakes from tottering balustrades. Only a few which express the cult of mountain-worship and the many-sided personality of the T'ai Shan need be mentioned here. Of such is the Temple of Hell, a weird uncanny place where his grim image presides as Judge; the Temple of the Three Sovereigns, Fu Hsi, Shên Nung and Huang Ti, mythical rulers of primeval China (see "The Hundred Gods"); the picturesque belvedere of Yü Huang, the Jade Emperor, who has another altar on the mountain summit; the adjoining Palace of the Great Bear Constellation, a popular shrine on certain festival days when its altar is piled with gifts of corn and cakes; the Hall of Lao Tzŭ, with engraved models of the best writings of the T'ang dynasty—beautiful inscriptions from China's Golden Age of Literature that smoulder here like jewels in the dark.

The winding course of the Long Stone Road, worn smooth and slippery by countless feet, is spanned by three stone archways called "Gates of Heaven". Each marks a definite stage in the ascent. The first, leading to a long flight of stone steps, is the beginning of the true climb, all below this portal being considered the foot of the mountain.

As the gradient stiffens, tired men and women make every temple an excuse to stop and rest. They burn incense, whatever their creed, at the Red Gate Palace, the chief Buddhist Monastery on the T'ai Shan. Here one notices a curious anomaly—a shrine to that modern Taoist goddess, the Princess of the Coloured Clouds, whose image in one of the halls is associated with two Bodhisattvas. A little further on comes the Shrine of Maitreya, the fat "Laughing Buddha." Carnally represented, the Mi Lei Fo caters to the physical needs of the pilgrims and, attached to his temple, is a charming open pavilion where people sit and sip tea while enjoying the view upwards towards the peak, and downwards, over the Peach Orchard Valley. In olden times, this was the "Raiment Changing Pavilion." The very name, like that of "Turning Back Horse Arch" that one reaches a little further on, calls up pictures of old-time dignitaries, who started on gayly caparisoned mounts and in rich robes of ceremony but, humbled by the difficulties of the Road to Perfection, quitted their silken garments to climb the last steep track on foot like simple folk.

A shrine to Kuan Yin, Goddess of Mercy, here called "The Great Personage from the Southern Seas," adjoins Maitreya's temple. But, strange to say, for once the Chinese Madonna seems neglected and eclipsed by the Princess of the Coloured Clouds.

On a hillock to the east of the road is a curious memorial—the Tomb of the White Mule, with an odd legend attached to it. The mediaeval chronicles relate "that, in the year A.D. 726, the great Emperor Ming Huang of the T'ang house (grandson of the notorious Empress Wu Tzê T'ien) planned to follow the ancient custom of paying his respects to the Eastern Peak in person. . . . The prefect of Yi Chou in Southern Shantung had a very strong white mule which he sent to T'ai Shan as a mount for his Emperor. Although the sages had inveighed against the Book Burner, Ch'in Shih Huang, for his disrespect to the peak by the use of a chariot with horses, they had nothing to say against the use of the mule for the royal body. The ascent and descent were made in perfect safety on the back of the remarkable beast, there being, of course, no stone stairs but only a rude trail to the summit . . . As the Imperial rider put his foot to earth . . . the strong white mule who had so faithfully performed the duties required of him, expired without warning and without apparent malady. The Emperor, convinced that the animal must have been of supernatural character, and especially interested in the house of T'ang, at once conferred upon the dead steed the posthumous title of military commander as 'General White Mule.' He also ordered a coffin of special dimensions, a splendid funeral and interment under a cairn of mountain stones." The last, of course, was complimentary, for T'ai Shan stones (as we have noted) are able to withstand all evil spirits.

Next comes the "Tower of Ten Thousand Fairies," once a prosperous monastery, now a ruin whose tumble-down cells are farmed out to bedraggled beggars. Each of these "want-food-persons" has his own beat, marked out by piles of rocks, and here he lies like Lazarus, with his sores exposed to the scorching rays of the sun, asking alms of the passers by: "Honourable Sir, do a good deed! Give to the poor that the gods may give to you." The appeal is irresistible, because the Chinese, so prone to listen to reason, find the argument sound.

A picturesque stretch of road meanders through groves of evergreens and scrub-oaks to the Monastery of the Mother Goddess of the Great Bear (whom scholars ask to intercede for them with her son, the God of Literature), passes the "Cave of the Veil of Spray" with its pretty waterfall, and then enters the Valley of the Stone Canons whose rocks are covered with quotations from the Diamond Sutra carved in bold characters, finally reaching the Second Heavenly Gate.

This is reckoned the middle point of the road. As a matter of fact, it is more than half-way, because the Chinese calculation of distance is inaccurate, and represents rather the amount of energy required for a given stage than the actual ground covered.

On a pleasant terrace everyone stops for rest and refreshment. The place is crowded with those two-bearer chairs peculiar to the T'ai Shan —quaint conveyances which are simply bags of netting with wooden back and foot-rests, attached to stout poles suspended from the shoulders of the bearers by leather straps. Wonderful men these coolies are, Mohammedans all, members of one guild, inhabitants of one village of mud-hovels near T'ai An Fu. Their feet seem made of steel, and the play of muscles on their bronze backs is like the working of springs. Trained from childhood to this mountain work, they can take the steepest bits of the road (when a single false step would mean a terrible accident) at a trot. But they never slip. Their fathers and grandfathers before them have been carriers on the peak, in the days when a misstep was punished by death, and they themselves, defying the normal capacity of human bone and muscle, are a constant miracle in their climbing.

After the Second Heavenly Gate, the road descends three flights of steps called the "Upside Down Stairs," and winds in a level stretch along the slopes of the Pleasant Mountains. At every turning are pinnacled rocks with inscriptions. Above towers "The Peak which Advances Threatening," and the romantic "Nine Maidens' Fort," a ruined refuge of nine amazons in the troublous times of the VIth century.

The hardest part of the whole climb begins at the Gateway of the Dragon, opening on tremendous flights of steps (of which there are in all seven thousand on the holy mountain), known as the "Sudden Eighteen Flights." These last rock-stairways are so steep that iron chains have been hung on either side to assist the weary and dizzy pilgrims to mount them. Imagination pictures this gorge as a kind of Jacob's Ladder, up whose stone rungs the faithful climb because nothing can quell the soaring of the human spirit, or its visions of dreamed-of divinity. But the figures that mount and descend, leaning heavily on pilgrim staves, hobbling painfully on bound feet, wear no shining wings. They are no glowing band of mystics but everyday men and women, and many we guess, from the look of them, have known more dinner-times than dinners.

Standing high above them, like a promise, is the last Heavenly Gate, framing a patch of turquoise sky. Though wilder and steeper, harder and rougher grows the road, the pilgrims, stopping to rest every few minutes, know that a few more efforts will bring their painfully performed act of piety to an end. Glimpses of Heaven's nearness lend them courage to go on, till they look upon the majesty of the setting sun, "and try to touch the near-by stars with outstretched hands."

The T'ai Shan, in reality, has several summits, including the "Peak Where One Contemplates the Sun," the "Peak Where One Contemplates the Moon," and the "Father-in-Law Peak," the highest. All are temple-crowned. Two of the shrines near the Heavenly Gate present a curious contrast —the "Tower Which Touches Upon Emptiness" (with its mystical symbolism) and "The Temple of the Military Sage," Kuan Ti. How strange that even here, so near to Heaven, the War God holds the strategic pass and the idea of human strife intrudes!

Modern worship centres largely in the Palace of the Princess of the Coloured Clouds which stands at the end of the Heavenly Street (*see* "The Fourth Moon"). The T'ai Shan's daughter began her connection with the mountain when a rude stone statue of her was discovered on the summit in A.D. 1008, and later replaced by one of jade. Now she has the grandest sanctuary on the peak, with bell tower and drum tower, a pagoda, called the Fire Pool, for burning offerings, and a holy of holies covered with copper tiles. This hall is closed by a wooden grille. None may enter, but the pious bowing before the bars catch a glimpse of the Princess sitting in state on her red throne between Our Lady of Good Sight and the Goddess of Fecundity. Offerings are thrown through the palings: cakes, cash, women's shoes, clay dolls and cardboard eyes. If the offerings reach the altar, good luck may be expected. If not, the blind man goes away as sightless as he came; the mother longing for a son feels that her prayers have fallen on deaf ears.

Near the big temple is a smaller shrine with a rare and curious representation of the Princess of the Coloured Clouds. On a bed hung with curtains in formal folds, she lies asleep for nine months of the year. But in the rainy season she is invited to get up, lest her robes be injured by the damp. The ceremony of arousing Her Highness takes place annually on the 18th of the fourth moon, when Taoist priests enter her chamber, recite prayers, play upon musical instruments, and finally place her image on a chair. The quilt which covers her while she sleeps, her robes and gilded head-dress with brass pendants, are all gifts of women-worshippers seeking her favours. Altogether, she is a very friendly human person, as Chavannes remarks, "more like an ordinary peasant-wife in holiday attire than a great lady or a royal princess."

Among the temples of the peak there is one dedicated to the God of the T'ai Shan but, contrary to our natural expectations, it is small, uninteresting and neglected, though once upon a time, before his grand-daughter stole his popularity, it seems to have been one of the favourite places visited by pilgrims. The sanctuary dedicated to the Green Emperor, "governing principle of growth and fertility", also attracts few visitors these days. Likewise the temple to the Jade Emperor, the topmost shrine of the mountain, is not over-popular. In the courtyard, the boulder which forms the actual summit of the Father-in-Law Peak is enclosed by a stone balustrade. It is regrettable that the little figure of the Supreme Lord is so unimposing. Indeed, there is something rather pathetic in the small image of the "Great and Honourable Sovereign of Jade, the Emperor on High," sitting abandoned in his gilded robes and mortar-board hat from which hang the thirteen red-beaded cords indicating his supreme rank, with a plaintive expression on his pear-shaped face, to which drooping moustaches of black hair lend a false look of fierceness.

More impressive than the little Lord of Heaven is the famous Tablet Without Inscription, a massive monolith just outside his gate. One single character and one only, *Ti*, "Lord," breaks its smooth polished surface, but the column is older than the ideograph. Indeed, tradition says it was placed here by order of the Emperor Ch'in Shih Huang Ti when he came as a pilgrim to the T'ai Shan in 219 B.C. This legend is doubtful. Historians now agree that the tablet probably dates from the visit of Wu Ti of the Han dynasty who came on an inspection of the Eastern Peak in A.D. 110.

Passing the "Summit where Confucius Contemplated the Kingdom of Wu" and where, naturally, a temple commemorates the Master's clarity of vision, lovers of historic associations, rather than pious pilgrims, will climb to the plateau on the "Peak Where One Contemplates the Sun." Marked by a single stone tablet on a ruined altar, this is the site of the infinitely old Imperial *Fêng* and *Shan* Sacrifices, and admirers of Chinese culture with its wonderful continuity feel this to be the holiest ground on the mountain. Here Chinese sovereigns bowed before Heaven in the days of Noah, when the first ceremonies were conducted by the Emperor alone. Later the ritual became more elaborate, and the *Fêng* sacrifices were offered to Heaven at the summit of T'ai Shan, while the *Shan* sacrifices to Earth took place at the Shê Shan hillock at its base. Chavannes suggests that the two places chosen were not merely fixed upon "because of a physical configuration which approximates the one to Heaven, the other to Earth . . . They intervene as divinities, playing the part of intermediaries between the Sovereign of Men and Heaven or Earth, as the case may be. Thus, the prayer addressed to Heaven is confided to the spirit of T'ai Shan, that to Earth to the spirit of Shê Shan, in order that these deities may deliver them to their respective destinations." So interpreted the antique ceremony becomes intelligible. The *Fêng* and *Shan* differed from other Chinese sacrifices in that none of the offerings were burned. The

Imperial petitions, instead of being sent by the usual channel of mounting flame, were inscribed on tablets of jade and officially placed in charge, as it were, of the divinities of the two sites. Some of these jade slabs, elaborately enclosed in a stone casket of geometrical design, were found on this very spot about 1480. The Ming Emperors reburied them, fearing to remove them, lest they disturb the spirit of the mountain. Ch'ien Lung, however, with a love of beauty stronger than a superstitious dread of evil consequences, later took them to his capital.

Not far from the antique Altar of Heaven is the "Cliff of the Love of Life," a sinister and forbidding precipice, place of suicide. With a single leap, a weary soul is freed from a tired body. But nobler motives have often inspired supreme sacrifice on this tragic spot—filial piety, for example, driving devoted sons and daughters to die in order that a parent may recover from illness or be spared misfortune. This Oriental notion that a life given can avert disaster from the family, is expressed in the older name of this sad place: "The Cliff of Atoning Death."

Yet "to normal-minded folks who climb in the footsteps of scholars and Emperors, the poets and peasants of China's great past," there is every reason to cling to life here. The soothing peace from T'ai Shan is spread below us—a splendid panorama of nature, a view unsurpassed over the homeland of Confucius, over the river Wên "twisting like a huge yellow dragon shedding his sandy scales on the near-by farming lands, over the city of T'ai An with its temple-roofs gleaming in the sun." Whatever his creed, the pilgrim who looks down while a last rich flood of sunshine pours over the earth from the glowing, everlasting urn resting on the far horizon, feels himself near to the Unseen. The Great God of Nature, call him T'ai Shan or what you will, seems to offer the gift of faith—that universal faith that often languishes in crowded cities, or steals shamefaced to hide itself in dim churches or dusty temples, but here, on the mountain-top, flourishes greatly, filling the soul with solemn joy.

FIVE HUNDRED LOHANS, PI YÜN SSÜ, PEKING.

Photo by Yamamoto, Peking.

BUDDHIST NUNS.

Photo by the Asiatic Photo Publishing Co.

CHAPTER XV.

THE ELEVENTH MOON, OR "THE WHITE MOON."

HE eleventh moon sees winter firmly entrenched in North China, not to relax his dominion till after the New Year and then reluctantly, inch by inch, fighting as he goes with bitter winds and cruel spasms of cold. Meanwhile, all the rivers and lakes and ponds are in the grip of his white fingers. The formal gardens are stripped to bare skeletons of rocks and twisted evergreens. Even in the south, verdant valleys become chess-boards of brown squares, spotted with the stubble of harvested rice which looks like the stunted tufts of bristles in a worn-out scrubbing brush.

In Peking, the first "white day" is fixed for "stuffing up windows," and in every house all cracks are tightly sealed with stout paper to prevent draughts, for ventilation—always dangerous in Chinese eyes—is entirely dispensed with in the cold weather. Under the monarchy, officials who had the right to wear martin-skins were expected to put them on at this date. The regulation of such details proves how the conduct of the individual in Chinese society was formerly bound by the pressure of common opinion. Politically much freer than we were in olden times, the Sons of Han were socially far less so. The individualist and the eccentric could not flourish where unusual behaviour was judged adversely as a departure from custom, and it was an accepted axiom that "the striving for individualisation could not lead to valuable results among average men . . . and that tradition was always wiser than the mediocre individual."

Nevertheless, though revolt in small matters was frowned upon because it might constitute too dangerous a precedent in larger ones, such petty restraints as when clothing should be changed from thin to thick, and vice versa, was no hardship to the Chinese because their wonderful calendar predicted the seasons so accurately. It could be relied upon even where food was concerned. For example, when hare-stew was the dish recommended for the tenth moon, we may be sure hare used to be plentiful at this season, for the regulations were always founded on wise economic considerations.

At the Period of Greatest Cold, men knew to a day when it was safe to launch their sleds on the ice of the T'ung Chou Canal—that continuation of the Grand Canal which, for centuries, served to transport the tribute rice from the southern provinces to the Court. Moreover, they could be sure of carrying heavy loads safely till the *Li Ch'un* ("calendar spring") when the ice, as per schedule, was no longer solid enough to hold them.

The *p'ai tzŭ*, or bed-sleds, so called because they resemble the ordinary wooden "sleeping platform" mounted on trestles, are fitted with iron runners and propelled by two men, one pulling and one pushing. When the canal is "sealed," according to the local expression, and the heavy barges which do a thriving trade all summer are hauled up on the banks and protected by clumps of thorn-branches stuck in lumps of mud, passengers and light freight are carried on these sleds from lock to lock. Huddled under their wadded quilts, the farmers come to town bringing basket-loads of chickens, or squealing pigs with feet tied together. Or a group of women, with flowers in their hair, and babies in bright wadded coats, lend a note of colour to the drab winter landscape as they bargain for a *p'ai tzŭ* to take them home to their village after a day's shopping in the city.

Along the banks, life seems suspended. The outdoor tea-houses are closed; their vines, which cover trellises that shade summer-customers, are shrivelled to the roots, the earthen benches and tables—empty and deserted. But the scene on the frozen highway itself is animated and picturesque. A solitary fisherman has bored a hole in the ice near the shore, and by a system of mysterious signals, incomprehensible to us, all the dogs of the neighbourhood have gathered for a drink. A duck-dealer breaks a space of clear water with his heavy pole, and jealously guards his fat white birds while they take their morning swim. Yonder, the ice-gatherers are at work cutting huge blocks for the city's summer-supply, and hauling them on to the bank where they are buried in a deep pit-store-house and covered with earth.[1]

None of these people, however, interfere with the fairway, where there is a busy traffic of sleds passing and re-passing, and where young men and boys indulge in crude winter sports. Some of the children have made themselves little low sleighs on which they stand upright, pushing themselves along at a good speed by means of a metal-tipped pole, and avoiding collisions by a miracle. Others are skating on blades roughly forged from old iron by the village blacksmith, and tied on to their cloth slippers by leather thongs. Most of these youths can only afford one skate, the other being lent to a small brother. Of course, figures-of-eight are not possible under such conditions, but it is remarkable what some of the boys can do on rough ice, sticky with dust and cut-up by sled-runners. Even tiny youngsters amuse themselves all day, pushing-off with one slippered foot for long slides on their single blunt blade. If they fall, a collection of wadded garments, worn one on top of another, saves them from hurt.

Formerly, Manchu lads were taught to skate on the Palace lakes, and the Emperor himself distributed prizes to the most proficient. Ch'ien Lung, by nature a good sport, seems to have taken a particular interest in the youngsters. He found

---

[1] This method of harvesting ice has been followed ever since the Chou dynasty, when a special official was appointed to superintend the operation. Although the ice taken from canals and ponds is often of doubtful cleanliness, it is considered fit for use until the "third nine" (*vide infra*) when, although to our eyes it appears as good as ever, custom forbids any further harvesting.

time, in the midst of his more serious compositions, to write about skating as follows: "The horse is the animal which covers the ground quickest. The boat travels fastest over water, the bird—through air. Now, skating, with the aid of special shoes, permits man to slide over the ice with the swiftness of the horse, and the boat, and the bird. Thus we may say it is indeed the sport of lightness and speed." Great men are not always above stating the obvious—dear to the Oriental heart!

The fifteenth of the eleventh moon people say: "The Moon is Above the Head." On this one night of the whole year, "the shadow of a pagoda is not pointed, and the shadow of a man grows very, very small." Children lie awake to see if this is really true because, though grown-ups do not seem to fear the long black grimacing shapes that follow them all their lives save only on this one evening every year, they frighten the little folk when they fall like inky fingers on their toys and smudge the pretty colours.

* * *

The chief event of the eleventh moon is the *Winter Solstice*, marking the end of the agricultural and astronomical—if not of the civil—year, because the sun has now completed his cycle and is about to be re-born, and begin a new period of production.

The Chinese are not alone in their observance of a festival at this season. Many races and many nations did likewise. It was only natural that among primitive peoples, who watched with awe the long encroachment of the night upon the day and suffered from the chill and gloom of dying nature,[2] the first sign of a new victory of light over darkness should be celebrated as an occasion of rejoicing. Indeed, our own Christmas, which falls about this time of the year, is a transformed survival of a much older feast observed by the Romans, the Britons, and various Teutonic tribes. The Chinese period of *Tung Chih*, which begins at the Winter Solstice and lasts twelve days, exactly corresponds to the Teutonic Yule-tide,[3] and the expression *Yang Tung Chih*—"*Yang*" meaning foreign—is sometimes used for Christ's birthday.

Before the abolition of monarchy in China, the Sovereign performed stately ceremonies in honour of the turning point of the year (*see* "Imperial Ceremonies"). So much importance was attached to this solemn thanksgiving to Heaven, Earth, and the Imperial Ancestors for benefits received during the past twelvemonth, that there was an axiom: "When no winter sacrifices are offered, the State has lost its independence."

Though ritual thanksgiving is considered undemocratic, the Republic still so far recognises

---

[2] Before the promise of the Solstice, when winter is at its darkest—in a word, at the natural season of death—it was the habit in old China to execute condemned criminals who were, if possible, "kept over" till this suitable time, no matter when they were convicted. Hence, the old name of the eleventh moon, "The Moon of the Condemned."

[3] The word "Yule" is supposed to be derived from the Sun symbol. According to the Rig Veda, the twelve days after the Winter Solstice were looked upon as the Sun's holiday, when the sun went into retirement, leaving another and lesser deity in charge. This lesser deity corresponds to the God of the Hearth of the Laps, who later became our Father Christmas, St. Nicholas, or Santa Claus. See *The China Journal of Science and Arts*, December 1925, p. 638.

the Winter Solstice as to close some of the public offices, whereas the people celebrate their simple home ceremonies as they always did. Dough-cakes called *hun tun* are the staple dish at the family feast, just as rice-cakes (*see* "The Fifth Moon"), and *vermicelli,* are the mainstays of the meal eaten in honour of the Summer Solstice. Speaking of these cakes, a very old custom in some provinces requires that if any son of the house has brought home a bride within the year, she, being still entitled to wear a red skirt, must prepare them. A newcomer in new clothes asking blessings for the first time on a new home, she is supposed thus to insure plenty to eat and wear for her relations. When the dough is kneaded to the required consistency, each member of the household takes a piece and rolls it into a ball to be cooked in water and offered to the Gods of Doors next morning. Some people limit the dough bullets to twelve, one for each month of abundant food asked or thanked for. Other families provide extra cakes which are stuck on the posts of outer doors, and on window-sills, much as American school-children stick lumps of chewing-gum under their desks. Farmers' wives, to make assurance doubly sure, often model rude figures of pigs and chickens for the home altar. This allusion to the twelve animal signs of the Chinese Zodiac (*see* "The Chinese Calendar") helps to bring a blessing on their livestock.

More important than any of these local superstitious practices are the general sacrifices before the ancestral tablets, for again, as always, the cult of the dead streams through the Moon Year like the recurrent phrase of a symphony. This time, they are not remembered at their graves but in the ancestral hall where their posthumous portraits are hung for the occasion. A rich feast of smoking dainties is spread, and the whole family assembles. Chairs for the spirits are placed on the north side of the altar-table, while the living head of the family stands reverently on the south, facing them. There is perfect silence as he invites the ancestors to "share the Winter Feast," and all present politely wait while the ghosts partake of the ethereal vapours before sitting down to the repast themselves. So simply is re-union accomplished, and the whole family, both on earth and in heaven, made one.

The sacrifices to the dead, at the Winter Solstice, were originally held in honour of military heroes, and may be traced back to those legendary times before the Chinese people had been welded into a nation but consisted of separate tribes. These communities, often warring against one another, all had to resist barbarian nomads less thrifty than themselves and, therefore, anxious to pillage the results of their labours. Moreover, the settled agriculturists must be specially on their guard after the autumn harvest, and thus winter became the natural season for campaigning, defensive or offensive. Idle farmers could then exchange the sword for the hoe; frozen roads made transport easy. Men not required for home-protection had the leisure—and, too often, the inclination—to take revenge on former enemies, or loot from their neighbours. Workers became warriors.

This was so well recognised that the Book of Rites formally commands the Chief of the State and his subordinates to inspect weapons, discipline recruits, strengthen city walls, and get the war-chariots ready towards the end of the year. Thus advised, what more natural than that the rulers should choose this season to heighten the patriotism of their subjects by offerings to the manes of warriors fallen in former wars, and rewards to living soldiers of distinction?

The twilight of history seems very far from our twentieth century, yet it is to these shadowy times that we must trace the sacrifices of dough-balls still made to the Gods of Doors, descendants of local protecting spirits on whom men called for success in arms, and from whom they now beseech peace and plenty.

* * *

Winter was also the natural season for celebrating marriages in China, as in all agricultural countries, because once the harvest is in and cold weather automatically stops field-work, people have money and leisure to attend to their family affairs. Moreover, in the East, the Winter Solstice is considered the time when the struggle between the *Yang* and the *Yin*, or male and female principles, is in abeyance just before the re-birth of Life. The mysterious union of Heaven and Earth then begins, in readiness to answer the eternal call of Spring for fecundity and birth. Mankind also conforms to the universal law, following the example of Nature. To this reproductive instinct de Groot traces the origin of the legend of the Cowherd and the Weaver, the most poetic and romantic of Chinese myths (*see* "The Seventh Moon"). Now the two stars, Lyra and Aquila, that symbolise the Eternal Lovers, used in olden days to reach their zenith simultaneously about the time of the Winter Solstice. Both were given attributes suitable to the season. Aquila, "the Herdsman," became patron of the cattle brought back to their pens at the first frosts; Lyra—patroness of women who devote themselves to the making and mending of clothes for the coming year, when bitter weather kept them indoors. Thus it came about that these stars presently assumed, in addition, the joint rôles of guardians of the sexes and, particularly, of lovers. As a result of the precession of the equinoxes in the course of milleniums, Lyra and Aquila now reach the highest point in the sky together not in winter, as once they did, but in the seventh moon, when their festival is celebrated as we already know. Nevertheless popular custom, harking back to the time when the immortal Star-Lovers were first observed together in the heavens at the Winter Solstice, decrees that the red chair of the Chinese bride shall make many a lucky journey at the period of the *Tung Chih*, since conservative families like to conform to custom even if it dates from a prehistoric period.

* * *

"The Nine Nines" After the Winter Solstice begin the "nine periods of nine days" which mark the end of the

cold weather. They are called the "Nine Nines." When they commence, the Chinese make up what they call "a chart of the lessening of cold." This chart consists of nine lines, on each of which they trace nine circles which are further divided into five parts by a smaller inner circle and four strokes connecting it with the outer one. The upper section serves to mark cloudy weather, the lower, fair weather; the left, wind; the right, rain; the centre, snow. One circle is scratched off each day after the atmospheric conditions have been noted in it. By this primitive method, accurate weather records are kept. Some people make their chart in the form of a branch with eighty-one plum blossoms, one of which is tinted each day, in imitation of the old tablet at Hsi An Fu, dating from A.D. 1488. This has an engraved vase upon it with nine branches, each showing nine blossoms. The extreme simplicity of such an almanac permits even illiterate country folk to know the proper date for the spring sowing, and the farmer goes to his fields when the last circle is scratched out, or the last blossom marked off. For then the *Yang* principle has once more gained the ascendant, and spring in its fulness and beauty celebrates his triumphant return.

Various popular proverbs concerning the "Nine Nines' Period" are often on the lips of country folk. For example: "During the first and second nines, the wind roars like a buffalo." "During the third and fourth, wise men keep their hands in their sleeves." (This, of course, is an allusion to the extreme cold when the Chinese, who use no gloves, keep their hands warm by hiding them in their long sleeves). "During the fifth and sixth nines, men lift their heads and look at the blush of green on the willow-trees." "During the seventh nine, the rivers open." "During the eighth, the wild geese arrive from the north." "After the 81 days, there is no longer a needle of ice, and the grass is higher than the hoofs of an ox".

Translated into terms of climate, in North China at least (the Chinese calendar was made by, and for, northerners), these wise laws mean simply that the greatest cold may be expected in the first two nines, but the third and fourth are scarcely milder, and sensible men still keep their hands in their sleeves. During the fifth and sixth, the temperature drops a little, and spring is scheduled to arrive every year at the beginning of the sixth nine. Let us not forget that the Chinese seasons are forty-five days ahead of ours, as our spring begins only with the vernal equinox. (*See* "The Chinese Calendar").

* * *

Buddha Amitabha One beloved personage sheds a golden glow over the winter darkness of the eleventh moon. This is the Buddha Amitabha, the most popular Buddha in China. He is enthroned in the temples between two attendant Bodhisats hardly less glorious than himself, Ta Shih Chih, "The Most Mighty," and Avalokiteshvara, the Chinese Kuan Yin, who act as guides and protectors to humanity journeying across the perilous ocean of life and death. Amida, the Well Beloved, was once a rich and powerful king who so loved his fellow-men that he gave up his throne and became an ascetic, thereby attaining Buddhahood in the presence of Gautama. It was on this occasion that he made a series of great prayer-vows "whereby he undertook to establish a heavenly kingdom of perfect blessedness, in which all living creatures might enjoy an age-long existence in a state of supreme happiness, sinlessness and wisdom." That is why he is often affectionately called *Tsie Yin Fo*—"He who guides Souls from this Earth to Paradise."

His promise is, in fact, unorthodox. Gautama, founder of Buddhism, taught that the Perfect State, only to be attained when man's weary chain of earthly re-births ends, is Nirvana, a condition of mind rather than a definite place. Indeed, Nirvana simply means "the attainment of perfect peace within the heart." The Master told his followers very little about another world, nor did he offer them the pleasures they longed for in their earthly lives. He only said that all suffering would cease, all human passions be stilled, and the fires of hatred and evil thought be extinguished.

For the mystically-minded, there is no heaven to look forward to, save such as they keep shining in their own souls as a pattern for the modelling of their lives, and they are satisfied with a hope of escape into calm and nothingness, everlasting and unchanging. Not so the average man who needs a paradise moulded nearer to the heart's desire.

In the doctrines of the Amidist, or "Pure Land" School, the people's craving for immortality dominates the philosopher's doctrine of a return to the absolute.[4] Founded in the IVth century A.D. by a native of Shansi, its teaching of salvation through faith in the Buddha Amitabha, and its promise of a future life of perfect happiness in the "Pure Land," instead of the legitimate Nirvana of Sakyamuni, or Gautama, made it the most flourishing sect in China, popular among all classes. Nowadays, nearly every Chinese Buddhist is more or less of an Amidist at heart, and even the priests manage to combine the comforting beliefs of this school with the more orthodox doctrines to which it is, strictly speaking, antagonistic.

Religious imagination paints marvellous pictures of the "Pure Land," or Western Paradise, as it is called, though "the supposed western position of Amitabha's heaven has no reference to mundane geography." In that blessed country the sands of the rivers are of pure gold, and the trees have trunks of coral, with branches of diamonds, leaves of shining precious metals, and fruits of glittering gems. A sevenfold screen of balustrades and a sevenfold row of silver nets surround a lake where lotus-flowers float. True believers, when life—the curtain hiding them from reality—is gently pushed aside, are re-born within the calyx of one of these holy flowers. The pure in heart, by reason of their purity, will find that their lotuses open immediately, so they can enter at once into all heavenly joys and contemplate the glory of the

---

[4] These doctrines differ so widely from the teachings of the earlier Buddhism that some people are inclined to classify them as a separate religion. An interesting theory has been put forward that Amidism was influenced by Christianity and Gnosticism through a Persian medium.

Lord Amitabha. Less virtuous souls remain imprisoned within the closed buds for a longer or shorter time, according to their earthly sins, and this "state may be regarded as a kind of painless purgatory, since they are in heaven and yet not of it". But none will be eternally kept waiting for the fulness of his joy; none will be forever excluded from the beatific vision of their Lord, because the Buddhist Law says that there is no eternity of things imperfect, and that the "whole universe will ultimately enter into Buddhahood. . . . This doctrine is equal and alike for all. There will be neither superior nor inferior, neither above nor below."

Further, it is said "that when anyone becomes a disciple of Amida Buddha BY INVOKING HIS NAME, a lotus-plant resembling that person makes its appearance in the Sacred Lake. If during his earthly career he is devout, virtuous and zealous, his lotus will thrive,—if he is irreligious, vicious or negligent, it will languish and shrivel up." Moreover, Kuan Yin the Merciful will come to the death-bed of the righteous man holding his lotus in her hand, and, as soon as his spirit is released, will place it in the heart of Buddha's Flower which will then be carried by angels to the Pure Land, there to re-open on the surface of the Sacred Lake at the appointed time,—a spirit released from the darkness of illusion, as the perfume of a blossom is set free at the breaking of a bud.

There is much beautiful religious symbolism associated with the lotus—a flower which may be said to occupy in the Buddhist imagination a place somewhat analogous to that occupied in Christian thought by the Cross. Gautama compared himself with the holy flower, using the following striking words: "Just as a lotus born in water, bred in water, overcomes water and is not defiled by water, so I, born in the world, and bred in the world, have overcome the world." This same beautiful blossom which he made his own is found wherever the *Sutras* are chanted, in all Buddhist temples, serving as a pedestal for the gods. Nor has the richness of Oriental imagination conceived any more beautiful picture than the Sacred Lake of the Western Heaven, starred with the dawn-pink calyxes wherein the radiant figures of the blessed are enshrined.

Now, to attain this Paradise, the Amidists teach that two things are needed, faith in the God Buddha Amitabha and, especially, constant repetitions of his holy name. Indeed the phrase "Na mo O mi to Fo," or simply "O mi to Fo," has in China the force of a quasi-magical talisman. It is inscribed on the walls of temples, and carved on the rocks of sacred mountains. It is used as an invocation alike by priest and layman. "Let them call upon me," says Amida, "and they shall receive exceeding bliss." Therefore, daily and hourly, believers in the Pure Land repeat this prayer, and devout persons buy sheets of paper, specially prepared with a figure of their Saviour, whose halo is starred with little circles, in order that they may keep track of the number of their invocations. After a hundred or a thousand repetitions, a circle is marked with a dot of red ink, and when one page is finished, another is begun. Thus, they continue filling sheet after sheet and carefully laying them away till, when Kuan Yin comes, lotus in hand, to call their souls away, the book of prayers is ceremoniously burned, in the hope that it will serve as a spiritual passport to the Better Land.

We may smile at such childish faith in the efficacy of the mechanical repetition of a sacred name but, as Johnston says in *Buddhist China,* it "has many parallels in other countries and in other religions. The Bengali Vaishnanvas, for example, believe that the mere utterance of the name, of Krishna is a religious act of great merit, even though such utterance is unaccompanied by any feeling of religious devotion. A European observer has defended the worshippers of Krishna against hostile critics of this practice by remarking that the mechanical repetition of the holy name is based on sound principles, inasmuch as the habit was originally prompted by a devotional intention, which intention is virtually continued as long as the act is in performance." Growse, a Catholic writer, speaking of similar Christian practices, remarks that "it is not necessary that the intention should be actual throughout, only a virtual intention is required—that is to say, an intention which has been actual and is supposed to continue, although, through inadvertence or distraction, we may have lost sight of it."

Needless to say, the enlightened Amidist does not accept "O mi to Fo" literally as a potent charm like the simple peasants do, but rather believes that this invocation is useful only to awaken the god that dwells within the depths of his own soul, and lead him at last, when egotism and individuality have faded, to become one "with the Buddha that is at the heart of the universe." Nor does he believe "in the literal truth of the tales of Amida's Lotus Lake, and in the personal and separate existences of its divine lords, any more than the educated Christian of to-day believes in the real existence of the winged cherubim, the golden crowns, the jewelled streets and glassy seas that characterise the bric-à-brac roccoco heaven, as George Tyrrell called it, 'of hymnal and Apocalypse.' 'These,' says the Christian priest, 'are symbols of divine truth.' 'Those,' says the Buddhist monk, 'are parables of Buddhahood.'"

The debt that Chinese Buddhism owes to the Amidist School is incalculable. The dogma of individual resurrection, though a late addition to the faith, was the most essential contribution to the success of Gautama's religion. Buddhism might have overspread the Middle Kingdom without many of the orthodox tenets strictly held in India, the land of its birth, but their lofty morality alone would never have sufficed to insure its lasting success in China. Without Amida, the Saviour, the popular heart could never have been won. For what the simple folk needed in the beginning, as indeed they still need and always will, was not merely a rule of life and a mirror to the heart, but a promise of Paradise where the weary shall lay down their burdens, and the unhappy shall at last find happiness.

BIBLIOGRAPHY

Ayscough, Florence,—*A Chinese Mirror*, London, 1925.
Ayscough, Florence,—"Cult of the Ch'êng Huang Lao Yeh," Shanghai, in the *Journal of the N.C.B.R.A. Society*, 1924.
Backhouse, E. & Bland, J. O. P.—*Annals and Memoirs of the Court of Peking*, London, 1914.
Backhouse, E. & Bland, J. O. P.—*China under the Empress Dowager*, London, 1914.
Baker, D. C.—*The T'ai Shan*, Shanghai, 1925.
Ball, J. Dyer,—*Things Chinese*, Shanghai, 1903.
Baranoff,—"A Visit to Chinese Temples" (Ajiho, Manchuria), Harbin, in the *Far Eastern Monitor*, 1926.
(Bichurin), Father Hyacinth,—*China's Civil and Moral Condition, I-IV*, St. Petersburg, 1848 (in Russian).
(Bichurin), Father Hyacinth,—*Description of the Religion of the Literati*, St. Petersburg, 1844 (in Russian).
Bouillard, G.—*Usages et Coutumes à Peking*, Peking, 1923.
Bredon, Juliet,—*Peking*, Shanghai, 1922.
Brunnert, H. & Hagelstrom, V.—*Contemporary Political Organisation of China*, Peking, 1910 (in Russian).
Carus, Paul,—*Lao Tze's Tao-Teh-King*, Chicago, 1898.
Chavannes, Edouard,—*Le Tai Chan*, etc., Paris, 1910.
Cordier, Henri,—*Bibliotheca Sinica*, 2nd Ed., Paris, 1904-1924.
Cormack, Mrs. J. G.—*Chinese Birthday, Wedding, Funeral and other Ceremonies*, Peking, 1923.
Cornaby, W. Arthur,—*A String of Chinese Peach-Stones*, London, 1895.

Couling, Samuel,—*Encyclopaedia Sinica*, Shanghai, 1917.
Doolittle, Rev. Justus,—*Social Life of the Chinese*, I-II, New York, 1867.
Doré, le Père Henri,—*Recherches sur les Superstitions en Chine*, I-XIV, Shanghai, 1911-1919.
Edkins, Rev. Joseph,—*Ancient Symbolism among the Chinese*, London, 1889.
Edkins, Rev. Joseph,—*Chinese Buddhism*, London.
Eitel, (Rev.) Ernest J.—*Handbook of Chinese Buddhism*, etc., Hongkong, 1888.
Fiske, John,—*Myths and Myth-makers*.
Forke, Alfred,—*World Conception of the Chinese*, London, 1925.
Frank, Harry A.—*Roving through Southern China*.
Frank, Harry A.—*Wanderings in Northern China*, New York, 1923.
Fraser, J. G.—*Folk-lore in the Old Testament*.
Frazer, J. G.—*The Golden Bough*.
Gailey, Charles Mills,—*Classic Myths*.
Geil, William Edgar,—*The Sacred Five of China*, London, 1925.
Giles, Herbert A.—*Adversaria Sinica*, Shanghai, 1914.
Giles, Herbert A.—*The Civilisation of China*, London, 1911.
Giles, Herbert A.—*A Chinese Biographical Dictionary*, London, 1898.
Giles, Herbert A.—*Confucianism and its Rivals*, London, 1915.
Giles, Herbert A.—*Chuang Tzŭ*, etc., London, 1889.
Giles, Herbert A.—*Strange Stories from a Chinese Studio*, Shanghai, 1908.

BIBLIOGRAPHY 511

Granet, Marcel,—*Fêtes et Chansons Anciennes de la Chine*, Paris, 1919.
Gray, John Henry,—*China*, etc., I-II, London, 1878.
Green, G. P.—*Some Aspects of Chinese Music*, London, 1913.
Groot, J. J. M. de,—*Les Fêtes Annuellement Célébrées à Émoui* (Amoy), I-II, Paris, 1886.
Groot, J. J. M. de,—*The Religious System of China*, I-VI, Leyden, 1892-1910.
Groot, J. J. M. de,—*Sectarianism and Religious Persecution in China*, Amsterdam, 1903.
Groot, J. J. M. de,—*Universismus*, Berlin, 1918 (in German).
Grube, Wilhelm,—*Fêng-Shêng-Yên-I*, (*Metamorphoses of the Gods*), Leyden, 1912 (in German).
Grube, Wilhelm,—*Zur Pekinger Volkskunde*, Berlin, 1901.
Grube, Wilhelm,—*The Religion of the Ancient Chinese*, Tübingen, 1911 (in German).
Gueorguievsky, S.—*Mythical Conceptions and Myths of the Chinese*, St. Petersburg, 1892 (in Russian).
Gueorguievsky, S.—*The Principles of the Life of China*, St. Petersburg, 1888 (in Russian).
Hastings, James—*Encyclopaedia of Religious Ethics*.
Hayes, Newton A.—"The Gods of the Chinese," Shanghai, in the *Journal of the N.C.B.R.A. Society*, 1924.
Headland, Isaac Taylor,—*Court Life in China*, New York, 1909.
Headland, Isaac Taylor,—*Home Life in China*, London, 1914.
Hearn, Lafcadio,—*Some Chinese Ghosts*, Boston, 1906.
Hearn, Lafcadio,—*Japan, an Attempt at Interpretation*, London, 1904.

Hoang, Father Peter,—*A Notice of the Chinese Calendar*, Zi-Ka-Wei, 1904.

Hoang, Tsen-yue,—*Étude Comparative sur les Philosophies de Lao Tseu, Khong Tseu, Mo Tseu*, Lyon, 1925.

Hu Shih,—*The Development of the Logical Method in Ancient China*, Shanghai, 1922.

Imbert, Henri,—*Le Nélombo de l'Orient (Lotus)*, Peking, 1922.

Imbert, Henri,—*La Pivoine*, etc., Peking, 1922.

Imbert, Henri,—*Poésies Chinoises*, etc., Peking, 1924.

Johnston, R. F.—*Buddhist China*, London, 1913.

Johnston, R. F.—*Lion and Dragon in Northern China*, London, 1910.

Johnston, R. F.—"The Cult of Military Heroes in China," Shanghai, in *The New China Review*, 1921.

Kayserling, Count,—*The Travel Diary of a Philosopher*, I-II.

(Kulchitsky), Father Alexander,—*Chinese Marriage*, Peking, 1908 (reprint—in Russian).

Kupfer, Carl F.—*Sacred Places in China*, Cincinnati, 1911.

Laufer, Berthold,—*Jade*, Chicago, 1912.

Legge, James,—*The Chinese Classics*.

Leong, Y. K. & Tao, L. K.—*Village and Town Life in China*, London, (1915).

Li Ung-bing,—*Outlines of Chinese History*, Shanghai, 1914.

Lowell, Percival,—*The Soul of the Far East*, London, 1888.

Macgowan, Rev. John,—*Chinese Folk-Lore Tales*, 1910.

Mackenzie, Donald A.—*Myths of China and Japan*, London.

Maybon, Pierre B.—*Essai sur les Associations en Chine*, Paris, 1925.
Montuclat,—(articles on Chinese astronomy in *La Chine*, Peking, 1923).
Obata, Shigeyoshi,—*The Works of Li Po*.
Plopper, C. H.—*Chinese Religion seen through the Proverb*, Shanghai, 1926.
Pokotiloff, D.—*Wu T'ai, past and present*, St. Petersburg, 1893 (in Russian).
Popoff, P. S.—*The Chinese Pantheon*, St. Petersburg, 1907 (in Russian).
Richard, Timothy,—*A Mission to Heaven*, etc., Shanghai, 1913.
Rockhill, W. Woodville,—*The Life of the Buddha*, etc., London.
Ross, John,—*The Original Religion of China*, London, 1909.
Russian Ecclesiastical Mission, *Works of the*, I-IV, Peking, 1909-1910 (reprint—in Russian).
Saussure, Léopold de,—"The Lunar Zodiac," Shanghai, in *The New China Review*, 1921.
Saussure, Léopold de,—*Le Système Astronomique des Chinois*, London, 1921.
Shkurkin, P.—*Chinese Legends*, Harbin, 1921 (in Russian).
Smith, Arthur H.—*Chinese Characteristics*, London, 1892.
Smith, Arthur H.—*Proverbs and Common Sayings from the Chinese*, etc., Shanghai, 1902.
Smith, Arthur H.—*Village Life in China*, New York, 1899.
Soothill, Rev. W. E.—*The Three Religions of China*, London, (1913).
Spencer, Herbert,—*Principles of Sociology*.

Staël-Holstein, Baron A.—(Lectures on Buddhism, Peking, 1927).

Suzuki, D. T.—*A Brief History of Chinese Philosophy*, London, 1914.

Vassilieff, V.—*Religions of the East: Confucianism, Buddhism, Taoism*, St. Petersburg, 1873 (in Russian).

Vitale, Baron Guido,—*Pekingese Rhymes*, etc., Peking, 1896.

Waidtlow, Rev. C.—"Ancient Religions of China," Shanghai, in *The New China Review*, 1922.

Werner, E. T. C.—*Descriptive Sociology*, etc., Chinese, London, 1910.

Werner, E. T. C.—*Myths and Legends of China*, London, 1922.

Wieger, le Père L.—*Folk-Lore Chinois Moderne*, Hsien hsien, 1909.

Wieger, le Père L.—*Morales et Usages.*

Wieger, le Père L.—*Histoire des Croyances Religieuses et des Opinions Philosophiques en Chine*, etc., Hsien hsien, 1917.

Wilhelm, R. & Martens, Frederick H.—*The Chinese Fairy Book*, New York, 1921.

Williams, Edward T.—*China Yesterday and To Day*, London, 1923.

Williams, S. Wells,—*The Middle Kingdom*, New York, 1883.

INDEX

A

                                                                                                                    PAGE

Almanac, Chinese 歷書.
Amida Buddha 阿彌陀佛   50, 456, 461-463, 497-498, 502-508
Ancestor Worship 祈禱祖先   30-32, 40, 92, 96-99, 104-105,
                                                                             219-229, 377
Arbor Day (*See* Ch'ing Ming)   218
Arhats (*See* Lo Hans).
Artemisia (mugwort) 艾蒿 314-315
Artillery, Gods of 火砲大將軍 (五虎神) .   .   364

B

Board of Mathematicians   5
Bodhidharma ("Ta Mo") 達摩 463-468
Buddha Sakyamuni 釋迦文佛 . 257-262, 271-272, 503, 504
Buddhism 佛教   49-51, 71, 180-196, 255-265, 269, 275-278,
                                  331, 386-390, 463-468, 481, 502-508
Buddhist Triads 271-273

C

Calendar, Chinese .   .   .   . Chapter 1, 500, 502
Calendar, Gods of the   23-28
Carpenters, God of (*See* Lu Pan).
Cassia Tree, Sacred 桂樹   .   .   409, 410, 411
Cattle, God of ("Niu Wang") 牛王 .   . .360-361
Ch'an, the Moon Toad (*See* Hêng O) 嫦.
Chang Fei 張飛 325-327
Chang Kuo, Immortal 張果 294-295
Chang Tao-Ling (*See* Chang T'ien Shih) 張道陵.
Chang T'ien Shih 張天師   . 46, 319-323, 342, 441, 475
Ch'eng Huangs (*See* "City Gods").
Chêng Wu (Chin Tu Fo) 眞武   .   . 138-140, 418-419
Chih Ma (posters) 紙禡   .   .   . 95-96, 146
Ch'in Shih Huang Ti (The First
    Emperor) 秦始皇帝   42-43
Ch'in Kuang, a King of Hell 秦廣   .   .   . 173
Ch'ing Ming Festival 淸明   .   . 63, 69, 218-221
Chiu Hua, Mt. 九華山 388-389, 475
Chou Wang (Hsi Shên) 紂王   .   . 37, 416-417
Chrysanthemums 菊花   .   .   .   426-429
Chuan Lun, a King of Hell 轉輪 179
Ch'u Chiang, a King of Hell 楚江 176

INDEX

	PAGE
Chu Hsi (Chu Fu Tzŭ) 朱熹	437-438
Chung Kuei 鍾馗	313
Ch'ü Yüan (*See* "Dragon-Boat Festival") 屈原	
City Gods ("Ch'eng Huangs") 城隍	36, 68, 82, 94, 438-455
Clouds, Gods of	57, 351, 358-359
Confucius ("K'ung Tzŭ") 孔子	40-45, 64, 203-213, 488
Constellations, Twenty Eight 二十八宿	13-17, 57
Cowherd and Weaver 牛郎織女	370-376, 500
Crickets 蟋蟀	368, 369, 433-435
Cyclical Signs 輪囘	9-10, 141

D

Dragon-Boat Festival ("Tuan Yang") 端陽	69, 80, 300-308
Dragon Boats 龍舟	303-308
Dragon Cult	67, 332-333, 334-351, 396-397

E

Earth Goddess	35, 163
Earth Worship	61-62, 93, 104, 129, 162-165, 391-392
Emperors, Perfect 五帝	29-30, 57, 308, 469, 480
Erh Lang 二郎	172-173
Exorcism (Exorcists) 驅邪術	122-124, 312, 322-323

F

Fasting 齋戒	56, 116-117
Feast of Lanterns ("Têng Chieh") 燈節	84, 133-141
Fêng Hsien Tien (Palace temple)	63
Fêng Shui 風水	53, 159
Festivals of the Dead (*Kuei Chieh*) 鬼節	69, 218-229, 376-386, 460-463
Festivals of the Living (*Jên Chieh*) 人節	69, Chapter V, 300-308, 397-404
Fire-crackers 爆竹	78
Fire God 火聖	67, 74, 149, 361-364
Five Planets 五行星	57
Five Poisonous Animals 五毒	308, 317-319
Five Sacred Mountains 五嶽	470, 474-490
Five Rulers ("Wu Yo Shên") 五嶽神	24, 469-473
Four Diamond Kings ("Chin Kang") 四金剛	266-268

INDEX   XV

	PAGE
Foxes, Spirits of 狐仙節	417
Frog Worship 青蛙神	165-167
Fu Hsi, Perfect Emperor 伏羲	29, 47, 308-309, 469, 480
Fungus 木耳	244-245
Fu Shên, God of Luck 福神	419

G

Gate Gods 門神	86-88, 94
Graves, Chinese 墳墓	221-229
Great Bear Constellation ("Northern Measure") 度毋	57, 420-423, 480-483

H

Han Chung-li, Immortal 漢鍾離	288, 293
Han Hsiang Tzu 韓湘子	295
Han Shih (Cold Food Feast) 寒食	216-218
Hariti, Goddess	182
Harvest Moon Festival 中秋節	69, 80
Hearth God (See "Kitchen God")	
Heaven, worship of	54-61, 93, 104
Hells, Chinese 地獄	136, 173-180, 379, 444-448, 480
Hêng and Ha, Marshals (鄭倫, 陳西)	268-269
Hêng O (The Moon Lady) 姮娥	412-414
Ho Hsien-ku, Immortal 何仙姑	296
Horses, God of ("Ma Wang") 馬王	360-361
Hou Chi 后稷	64
Hours, Chinese	23
Household Gods 家神	36, 90, 92, 94-95, 105-106, 455
Hou T'u 后土	35-36, 64, 164
Hou Yi (The Divine Archer)	412-414
Hsien Nung (See Shên Nung) 先農	65
Hsien Ts'an T'an (Altar of Silkworms) 先蠶壇	66
Hsi Shên (God of Joy) (See Chou Wang) 喜神	416-417
Hsi Wang Mu 西王母	47, 184, 231-234, 413
Hua T'o, physician 華陀	311-312
Hundred Gods, Gathering of	144-147
Huo Shêng (See Fire God)	

I

Idols	49
Immortals, Eight 八仙	288-298

	PAGE
Imperial Ancestors	57, 62-63
Incense 香	111
Intercalary Moons 閏月	7

J

Jade 玉	220
Jade Emperor ("Yü Huang") 玉皇	47, 77, 145, 147-148, 173, 189, 441, 476, 480, 487
Jade Rabbit (*See* "Moon Hare") 玉兔	
"Joints and Breaths" of the year 中氣, 節氣	18-22
Jupiter, planet 太歲	26-27, 67

K

Kites 紙鳶	430-422
Kitchen God 竈君 (竈王)	73-78, 94, 106, 371
K'o t'ou 磕頭	102
Kuan Ti (God of War) 關帝	67, 176, 323-332
Kuan Yin (Goddess of Mercy) 觀音	71, 180-196, 481, 502
K'uei Hsing (God of Literature) 魁星	238-239

L

Land and Grain, Gods of 社稷	63, 64
Lan Ts'ai-ho, Immortal 藍采和	296
La pa ch'ou 臘八粥	71
Lao Tzŭ 老子	38-39, 40, 47, 201-203, 480
Lei Kung (*See* Thunder, god of)	
Lightning, Goddess of 電母	354
Lion Dancers 獅舞	395
Li Pu (Board of Rites) 禮部	54
Literature, Gods of ("Wen Ch'ang") 文昌	67, 237-240, 331
Li T'ieh-kuai, Immortal 李鐵枴	293-294
Liu Hai, Immortal 劉海	166-167
Liu Mên Chiang Chün (*See* Locusts, God of) 劉猛將軍	
Liu Pei 劉備	325-327
Liu Tung-pin, Immortal 呂洞賓	240, 288, 292-293
Local Gods 土地	36
Locusts, God of ("Liu Mên Chiang Chün")	165
Lo Hans (Arhats) 羅漢	273-275
Lotus 蓮花	355-358

INDEX   xvii

	PAGE
Luck Posters	28-86
Lung T'ai T'ou 龍抬頭	170-172
Lung Wang (*See* "Dragon Cult") 龍王	
Lu Pan (Patron of Carpenters) 魯班	364-367

M

Ma Chu, Maritime Goddess	196-197
Maitreya ("Mi Lo Fo," *see* Pu Tai) 彌勒佛, 布袋	
Manicheism 摩尼教	182
Manjusri (*See* Wên Shu Pusa) 文殊菩薩	
Maritime Goddesses	196-200
Marriage, Gods of	414-419
Ma T'ou Niang (*See* Patroness of Silkworms) 馬頭娘	
Ma Tsŭ P'o, Maritime Goddess	198, 337
Ma Wang (*See* Horses, God of)	
Medicine, Gods of 藥王	67, 308-312
Metonic Cycle	8
"Ministry of Earth"	62-67
"Ministry of Heaven"	57, 67
Monkey, Heavenly 齊天大聖	172-173
Moon cakes 月餅	399-400
Moon Hare 月兔	400-401, 404-410, 414
Moon Worship	57, 65, 397-416
Mo Tzŭ, philosopher 墨子	39
Mountain Cult	34, 469-490
Mugwort (*See Artemisia*)	
Mu Kung (*See* Tung Wang Kung)	
Mu Lien ("Ti Tsang Wang")	387-389
Music, Chinese 音樂	206-208

N

Nature Cult	32
Neo-Confucian School	44-45
New Year 新年	69, 80, Chapter V
Niang Niangs 娘娘	281-311, 417-418

O

O Mei, Mt. 峨眉山	277-278, 475

INDEX

P

	PAGE
P'an Ku 盤古	33, 47, 162
P'an Kuan 判官	312
Paper money 冥錢	95
Peach, symbolism of the 桃子	84, 318
Peonies 牡丹	251-255
Pien Ch'êng, a King of Hell 卞城	179
Pi Hsia Yüan Chün (Princess of the Coloured Clouds) 碧霞元君	281, 479, 485-486
Pilgrimages 聖地遊	279-288
P'ing Têng, a King of Hell 平等	179
Ploughing Ceremony	65, 68
Polar Star 紫微星	26, 68
P'u Hsien Pusa 普賢菩薩	277-278
Pusa (Bodhisattva) 菩薩	181
Pu Tai (The Laughing Buddha) 布袋	269, 481
P'u T'o, Island of 普陀	188, 191-196, 475

R

Rain Gods (*See* "Dragon Cult")	57, 353-354
Rain prayers and processions 祁雨	340-351
Rats' Wedding Day	143-144
Riches, God of (*See* Tsai Shên)	
River Spirits 水神	34

S

Sacred Mountains	62
Sacred Rivers	62
Sacrifices: Great, Medium, Small 祭祀	Chapter III
San Ch'ing (The Three Pure Ones) 三清	47
San Kuan 三官	168
Sao Ch'ing Niang Niang 掃晴娘娘	358-359
Shang Ti	32-33, 35-36, 57, 60
Shê Chi T'an (Altar of Land and Grain) 社稷壇	63
Shên Nung, Patron of Agriculture 神農	64-66, 132, 163, 308-309, 469, 480
Shou Huang Tien, Palace hall 壽皇殿	63
Shou Hsing (God of Longevity) 壽星	420-423
Shui Kuan 水官	163
Silkworms, Patroness of 蠶女	66
Smallpox, Gods of 痘神娘	281, 311
Snakes, worship of 拜蛇	336, 340, 344, 348-349

INDEX

	PAGE
Spring, Ox of	129, 133
Star Festival	68, 141-143
Stilt-walkers 踏高蹻	395-396
Story-tellers 說書者	151-153
Sung Ti, a King of Hell	176
Sun Worship	57, 65, 158-162, 230, 414, 415

T

T'ai Miao 太廟	62
T'ai Shan, Mt. 泰山	68, 179, 234, 243, 476-490
T'ai Sui (See Jupiter, Planet)	26, 67
T'ai Yi (T'ai Chi) 太乙, 太極	26, 43, 133
T'ai Chê Ssŭ, temple of	240, 241, 244
Talismans 驅邪符	314-319
Tao, the 道	38, 43-44, 53
Taoism 道教	45-49, 289-290, 319-323, 339
Ta Shih Chih (Mahasthamaprapta) 大勢至	502
Temple of Agriculture 先農壇	65
Temple of Earth 地壇	61
Temple of Heaven 天壇	55
Têng Kao Festival 登高 (Chung Yang Chieh) 重陽節 427-430	
Theatre, Chinese 戲園	155-157, 392-394, 412
Thieves, God of	423-424
Thunder, God of 雷公	57, 242, 352-353
T'ien Fei, Maritime Goddess 天妃	198-200
T'ien Kuan 天官	168
Tiger, symbolism of 虎	316
Ti Kuan 地官	168
Ti Tsang Pusa 地藏菩薩	386-390
Ti Wang Miao, "Temple of Emperors and Kings" 帝王廟 65	
Tree Worship	229-231, 241-249
Tsai Shên (God of Riches) 財神	108-113, 170
Tsao Chün ⎫ (See Kitchen God)	
Tsao Wang ⎭	
Ts'ao Kuo-ch'iu, Immortal 曹國舅	295
Tuan Wu Festival (See "Dragon Boat Festival") 端午節	
T'u Kung, Earth God 土公	164
T'u Mu, Earth Goddess 土母	164
Tung Wang Kung 東王公	47, 231-232
Tung Yo Miao, temple 東嶽廟	174, 234-235
Tu Shih, a King of Hell 都市	179
T'u Ti (Local God) 土地	36, 94, 455-458
Tz'ŭ Hsi, Empress 慈懿太后	54
Tzŭ Ssŭ 子思	212

V

Verbiest, Father 南懷仁	6

W

	Page
Wealth, God of (*See* Tsai Shên)	330
Wei T'o 韋陀	269-270
Wên Ch'ang, God of Literature 文昌	237-238
Wên Shu Pusa 文殊菩薩	275-276
Willows, symbolism of 柳樹	133, 229-231, 245-247, 318, 344
Wind, God of 風伯	57, 354-355
Winter Solstice 冬至	495-499
Wu Kuan, a King of Hell 五官	177
Wu Shêng (Five Saints) 五聖	111, 168-170
Wu T'ai Shan (Mt.) 五台山	191, 275-276, 475
Wu Yüeh Chieh (*See* "Dragon-Boat Festival") 五月節	

Y

Yama (*See* Yen Lo Wang).	
Yang and Yin principles 陽陰	35-43, 47, 60-61, 65, 70, 133, 231, 338, 397, 398, 414, 427, 435-436, 499
Yang Chu, philosopher 楊朱	39
Yao Wang, God of Medicine 藥王	310
Yen Lo Wang 閻羅王	136, 175, 177-178, 188, 382, 386, 444
Yen Tzŭ (Yen Hui) 顏子 (顏回)	212, 436-437
Yo Shih Fo, the Buddha of Healing 藥師佛	308
Yü Huang (*See* Jade Emperor) 玉皇	
Yü lan pên (Magnolia Festival) 盂蘭盆會	69, 376-386
Yü Shih (*See* Rain, Gods of) 雨師.	

Z

Zodiac, Chinese Lunar 黃道	11-12, 147
Zodiac, Chinese Solar 日道	10, 492

www.ingramcontent.com/pod-product-compliance
Lightning Source LLC
Chambersburg PA
CBHW051415290426
44109CB00016B/1301